CONTEMPORARY URBAN ECOLOGY

CONTEMPORARY URBAN ECOLOGY

Brian J. L. Berry
Harvard University

John D. Kasarda
The University of North Carolina

Macmillan Publishing Co., Inc.
New York

Collier Macmillan Publishers
London

Copyright © 1977, Macmillan Publishing Co., Inc.

Printed in the United States of America

All rights reserved. No part of this book may be reproduced or
transmitted in any form or by any means, electronic or mechanical,
including photocopying, recording, or any information storage and
retrieval system, without permission in writing from the Publisher.

Macmillan Publishing Co., Inc.
866 Third Avenue, New York, New York 10022

Collier Macmillan Canada, Ltd.

Library of Congress Cataloging in Publication Data

Berry, Brian Joe Lobley, (date)
  Contemporary urban ecology.

  Bibliography: p.
  Includes index.
  1. Sociology, Urban. 2. Cities and towns—
Growth. 3. Urbanization. 4. Metropolitan areas.
I. Kasarda, John D., joint author. II. Title.
HT151.B4338   301.36'3   76-6862
ISBN 0-02-309050-2

Printing: 1 2 3 4 5 6 7 8   Year: 7 8 9 0 1 2 3

To Jan and Mary Ann

Preface

The idea to present an interdisciplinary approach to urban ecological issues emerged from a series of informal exchanges between us while we were colleagues at the University of Chicago. These discussions, oriented to current urban problems, served to underscore numerous theoretical and substantive convergences taking place among the social sciences on issues in urban ecology and, likewise, to strengthen our conviction regarding the need to span traditional disciplinary boundaries if we are to achieve a fuller understanding of the patterns, processes, and problems of contemporary urban organization and change. In this regard, both authors share a lively interest in urban economics and city and regional planning which, when integrated with our individual backgrounds in urban geography and social ecology, has given rise to the form and content of *Contemporary Urban Ecology*.

The book, intended for use in upper-division and graduate-level courses in urban sociology, urban geography, human ecology, urban studies, and city and regional planning, contains seven sections. Part I introduces the student to fundamental assumptions, principles, propositions, and paradigms in modern ecological inquiry, and traces their evolution and convergence from the separate research traditions of sociology and urban geography. Parts II through VII are organized into successive levels of a sociospatial hierarchy (neighborhoods, cities, metropolitan areas, regional and urban systems, and total societies) proceeding, in turn, from questions about social behavior within local community units at one extreme to global concerns about comparative urban structure and planned societal change at the other.

Our objective is not to provide an overview of all possible issues and problems of concern in modern ecological inquiry. Rather, we have selected at each level of the sociospatial hierarchy a number of issues and problems that we consider among the most important and have attempted to provide some depth to their treatment. Thus, in Part II, the social construction of local communities is explored: first by examining organized neighborhood responses to racial transition (illustrating the processes by which residents of local communities resist changes in their residential environments which they perceive as detrimental to their neighborhood's status); then from the perspective of changing community attachments and local social bonds under conditions of increased urbanization; and finally, from the vantage of an alternative theory of the bases of local community structure in mass society.

Parts III and IV proceed beyond the neighborhood and local community level to issues of city and metropolitan structure. Part III analyzes intraurban form and structure in terms of its nodes and hierarchies, networks and gradients, and social and physical subareas. Part IV examines metropolitan expansion and structural changes in central cities and suburban areas and assesses their implications for current economic and fiscal problems plaguing U.S. central cities.

Moving further up the sociospatial hierarchy, Part V considers regional growth dynamics, systems of cities, and the economic integration of a national urban system, along with the structural effects of size for different types and levels of systems in the hierarchy. Part VI addresses a range of comparative issues and cross-cultural urban problems, with special emphasis on the role of urban public policy and planning in Western Europe, North America, the Soviet Union, and the Third World.

We conclude, in Part VII, with an agenda for future ecological inquiry which brings together concepts and variables from various social science disciplines in the form of a synthetic model for planning and deliberately directing change in modern sociospatial systems. The mutuality and complementarity of these concepts and variables highlight the interdisciplinary nature of modern ecological inquiry.

As is typical with most books of this nature, we are deeply indebted to numerous persons who contributed either directly or indirectly to the volume. We would especially like to acknowledge those who collaborated with us in the preparation of a number of individual chapters: Carole Goodwin, Robert Lake, and Katherine Smith (Chapter 2); Morris Janowitz (Chapter 3); James Simmons and Robert Tennant (Chapter 5); Philip Rees (Chapter 7); George Redfearn (Chapters 8 and 9); and Yehoshua Cohen (Chapter 13).

We would also like to acknowledge publishers of our previous works who generously granted us permission to incorporate these materials in the book. Such publishers include the *American Journal of Sociology* (Chapters 2, 7, 11, and 12), the *American Sociological Review* (Chapters 3 and 16), St. Martins Press (Chapters 4, 17, and 18), *Law and Contemporary Society* (Chapter 5), *The Geographic Review* (Chapter 6), the *Journal of Urban History* (Chapter 8), *Social Forces* (Chapter 10), Sage Publications (Chapter

13), Ballinger Publishing Company (Chapter 14), John Wiley and Sons (Chapter 15), and the *South African Geographical Journal* (Chapter 21).

Finally we would like to express our gratitude to those who aided in the production process, especially Linda Zouzoulas and Perdita Mixon, who typed the manuscript.

B. J. L. B.
J. D. K.

Contents

VI. COMPARATIVE URBAN STRUCTURE AND PLANNED CHANGE

VII. TOWARD AN ECOLOGY OF THE FUTURE

PART I
INTRODUCTION

Chapter 1
The ecological approach

The ecological approach to the study of social and spatial organization has had a substantial impact on the development of both sociology and geography in the United States. Its benchmark in sociology was a paper published in 1916 by Robert E. Park titled "The City: Suggestions for the Investigation of Human Behavior in an Urban Environment." This paper laid the foundation for an impressive series of articles and monographs on the ecology of urban life published over the following two decades. These publications served to place the "ecological school" into the mainstream of sociological research and created what Milla Alihan termed in 1938 "one of the most definitive and influential schools in American sociology."

The period of eminence of the ecological approach was relatively brief, however. Between 1938 and 1945, both its theoretical framework and its empirical generalizations came under increasingly sharp criticism. By 1950, the ecological approach as developed by Park, his colleagues, and students at the University of Chicago was virtually dead. It was at this date that Amos Hawley's treatise *Human Ecology: A Theory of Community Structure* appeared. Hawley reformulated the ecological approach and initiated its present revival within the field of sociology. It was at this time, too, that the rapid growth and intellectual transformation of urban geography provided a parallel stream of research findings and, ultimately, a healthy intellectual discourse between the two disciplines within the broader interdisciplinary context of urban studies. Today, the ecological approach is generally recognized as an important mode of inquiry, but, because of its history of shifting substantive objectives and the diversity of its methodological techniques, confusion as well as debate continue to surround its role not only in sociology, but in geography as well. The purpose of this chapter is first, therefore, to review the development of the ecological approach in sociology and, in so doing, to shed some light on the reasons for the confusion over the present meaning of that approach. The separate research traditions in urban geography are also examined, and attention is given to contemporary ecological principles and their application to problems of social and spatial organization.

Traditional human ecology

The theoretical origins of the ecological approach may be traced to nineteenth-century concepts and principles formulated by plant and animal ecologists, and to related work by geographers, economists, and social theorists. Park, whose previous experience included nine years as a newspaper reporter, was able to witness at first hand the conflict and adjustment between the waves of immigrants who poured into midwestern communities at the turn of the century (Alihan, 1938:5). He was particularly impressed how similar their "struggle for existence" was to the Darwinian conception, and how, over time, each community seemed to progress toward a state of economic and social equilibrium based on competitive cooperation among the various immigrant groups. Just as in plant and animal communities, Park concluded, order emerged in the human community through the operation of "natural" (that is, unplanned) processes such as competition, dominance, succession, and segregation. With this conception of community organization, it was a logical step for Park to begin to formulate an ecological theory of the human community analogous to the existing theories of plant and animal communities.

The starting point of Park's ecological approach was the division of community organization into two levels—the biotic and the cultural. The former, based on competition and the division of labor, was derived from Darwin's concept of "the web of life." The latter, based on communication and consensus, followed from Comte's notion of the moral order. Park was careful to point out that the two levels were separable only analytically and that every human community actually was organized simultaneously on the biotic and cultural levels. He contended, however, that the proper focus for human ecology was the biotic level. Analysis of the cultural level was considered to be a problem for social psychology. This early decision to exclude ideational factors from the scope of human ecology eventually led to a substantial amount of polemics, and to a split among sociologists investigating ecological problems.

To a great extent, traditional human ecologists accepted the analytical separability of the biotic and cultural levels. Their approach was to view the human community as a dynamic adaptive system in which competition served as the primary organizing agent. Under the pressure of competition, each individual and group were said to carve out both residential and functional niches in which they could best survive and prosper. The effect was to segregate people and their businesses into relatively homogeneous residential and functional subareas within the community. These subareas were called "natural areas" because they purportedly resulted not from conscious design, but from the spontaneous operation of the market. Examples of such natural areas in the urban community included the central business district, the slum, the ethnic ghetto, and the rooming-house area.

The position of traditional ecological theory was that the overall spatial pattern of the community, including the location of specific natural areas, was also regulated by competition. In this the urban geographers and urban land economists concurred, maintaining that industries and commercial institutions competed for strategic locations which, once occupied, enabled them to exercise control (dominance) over the functional use of land in other parts of the community. The most strategic location, or area of dominance, was found at the point of highest accessibility, usually the center of the community. It was here that the largest number of people converged daily to work, shop, and conduct their business. Consequently, land values were shown to be much higher in the central business district (CBD) than in the surrounding area. The early ecological literature described how speculators acquired the land immediately adjacent to the central business district with the expectation of reaping profits when the central business district expanded. Because land held for speculative purposes was usually neglected, the area surrounding the CBD often deteriorated into a slum, the "zone in transition."

Ecologists pointed out, however, that the slum performed an important function for the community. Its relatively inexpensive housing and propinquity to expanding industries provided a convenient locus of first settlement for newly arriving migrant groups. Many ecological studies documented the invasion-succession sequences that followed the rapid influx of ethnic immigrants (Burgess, 1925; Cressey, 1938; McKenzie, 1929). The studies showed how each immigrant group initially concentrated in a highly segregated ethnic enclave within the deteriorating zone, and how, with the passage of time, they were able to climb the socioeconomic ladder and escape to better residences further removed from the slums, only to be replaced by another wave of newly arriving immigrants. These successionlike movements were responsible for the positive gradients between socioeconomic status and distance from the urban center so often referred to in the classic ecological literature.

Looking back at the traditional ecological studies, it appears that most can be classified into three basic types. First were those studies that applied concepts and principles derived from plant and animal ecology to the analysis of the human community. These studies focused primarily on the processes of competition, dominance, and succession, and their consequences for the distribution of population groups and commercial activities throughout the community. The early works of Robert Park, Ernest Burgess, and R. D.

McKenzie best represent the theoretical and empirical application of this approach.

The second type of study provided detailed descriptions of the physical features of specific natural areas along with the social, economic, and demographic characteristics of their inhabitants. Community studies of this nature, of which Harvey Zorbaugh's *The Gold Coast and the Slum* and Louis Wirth's *The Ghetto* are well-known examples, freely mixed the so-called biotic and cultural elements. They were not, as Milla Alihan (1938) noted, strictly human ecological studies but rather general sociological studies in which territoriality was taken into account. The recent work of Gerald Suttles (1968, 1972b) provides contemporary examples of this second type of study.

The third type of study took as its objective the coding of observable social phenomena such as crime rates or mental disorders, plotting their frequencies on maps or graphing their gradients with respect to distance from the city center. These studies, which gave rise to the considerable literature dealing with problems of "ecological correlation," included Shaw et al., *Delinquency Areas*, and Faris and Dunham, *Mental Disorders in Urban Areas*. Although they were not definitive, ecological studies of the third type did uncover a number of empirical relationships between social and spatial variables that provided the stimulus for a number of important sociological analyses in recent years (see, for example, Voss and Peterson, 1971), as well as providing the initial basis for the postwar dialogue with a transformed urban geography.

As the above typology indicates, ecological research did not follow one single path. The diffuseness resulted because the substantive boundaries of human ecology were never formally delimited. Park's (1950:6) own attempt to define the scope of human ecology is illustrative: "There remain the studies of the city, or urban and rural communities, what R. D. McKenzie and I call, quite properly I believe, 'Human ecology.'" With such a capacious definition of human ecology, it is no wonder that so many studies that originated at the University of Chicago during the 1920s and 1930s were labeled "ecological studies," whether or not they were based on concepts or principles derived from ecological theory.

The question now arises: What led to the eventual downfall of traditional human ecology? The answer appears to lie in the fundamental shortcomings in the approach's basic framework. A series of theoretical and empirical critiques by Alihan (1938), Davie (1937), Gettys (1940), Firey (1945), and Hatt (1946) revealed a number of weaknesses of traditional ecological theory: its muddled distinction between biotic and cultural elements, its excessive reliance on competition as the basis of human organization, its total exclusion of cultural and motivational factors in explaining land use patterns, and the failure of its general structural concepts, such as concentric zonation and natural area, to hold up under comparative examination. Taken together, these criticisms served to call into question the overall validity of the traditional ecological approach in sociology.

In the face of mounting skepticism during the late 1930s and the 1940s, human ecologists all but abandoned their broad theoretical and empirical studies of community structure. A majority shifted their attention to the narrow and somewhat unimaginative graphing and mapping of social data to uncover spatial (ecological) correlations. The unfortunate consequence of this substantive shift has been to create, among some, an image of the ecological approach as merely a methodological technique in which inventories of social data are collected for various subareas of cities such as census tracts, then are plotted on maps, which are in turn compared to elicit their "ecological" correlations (the latter contrasted with "individual" correlations among the characteristics of people or families).

A smaller number of ecologists shifted their focus to the analysis of social and cultural determinants of man's location in space. This group, which became known as the "sociocultural school" of human ecology, exhibited a preoccupation with land use patterns and spatial distributions of population groups. Their general restriction of the scope of human ecology to the study of areal patterns helped promote another widely held misconception that the ecological approach is more concerned with geographic space than with social organization, and with patterns rather than the processes giving rise to these patterns.

Traditional urban geography

This shift of human ecology from its original concern with symbiotic relations to that of mapmaking and analyzing land use patterns roughly paralleled the shift of urban geography from a philosophy of environmentalism to the descriptive approaches of "areal differentiation," yet the ecological approach and traditional urban geography converged only belatedly. That this convergence came so late is perhaps the most startling feature of the history of the two fields; as early as 1841, J. G. Kohl had written a provocative theoretical treatment of the location, growth, and spatial distribution of urban settlements in which he postulated explicit relationships between the social and spatial structure of cities (see Chapter 7). His work was, unfortunately, largely ignored by subsequent generations of geographers, so that in 1913 the leading urban geographer of the time complained that

> The goals and methods of current research in settlement geography are still insufficiently clarified. We are particularly far away from having available universally applicable methods. One...who takes his research seriously, is consequently forced to develop methods on his own (Gradmann, 1913a, p. 5).

Why was it that the small community of interested geographers had been unable to institute the study of urban settlements as an independent yet integral element of geographic research? A provocative essay by Müller-Wille (1976) provides the answers, and, in doing so, reveals the broad parallelism that existed between ecological and geographical research long before any convergence occurred.

Müller-Wille points out that early urban geographers frequently noted the necessity for a general, conceptual framework which was to provide the basis for more independent and systematic research work dealing with human activities, their material manifestations, and their temporal distribution over space. The most comprehensive effort to clarify the particular objectives of such a research program was made by Ratzel, in his monumental two-volume work *Anthropogeographie* ("The Geography of Man," 1882). Ratzel's basic plan was to define "the human element" in geographic research and to bring its evaluation in closer correspondence with the recognized system of geographic thought and explanation, that of environmental determinism. He proposed a catalogue of pertinent problems which geographers should be able to address, knowing that their solutions would be consistent with the principal goals of geographic research. That the study of "anthropogeography" was not yet considered an essential part of geography, but rather the intellectual link between (physical) geography and history, is clearly demonstrated by the subtitle of Ratzel's work, "The Principles of Geography's Application to History."

Such a definition of the association of the two disciplines obviously determined the scope of anthropogeographic studies, including the methods to be applied. The subject matter was an investigation of the causal relationships between the natural environment—categorized as *Lage, Raum,* and *Boden*—and human activities and movements in an evolutionary perspective. In order to attain any geographically relevant results such studies had to employ methods that had been instituted by physical geographers, so that the "anthropogeographic laws" obtained might reflect the varying modes of man's dependency on environmental conditions.

It is not surprising, then, that geographic studies dealing with urban settlements analyzed the location, historical development, and spatial distribution of cities in terms of their dependency on environmental conditions. A distinction was made between local and regional physiogeographic determinants—that is, "topographical" conditions (site) vis-à-vis "geographical" conditions (situation). A prime example of this research orientation seems to be Hettner's discussion (1895) of the origin and development of urban geography and its contemporary objectives and methods. Referring to Ratzel's work, the author flatly declared that to be in accordance with conventional procedures, urban geography had to adopt the specific methods of inquiry and interpretation successfully exercised by physical geographers, especially plant geographers. Only a thorough understanding of the areal variation of environmental conditions on both the local and the regional level permitted the geographer to classify urban settlements, compare them, and finally arrive at the formulation of generally valid laws explaining the location, size, and historic evolution of single cities and groups of cities. Thus, the urban settlement was compared to an organism which, much like a plant, had to adjust to the natural environment and compete with others in a constant struggle for survival and supremacy.

With this organismic approach Hettner supported the rationalist's argument that nature represents the absolute standard of order and purpose. This required, as in early human ecology, the strict application of the Darwinian concept of survival by competition and adjustment. Emerging out of nineteenth-century natural philosophy, the first twentieth-century paradigm in urban geography thus was that of environmental determinism. "A statement has geographic quality," said the first president of the Association of American Geographers in his address to the Association, "if it concerns some relationship between inorganic control and organic response."

The paradigm did not persist for long past World War I, however. A growing number of geographers disapproved the absolutism of environmental conditions. In reviewing developments in human geography and assessing their significance for the study of urban settlements Schlüter, in particular, argued that

> it is a serious misconception to believe that...anthropogeography is able to deal with man's dependency on nature or nature's influence on man....The objective of geographic research is to understand the form, shape, and spatial arrangement or association of geophenomena which are visible on the earth's surface and thus observable....To solve this problem geography has to have complete freedom as far as the explanation of these phenomena is concerned. All kinds of influential factors have to be taken into consideration may they relate to either or both environmental conditions and/or man's free intellect (1899, pp. 66–67).

Schlüter's statement signified the beginning of a new, promising era during which human geography was to develop into an equally respected, if not dominant, subdivision of geographic research as compared to physical geography, but one which, unfortunately, collapsed into atheoretical mapmaking and description, just as traditional human ecology did.

Schlüter proposed in his influential programmatic statement that the unifying viewpoint of geography had to rest on careful description of "factual forms on the earth's surface." Consequently, scientific scrutiny should be applied to mapping and describing the visible product of a yet unknown number of independent variables, rather than to a selected number of presupposed causal factors. In any case, the process of identifying the latter seemed to depart from premature conclusions, if not preconceived theory. Thus Schlüter stressed the vital necessity of unprejudiced, inductive field work:

> The only profitable approach (to geographic problems) depends on the most accurate and methodical researching of facts. This alone can give the addressed problem its tangible form and represents the indispensable foundation for any theoretical construct (1899, p. 75).

Two alternate routes of practical research in urban geography were proposed: (1) the analysis of the areal extension of urban influence and its material expression; and (2) the description of the internal structural differentiation of cities. The conventionally defined city was argued to constitute, in reality, the center of a larger urbanized region. With increasing distance from the "central city," urban influence would diminish in a regular fashion, creating under ideal circumstances a graded sequence of concentric zones. Schlüter tentatively listed them as: (1) The city proper or "central city"; (2) the inner ring of urban-influenced and -oriented settlements; (3) the zone of commercial interdependence; (4) the outer ring of long-distance trade and traffic routes and the sphere of cultural dependency and institutional dominance.

It was suggested that any settlement, ranging in size from hamlet to metropolis, generates a sphere of influence which extends beyond its immediate boundaries. The range of influence is determined by a settlement's size and proximity to others. This basic notion of a rural-urban continuum was subsequently modified insofar as only those settlements which developed and maintained a complete system containing all four rings were classified as urban.

The distinction between these zones was based on settlement-morphological differences, principally systematic variations in the intensity of land use. Descriptive indices were formulated relating to the size of real estate, the height of buildings, and the ratio of built-up area to open, unused space. In this context, Schlüter called for substantial improvements of the available, yet scant, census materials.

Schlüter interpreted the "central city" per se as representing the most appropriate object of study in urban geography. This allowed him to apply two standard procedures of geographic inquiry: (1) To investigate the historical development of a city, thus defining the factors of its growth and partially explaining its structural diversification; and (2) to discuss the contemporary distributional patterns of selected urban characteristics, individual or grouped.

He insisted that the complex question of what agents or causal factors had been most dominant in the development of cities could be answered on an individual basis only. In accordance with the current *Zeitgeist*, the influence of historical events, or more precisely the visible material consequences of human activity and decision making, was thought to be more significant than physicogeographic conditions in explaining the physical character of cities. By relying mainly on historical accounts that reported on changes in a region's sociocultural, political, and economic situation, the researcher was equipped with the necessary information which let him determine some of the reasons for the diversified structure of present-day urban morphologies as manifested by the distinctive differences between urban districts in terms of their architectural individuality and geometry of street plans.

The primary objective of urban geography, however, was to remain the detailed mapping and description of the contemporary urban landscape well

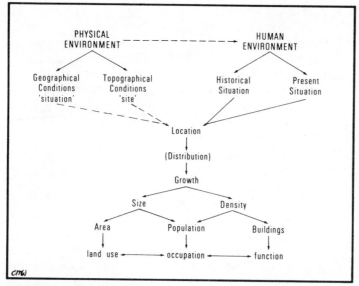

Figure 1.1

Topical organization of urban geography. Proposed by Schlüter.

past the middle of the twentieth century. Schlüter proposed that such investigations were best organized according to three elemental categories that formed a substantial part of the conventional definition of cities. They were *area, population,* and *buildings.* The raw data available on any of these categories were frequently presented as size or density calculations. In this form they already provided essential information about the internal differentiation of individual urban landscapes. A more comprehensive approach aimed at the identification of descriptive attributes which were commonly listed under three subgroups: land use, function of buildings, and population characteristics. Although such lengthy catalogues, if researched carefully, seemed to preclude extensive investigations of larger cities or groups of cities, such as in regional, comparative studies, Schlüter definitely encouraged this mode of research so as to establish a sound foundation for the classification of urban settlements. Figure 1.1 delineates the topical organization of urban geography as outlined by Schlüter. Broken lines indicate dependencies between variables whose validity had been questioned.

Transitional contributions

It was not until 1950, when Amos Hawley's treatise *Human Ecology* appeared, that a serious attempt was made in sociology to subordinate the emphasis on space to that of examining functional interdependencies of man's collective life. In his book, Hawley presented the ecological approach as the study of the form and development of community structure. Community structure was construed as the territorially based system of functional interdependencies that results from the collective adaptation of a population to its environment. Treating the human community as the smallest social system in which all properties of society are found, Hawley provided numerous hypotheses and propositions concerning the morphology and evolution of community structure that form the framework for much of contemporary ecological theory in sociology.

Paralleling the sociological experience, the 1950s saw a revolution in geography sparked by change in urban geography (Berry, 1965b). The dual bases of this transformation were the adoption and rapid development of a body of location theory, beginning with Christaller's central-place concepts, and a technical revolution sparked by the adoption of quantitative research methods, the emergence of regional science, and the rapidly evolving analytic capabilities of the scientific computer. Geography was dragged, frequently screaming, to the frontiers of modern social science and into broadly based interdisciplinary ventures, one consequence of which is a growing overlap between contemporary human ecology and modern urban geography.

Contemporary ecological inquiry

The central problem of contemporary ecological inquiry is understanding how a population organizes itself in adapting to a constantly changing yet restricting environment. Adaptation is considered to be a collective phenomenon, resulting from the population developing a functionally integrated organization through the accumulative and frequently repetitive actions of large numbers of individuals. The basic premise of the ecological approach is that as a population develops an effective organization, it improves its chances of survival in its environment (Hawley, 1971:11).

Unlike traditional models of human ecology which emphasized competition, contemporary ecological inquiry is primarily concerned with interdependence. Interdependence is seen as developing along two axes: the *symbiotic* and the *commensalistic*. Symbiotic interdependence arises from structural differentiation and integration of specialized roles and functions within a system. It varies with the degree of differentiation and frequency of exchange among the specialized parts engaged in complementary activities. Symbiotic interdependence is best illustrated by the division of labor which generally exists at all levels of social groupings from families to total societies. Commensalistic interdependence is based on supplementary similarities and arises whenever a specific system task is greater than one person can manage. Through the co-action of like units, enormous power may be assembled and huge undertakings can be accomplished. Labor unions, neighborhood clubs, and common interest groups represent modern social units which have developed on the basis of commensalism. From the ecological standpoint, symbiotic and commensalistic interdependencies form the basis of social system structure. As such, they are principal analytical elements in the ecological approach.

Another feature of contemporary ecological inquiry is that it is dominantly, though not exclusively, macrosocial in its perspective. A fundamental assumption of the ecological approach is that social systems exist as entities *sui generis* and exhibit structural properties that can be examined apart from the personal characteristics of their individual members. These structural properties are determined by the pattern of organized activities that arise from the routine interaction of various functions and roles performed within the system.

In line with its macrosocial perspective, contemporary ecology frequently analyzes the structure of organized activity without respect to the attitudes and beliefs that individuals may entertain in their roles. As a property of the aggregate, organizational structure is viewed as transcending the personal characteristics of the individual actors. This may be shown empirically in bureaucracies, communities, and societies where individuals of quite different personal characteristics periodically replace others in particular roles without disrupting the observable pattern of organized activity. It is this important property of social systems that provides for their stability and continuity in development over time.

A third feature of the ecological approach is its assumption that through the continuous interaction of their component parts, social systems tend to move toward a state of equilibrium. According to Hawley (1968:334) equilibrium will be attained:

1. *Functionally,* when the various functions that impinge on one another are complementary and when they collectively provide the conditions essential to the continuation of each.
2. *Demographically,* when the number of units-individuals (or man-hours) engaged in each function is just sufficient to maintain the relations of the function to each or all other functions.
3. *Distributively,* when the units are arranged in space and time in such a way that the accessibility of one to others bears a direct relation to the frequency of exchanges between them.

It should be emphasized, however, that equilibrium is more of an analytical construct than a social reality. It occurs only where there is a closed social system existing within a stable environment. Under these rare circumstances, the system will adjust its component parts until it has reached maximum efficiency in coping with prevailing environmental conditions. When complete adjustment occurs, the equilibrium stage is reached, and all demographic and structural change ceases. The population size and organizational structure of the system will remain in a stable state until disrupted by a change in the system's social or physical environment. Anthropologists have provided examples in which isolated societies have existed in an equilibrium state for hundreds of years without apparent changes in either their population size or social organization. But as soon as their isolation was broken, allowing new environmental inputs, immediate and extensive changes began to occur.

In analyzing stability and change within social systems, contemporary ecologists rely principally on four reference variables: *population, organization, environment,* and *technology.* These four variables, which are reciprocally causal and functionally interdependent, constitute what Duncan (1959) has labeled the "ecological complex." Predicated on the way a particular ecological problem is stated, each of the four variables may serve as either a dependent or an independent variable.

Ecologists tend to conceptualize their four reference variables rather broadly. *Population* refers to any internally structured collectivity of human beings that routinely functions as a coherent entity. As the common denominator of ecological analysis, a population exhibits a number of properties that are not shared by its individual members. These include independent mobility of its component parts (the individual human beings); no intrinsic limit on size, replaceability and interchangability of parts, and an indefinite life span (Hawley, 1968). Such properties have provided human populations with considerable resilience in adapting to changing environmental conditions.

Organization, the second key variable in the ecological complex, is synonomous with what has previously been described as system structure. It refers to the entire network of symbiotic and commensalistic relationships that enable a population to sustain itself in its environment. Since organization is an attribute of the collectivity, it is only analytically distinguishable from population. People are the bearers of the organizational parts (that is, of its roles and functions). Indeed, when ecologists talk of organization they often resort to population terms to explain it. Similarly, the ecologist tends to define his population in organizational terms. Only to the extent that a population exhibits an internal structure is it analyzable as a coherent entity.

The third principal variable, *environment,* is the least well conceptualized of the variables constituting the ecological complex. It has been broadly defined as all phenomena, including other social systems, that are external to and have influence upon the population under study. Like *population, environment* is a generic term and must be empirically redefined for each separate unit of investigation.

Despite its loose conceptualization, environment is a variable of utmost importance in ecological analysis. As the sole source of sustenance materials, it either directly or indirectly sets limits on both the size and organizational structure that a population may attain. Under primitive circumstances, the immediate physical environment constitutes the primary external influence. However, as Hawley (1970:6) states:

> With the multiplication and spread of interdependences among residence groups the immediately local physical environment recedes in importance. The social environment assumes progressively greater significance.... Yet a population is never actually emancipated from the influence of physical or natural events. Rather its physical environment is broadened and rendered diffuse. It is increasingly mediated through social mechanisms of various kinds. And in becoming diffuse, accurate description and measurement of the physical environment of any particular group is made exceedingly difficult.

The fourth variable, *technology*, refers to the set of artifacts, tools, and techniques employed by a population to obtain sustenance from its environment and to facilitate the organization of sustenance producing activities (Duncan: 1959:682). Through the application of technology, populations are not only better able to adapt to their environment, but they are often capable of substantially modifying it. The epitome of environmental modification may be observed in the modern industrialized city, where an artificial environment has essentially been created by man's application of his advanced technology.

Because population, organization, environment, and technology are functionally interdependent, any permanent alteration in one of the four variables will result in repercussions in the other three. This characteristic gives the ecological complex substantial heuristic value for interpreting change and feedback within social systems. It is not surprising, therefore, that the majority of recent ecological studies have focused on explicating interrelationships among demographic, environmental, technological, and organizational variables within evolving social systems (see, for example, Bailey and Mulcahy, 1972; Bidwell and Kasarda, 1975; Frisbie and Poston, 1975, 1977; Kasarda, 1971, 1972a, 1972b, 1974; Micklin, 1973; Schwirian, 1974; Sly, 1972).

The concern of the ecological approach with social system growth and development may be seen in the contributions contemporary ecologists have made toward understanding the process of expansion. Briefly stated, expansion is a process of cumulative change whereby growth of a social system is matched by a development of organizational functions to insure integration and coordination of activities and relationships throughout the expanded system. The process involves an enlargement of the areal scope of the system; an accumulation of cultural elements (including information and technology); an increase in members of the population who apply the accumulated elements of culture; an elaboration of territorial division of labor; and a centralization of administrative and control functions. If system change is to be cumulative, it is necessary that culture, population, territory, and organization all advance together.

Expansion results in a territorially patterned system in which control (dominance) is exercised downward from the organizational nucleus through progressively smaller and more distant subcenters. The limits to which the system may expand are determined by its own productivity; its facilities for moving men, information, and materials; and by other social systems that prevent further extension of dominant relationships (Duncan, 1964; Hawley, 1971).

The emphasis that modern ecologists have placed on cumulative system change points out a very important aspect of the contemporary ecological approach: it is basically evolutionary in its perspective. Contemporary ecological inquiry seeks to understand the developmental processes of man's adjustment to his environment as well as the form it takes.

Modern urban geography

Just as the central problem of contemporary human ecology is understanding how a population organizes itself in adapting to a constantly changing yet restricting environment, the key problem addressed by modern urban geography is how and why perceptions of the urban environment produce locational and other choices that reinforce (or change) the spatial processes that are responsible for maintaining (or changing) the urban environment. Generalizing a sociological truism, the key ideas are that spatial structure operates on spatial behavior through screens of cognition, and is in turn created by the spatial processes that are made up of aggregates of spatial behaviors. Thus, just as human ecologists are concerned with cumulative system development, maintenance (equilibrium), and change, so are modern urban geographers.

In the contemporary geographic paradigm the environment is understood in the sense of an ecosystem—a functioning interaction system of living organisms and their effective environment, physical, biological, and cultural. Ecosystems, in turn, are viewed not only from the static perspective of their spatial form and areal differentiation, including the characteristics of places, spaces, and regions, but also from the perspective of spatial interaction and organization, and especially with respect to their structuring with respect to nodes and hierarchies, networks and gradients, cores and peripheries, boundaries and barriers.

Specific ideational components are embraced by means of the role that cognition plays in forming the mental maps of the environment that are the basis of environmental and locational decision making, thus permitting an intimate interweaving of social and geographic theory.

Finally, spatial behavior—the actions taken to adjust to or change the environment—contributes to space-contingent processes that maintain a given steady state, to space-forming processes producing evolutionary change, or to space-transforming (revolutionary) processes that, to complete the circle, radically transform the environment. The bases of convergence of sociological, ecological and modern geographic theory thus are both clear and straightforward.

Concluding remarks

From this it should be apparent that although human ecology focuses primarily on the form and development of system structure, contemporary ecologists (both sociologists and geographers) are well aware that formal structure constitutes but one dimension of social organization. Social organization also contains economic, cultural, spatial, and behavioral dimensions that must be incorporated into the ecological framework if a fuller understanding of man's adjustment to his environment is to be obtained.

Second, it is worth reiterating that the ecological approach does not deny that values, sentiments, motivations, and other ideational factors are associated with patterns of observable social activities. These ideational factors are important and should be studied in their own right as well as integrated with basic ecological variables when they are substantively significant. What earlier human ecologists have argued is simply that one can study many observable patterns of organized activities without *necessarily* referring to the subjective feelings of the individual actors. This, of course, is the essence of a macrosocial approach.

Finally, one might be wondering why publications proselytizing the ecological approach in sociology have not appeared for more than a decade. The answer is that their raison d'être no longer prevails. The polemical papers published during the late 1950s and early 1960s by Duncan (1959; 1961) and Schnore (1958; 1961) may be viewed as a response to a powerful social-psychological trend which had been occurring in the discipline of sociology. A perusal of the sociological literature of the 1950s indicates that it was during this decade that refined scaling and other psychometric techniques were having a substantial impact on social research; George Homans, Robert Bales, and others were making their significant contributions on interpersonal dynamics of small groups; and even the leading sociologist of the period, Talcott Parsons, had moved into psychoanalytic and personality studies.

It was thus almost overdetermined that a series of papers would be written by students of human ecology to remind the sociological audience that all principles of social organization could not be reduced to individualistic concepts. These papers also served to re-emphasize the importance of long-neglected "ecological" variables such as population growth, environment, and technology in explaining patterns and processes of social and spatial organization.

Since the early 1960s, sociologists have become more cognizant of the value of ecological variables and have widely applied them as both *explanans* and *explanandum* in their studies. At the same time, both human ecologists and urban geographers have become more aware of the limitations of their own approaches, and there appears to be an emerging interdisciplinary agreement that there is a continuing—indeed, growing—need for enriched studies that span disciplinary and subdisciplinary concerns for understanding social and spatial organization. It is to that bridging function that this book is addressed.

PART II
THE SOCIAL
CONSTRUCTION OF
LOCAL COMMUNITIES

Chapter 2
The congruence of
social and spatial structure:
neighborhood status and
white resistance to residential
integration as an example

The notion of a direct relationship between a city's residential structure and its status hierarchy is a venerable ecological principle, as Paul Hatt (1945) noted more than thirty years ago. The ecologist's traditional concern with this relationship stems from the observation first made by Robert Park that residential mobility matches social mobility: those climbing the socio-economic ladder improve their "residential status" to keep pace with improvements in their social status. Residential mobility thus operates to sort like-statused individuals into socially distinguishable subareas of the city. A sociospatial structure evolves in which urban residents and urban neighborhoods become ranked in congruent hierarchical systems with status serving to relate the "vertical" structure of social stratification, so central to sociological inquiry, to the "horizontal" structure of residential areas ranked by "exclusiveness" or "desirability," which has been so central to research in urban geography.

The ecological processes which produce and preserve the sociospatial structure of the city involve the dual opposing forces of 1) mobility and change, and 2) stability and resistance to change. In contrast to the emphasis placed on locational shifts associated with improved socioeconomic status, however, relatively little attention has been paid to the corollary process by which residents of local communities *resist* changes in their residential environments when they perceive these changes to be detrimental to their neighborhood's status.

Black penetration into previously white urban neighborhoods encom-passes both of these ecological processes, of course. Upwardly mobile blacks view the contested areas as compatible to their housing needs and rising socioeconomic status. The resident white population, having previously identified the area as congruent with their own social standing, tends to oppose black entry, which they view as leading to neighborhood status deterioration.

We can come to a better understanding of white resistance as a socio-spatial process by focusing renewed attention on the explicit relationship between the social phenomenon of status and the spatial dynamics of racial transition. This, in turn, should highlight the interdependencies between sociology and modern geography that come together in urban ecology.

Racial composition and neighborhood status

Race, in Everett Hughes's terms (1945:357), remains a "master status determining trait." This may be a sociological truism, but it is often over-looked by contemporary scholars, who tend to measure status primarily in terms of education, income, or occupational prestige. For the bulk of white American society, race continues to play an important role in ascribing group as well as neighborhood status, even after education, income and occupational levels are considered. Since blacks as a group are considered of lower status by many whites and, in large concentrations, are associated with residentially undesirable areas, the arrival of large numbers of blacks reduces a neighbor-hood's rank in residential status hierarchy for whites; the lower status of the neighborhood deters other white families from entering, and remaining whites suffer status loss since, prima facie, they live in a less desirable area.

Because neighborhood status is so affected by racial change, areas that attract large concentrations of blacks are typically unable to retain white residents regardless of positive physical features such as environmental amenities, accessibility to place of work, or high-quality housing. Racial composition inexorably dominates physical standards in white residential decision making. Wolf and Labeaux (1967:101–102), for example, conclude that

> the physical standards of an area seem to play a minor role in the residential decision of white households, at least in comparison with the importance of social characteristics of the area's population. There are, of course, a number of largely Negro-occupied areas in many cities which present an excellent appearance. However, this was not sufficient to maintain the racial mixtures which such areas did possess at one time in the past.

And Abrams (1955:139) has noted: "It is no longer the type of house but the type of neighborhood which reflects social standing.... Fine looking homes in Chicago may be still fine looking but are considered blighted when Negroes or other minorities live in the neighborhood." In other words, having the "right kind of neighbors" (i.e., not of lower social status) is a factor of major importance in white residential choice. But, as we pointed out, the racial composition of the neighborhood is a powerful element in perceived social standing. Thus, dwelling units once occupied by black families rarely revert to white occupancy (Taeuber and Taeuber, 1965), and whites tend to avoid purchasing homes adjacent to black households (Rapkin and Grisby, 1960). When an "integrated neighborhood" was defined as attracting *both* blacks and whites, a national survey found a median black population of only 3 per cent in such neighborhoods (Bradburn, Sudman, and Gockel, 1970). This finding not only calls into question much of the survey's subsequent analysis of "integrated" neighborhoods but also, and more importantly, graphically portrays the myriad of white decisions to shun residential integration on other than a token basis. It has become a fundamental ecological principle that, *ceteris paribus*, residential desegregation of urban neighborhoods typically leads to white avoidance and abandonment, racial transition, and eventual resegregation.

As the tide of residential resegregation rolls across urban areas with blacks and other minorities replacing whites, it has elicited a mixture of responses from residents of "threatened" neighborhoods, however. Whereas numerous experiments in integration maintenance are being conducted by liberally-oriented local community organizations, new forms of black containment are simultaneously emerging. It is with these processes that this chapter is concerned.

We begin with an overview of residential movements in the Chicago region both to dispel some myths about what is happening in metropolitan housing markets and to provide a framework for interpreting white resistance to racial change. A series of local community case studies are then reported to document organized white responses to the arrival of blacks in Chicago and its suburbs. These case studies cover a broad spectrum of neighborhoods with widely divergent populations and varying conditions of black entry. Finally, the implications are evaluated in terms of the extent to which national housing goals regarding increased residential options for minorities are being achieved.

The Chicago experience

Between 1960 and 1970, 481,553 new housing units were built in the Chicago metropolitan area, 129,496 in the central city and 352,057 in the six-county suburban region. Chicago's suburban population growth of 941,000 during that decade was the greatest in the nation, yet the number of households in the metropolitan area increased by only 263,609—168,971 white and 94,638 black—so that the ratio of new housing units to new families was 1.8:1.0. The result was a growing housing surplus that enabled a massive chain of successive housing moves to take place, culminating in abandonment in the oldest inner-city neighborhoods. Homes vacated by occupants of the new units were purchased or rented by other families seeking better housing, and so on in echelon—the chains of successive relocations filtering simultaneously down the scale of housing values and inward from the suburbs to the core of the city. One consequence was that many families improved their housing conditions dramatically during the decade, while downward pressure was exerted on the prices of older housing units (prices of older units increased more slowly than those for newer housing), and discriminatory pricing of identical units—blacks paying more than whites—was eliminated. A substantial improvement resulted in the housing condition of many among Chicago's central-city minorities; for example, over 128,000 units were transferred from white to black occupancy (these, plus 33,000 new units in the central city occupied by blacks, exceeded the growth in black families in the central city in the decade by a ratio of 2.0:1.0, which compares with 1.75:1.0 for the white community). Moreover, 63,000 of the worst units in the city could be demolished at the same time that thousands of additional undesirable units were being abandoned. The Chicago region thus provides a classic example of filtering mechanisms at work because of an important and frequently unrecognized fact: *surpluses* rather than deficits characterized Chicago's housing market in the 1960s.

However, this dramatic improvement in housing access still took place within the context of a dual housing market, as Tables 2.1–2.4 show. The massive growth of the suburban housing stock is presented in Table 2.1. In the decade, 352,057 new suburban housing units were constructed. But in spite of this, in the entire six-county metropolitan area, only 4,188 out of 223,845 new homes and 3,712 out of 11,290 new apartments were obtained by blacks, whereas an additional 3,208 black families purchased homes previously owned by whites, and 3,153 black families moved into apartments

Table 2.1

Changes in the suburban housing stock, 1960–1970

	1960 stock	New construction 1960–1970	Withdrawn from stock 1960–1970	Net change in stock 1960–1970	1970 stock
Total housing units	812,652	352,057	83,980	+268,077	1,080,729
Occupied housing units	740,508	335,135	29,841	+305,294	1,045,802
Owner-occupied units	561,170	223,845	26,275	+197,570	758,740
White owners	555,480	219,657	34,998	+184,669	740,179
Black owners	8,690	4,188	(3,208)*	+7,396	16,086
Renter-occupied units	176,338	111,290	576	+110,714	287,052
White renters	167,964	107,578	4,859	+102,719	270,683
Black renters	8,374	3,712	(2,153)*	+5,865	14,239

* Net increase over new construction caused by transfer of units from white to black occupancy.

previously rented by whites. Thus, in contrast to the net increase of 287,000 white families in suburban Chicago, only 13,261 new black families were able to obtain residences in the rest of the six-county area, most of these located in or contiguous to "mini-ghettos" located in the crescent of industrial satellites ringing the metropolis.

Table 2.2 indicates this fact clearly enough. Some 12,168 of the 13,261 new black families in suburban Chicago moved into or adjacent to the traditional ghettos of the crescent of industrial satellite communities ringing the metropolis (Waukegan, Elgin, Aurora, Joliet, Harvey, and Chicago Heights) or into those suburbs with long-standing black enclaves such as Evanston and Wheaton. The increase in black families elsewhere in suburbia was only 1,093, bringing the total from 1,217 to 2,310.

But even these numbers are illusory. Of this increase, 20 per cent was in two communities, Park Forest and Oak Park (which together absorbed an additional 976 black families in the next four years, 1970–1974). Only 569 of the remaining 2,084 black families lived in suburbs with over 2,500 residents in 1970; the remaining 1,515 families lived in smaller places or in isolated rural backwaters that have had black populations for many years. In 1970, 68 of 148 suburbs with more than 2,500 population had no black families in residence, 54 had a token 4 or less, 15 had between 5 and 9, 9 had 10 to 24, and only 2 had more than 50. There are so few blacks in suburban DuPage County that they have formed their own organization, the Black Suburbanites Club, to provide a social outlet and a basis for mutual aid amidst white indifference or enmity. Thus, the civil rights advocates and fair housing groups who, coordinated by the Leadership Council for Metropolitan Open Communities, concluded in their monitoring efforts that only 411 black families were able to find homes in otherwise white suburbs in metropolitan Chicago in the years 1959–1968—180 of these in Park Forest—were not far short of the mark. These 411 in the entire period 1959–1968 should be compared with the 976 these same groups estimate to have moved into Park Forest and Oak Park alone in the years 1970–1974, years in which the developing new town of Park Forest South became 20 per cent black. We will have more to say about this numeric change as the chapter progresses, for the question it raises is whether Park Forest and Oak Park represent integration breakthroughs or simply the beginnings of ghetto extension into suburbia.

Contrast the picture of suburban constraints with that of the central city shown in Tables 2.3 and 2.4. There was a net decline of 41,500 white homeowners and 76,900 white renters in the central city in the 1960–1970 decade. Net increases in the central city black population consisted of 37,669 new homeowners (more than doubling black home ownership in the decade) and 43,708 new renters.

The complex dynamics of white-to-black filtering that maintained racially segregated living in the city were as follows: some 128,829 units were

Table 2.2
Estimated change in black population in metropolitan Chicago, 1960–1970

	1960 Population		1970 Population		1960–1970 Black population change
	Total	Black	Total	Black	
SMSA	6,230,913	890,154	6,978,947	1,230,919	+340,765
Chicago	3,550,404	812,637	3,366,957	1,102,620	+289,983
Suburbs	2,670,509	77,517	3,611,990	128,299	+50,782
Mini-ghettos[2]	658,730	63,387	774,747	112,501	+49,114
Park Forest	29,993	8	30,638	694	+686
Oak Park	61,093	57	62,511	132	+75
Rest of suburban area[5]	1,920,693	14,065	2,744,094	14,972	+907

	1960 Households		1970 Households		1960–1970 Black household change
	Total	Black	Total	Black	
SMSA	1,897,917	250,327	2,183,046	344,965	+94,638
Chicago	1,157,409	233,263	1,137,854	314,640	+81,377
Suburbs	740,508	17,074	1,045,792	30,325	+13,261
Mini-ghettos[2]	185,675	15,847[3]	229,850	28,051[4]	+12,168
Park Forest	7,522	2[3]	8,440	196	+194
Oak Park	20,897	13[3]	22,620	30	+17
Rest of suburban area[5]	524,414	1,202	784,882	2,084[6]	+882

	Black households, estimated 1974	Net monitored black family unit move-ins 1959–1968[1]
Park Forest	526[7]	180
Oak Park	450[7]	24
Rest of suburban area[5]		207

[1] The first reasonably accurate actual counts of the *black* population of suburban Chicago were developed through 1966 by the Commission on Religion and Race of the Presbytery of Chicago, in cooperation with the Illinois Commission on Human Relations and Home Opportunities Made Equal, Inc., an affiliate of the American Friends Service Committee. In 1967 the Leadership Council for Metropolitan Open Communities took over the job of estimation. The counts (not a census) were made by contacting several reliable sources (both public and private) in each community in the metropolitan area to obtain and verify the information. They refer to black move-ins into predominantly white suburban neighborhoods, and they exclude peripheral expansion of the suburban ghettos as well as the city of Chicago.

[2] Includes Aurora, Chicago Heights, Batavia, Blue Island, Broadview, Crestwood, Dixmoor, East Chicago Heights, Elgin, Evanston, Geneva, Harvey, Highwood, Joliet, LaGrange, Lake Forest, Lockport, Markham, Maywood, Melrose Park, Midlothian, Montgomery, North Chicago, Phoenix, Robbins, Summit, Waukegan, Wheaton, Zion.

[3] Imputed figures derived by dividing the 1960 black population by the ratio of black population to black households for 1970 for each category.

transferred from white to black occupancy, allowing net increases over new construction of 28,008 in black home ownership and 20,267 in black rental of good-quality apartments. In total, the new housing inventory available to black families increased by 81,377 units, in spite of the fact that 63,000 units within the area of the 1960 ghetto were demolished in the decade, and finally, there was a net increase by 1970 of 17,554 units vacant in the expanded black residential area of 1970, many of these the precursors to abandonment.

What these figures show is important. As white families withdrew to the suburbs, black families gained access to large numbers of good-quality housing units in previously white areas. There, population increased and

---

[4] Published census data do not give data on number of black households for places of 2,500–10,000 population. Estimates used were black-occupied housing units for Broadview, Robbins, Phoenix, Highwood, East Chicago Heights, Dixmoor, Crestwood, and black population divided by persons per household for Lockport, Batavia, and Montgomery.

[5] Includes Addison, Algonquin, Alsip, Antioch, Arlington Heights, Barrington, Barrington Hills, Bartlett, Bensonville, Berkeley, Berwyn, Bloomingdale, Bridge View, Brookfield, Buffalo Grove, Burnham, Calumet City, Calumet Park, Carol Stream, Carpentersville, Cary, Chicago Ridge, Cicero, Clarendon Hills, Country Club Hills, Countryside, Crest Hill, Crete, Crystal Lake, Darien, Deerfield, Des Plaines, Dolton, Downers Grove, East Dundee, Elk Grove Village, Elmhurst, Elmwood Park, Evergreen Park, Flossmoor, Forest Park, Fox Lake, Franklin Park, Glencoe, Glendale Heights, Glen Ellyn, Glenview, Glenwood, Grays Lake, Gurnee, Hanover Park, Harvard, Harwood Heights, Hazel Crest, Hickory Hills, Highland Park, Hillside, Hinsdale, Hoffman Estates, Hometown, Homewood, Itasca, Justice, Kenilworth, La Grange Park, Lake Bluff, Lake in the Hills, Lake Zurich, Lansing, Lemont, Libertyville, Lincolnwood, Lincolnshire, Lindenhurst, Lisle, Lombard, Lyons, McHenry, Matteson, Marengo, Morton Grove, Mount Prospect, Mundelein, Naperville, New Lenox, Niles, Norridge, North Aurora, Northbrook, Northfield, North Lake, North Riverside, Oak Brook, Oak Forest, Oak Lawn, Olympia Fields, Orland Park, Palatine, Palos Heights, Palos Hills, Palos Park, Park City, Park Ridge, Plainfield, Posen, Richton Park, Riverdale, River Forest, River Grove, Riverside, Rolling Meadows, Romcoville, Roselle, Rosemont, Round Lake Beach, Round Lake Park, St. Charles, Sauk Village, Schaumberg, Schiller Park, Skokie, South Wauccnia, Westchester, West Chicago, West Dundee, Western Springs, Westmont, Wheeling, Willow Springs, Wilmington, Wilmette, Winfield, Winnetka, Winthrop Harbor, Wood Dale, Woodridge, Woodstock, Worth.

[6] There were 569 black households in suburbs of 2,500 population or more in this category. (See Footnote 4 for method of estimating black households in suburbs of 2,500–10,000 population.) The remaining 1,515 are assumed to be in smaller places and unincorporated areas. The suburbs of 2,500 population or more can be categorized by number of black households as shown in the following table:

Number of Black Households	Number of Suburbs
50+	2[a]
25–49	0
10–24	9[b]
5–9	15
1–4	54
0	68

a. Glencoe (125) and Highland Park (89)—both with longstanding black populations.
b. Downers Grove (21); Hinsdale (21); Glen Ellyn (18); Posen (15); Skokie (15); Des Plaines (13); Carol Stream (12); Wilmette (12); Elmhurst (11).

[7] Estimates of the Park Forest Human Relations Commission and the Oak Park Community Relations Department.

Table 2.3
Changes in the central city's housing stock, 1960–1970

	1960 Stock	New construction 1960–1970	Withdrawn from stock 1960–1970	Net change in stock 1960–1970	1970 Stock
Total housing units	1,214,598	129,496	134,988	−5,492	1,209,106
Occupied housing units	1,157,409	118,484	138,039[a]	−19,555	1,137,854
Owner-occupied units	396,727	33,745	34,115[a]	−370	396,357
White owners	360,117	24,084	65,609	−41,525	318,592
Black owners	36,610	9,661	(28,008)*	+37,669	74,279
Renter-occupied units	760,682	84,739	103,930[a]	−19,191	741,491
White renters	564,029	61,298	138,220	−76,922	487,107
Black renters	196,653	23,441	(20,267)*	+43,708	240,361

* Net increase over new construction due to transfer of units from white to black occupancy.
[a] These 138,039 units were demolished in the decade. Of the demolitions, 63,000 were in areas occupied by black residents in 1960, and 75,000 in white areas.

Table 2.4
Dynamics of Chicago's dual housing market

	White	Black
Occupied housing units in 1960	924,146	233,263
New construction 1960–1970	85,382	33,102
Demolitions 1960–1970	75,000	63,000
1970 Housing stock in 1960 market areas	934,528	203,365
1970 Housing stock in 1970 market areas	805,699	314,640
Transfers from white to black market 1960–1970	−128,829[a]	+111,275[a]

[a] Difference between these figures represents a net increase in vacancies in black residential areas of 17,554 units by 1970, a growing surplus of property associated with abandonment.

school enrollments escalated as black childrearing families replaced whites at later stages in the life cycle. In turn, the traditional south- and west-side ghettos were depopulated. The population in the 1960 area of the ghetto declined 19 per cent by 1970, and the communities in which population declines in the decade were greatest were also the communities with the greatest amounts of abandoned housing: Garfield Park, Lawndale, Oakland, and Woodlawn. Increasing the pressures on substandard private market housing that would otherwise have to be occupied by the welfare poor, 13,250 of 19,000 units of public housing built in the decade were located within the 1960 ghetto. Also manifesting the changed market conditions, Chicago's overall vacancy rate increased from 3.7 per cent to 4.6 per cent, with the rate exceeding 6 per cent within the 1960 ghetto area.

These numbers suggest several important conclusions. The stock of housing available to blacks increased rapidly after 1960, enabling them to upgrade their housing conditions and also causing large numbers of less desirable units in the worst central-city neighborhoods to drop out of the market. At the same time, the dual housing market has continued to restrict access of blacks to the total housing supply, and particularly to that of suburbia. On the one hand, filtering processes appear to have been working well (that is, except for the financially afflicted—the elderly poor, the welfare poor, the sick and infirm poor—who, if they do not or cannot occupy public housing, are unable to participate in normal housing market channels on any basis and are thus left the dregs, unwanted by any market participant). On the other hand, subdivision of the housing supply into racially segregated submarkets is as profound as it ever was. Although racial relationships are improving in other spheres of social and economic life, residential patterns in city and suburban neighborhoods reflect a continuing white reluctance to share residential space with blacks.

Yet white resistance to residential integration often entails severe personal and neighborhood upheaval, and the decision to move away often involves considerable financial strain. At the metropolitan level, the dysfunctional consequences of this pattern of behavior include continued racial tension, the abandonment of the cities by the white middle class, growing city-suburban disparities in resources and income, and continued residential segregation by race. The search for an understanding of white resistance to residential integration, then, is a challenge to explain a process that is flourishing despite severe dysfunctional consequences. In light of these consequences, what is, in fact, gained by such behavior? What is being protected or defended in attempts to exclude blacks from a neighborhood? What is at stake? What is feared?

As suggested earlier, we believe that neighborhood status is the basic factor. Since blacks continue to be labeled with the imputation of status inferiority and, in more than token numbers, are considered detrimental to an area's residential status, most whites oppose growing concentrations of blacks in their neighborhoods. In fact, for those whites concerned with preserving the status of their neighborhood, shared opposition to the intrusion of blacks is often the strongest communal bond in an otherwise unorganized locality.

To explicate white opposition to sharing residential space with blacks we next offer six case studies of white responses to the arrival and expansion of black populations in the Chicago metropolitan region. Emphasis in the case studies is on the extent and direction of *organized* response as a generalized expression of community sentiment. Three of the case studies relate to heavily ethnic central-city communities: West Englewood, in which Francis X. Lawlor, a Catholic priest and Chicago alderman, has organized a defensive network of block clubs; Garfield Ridge, where the community

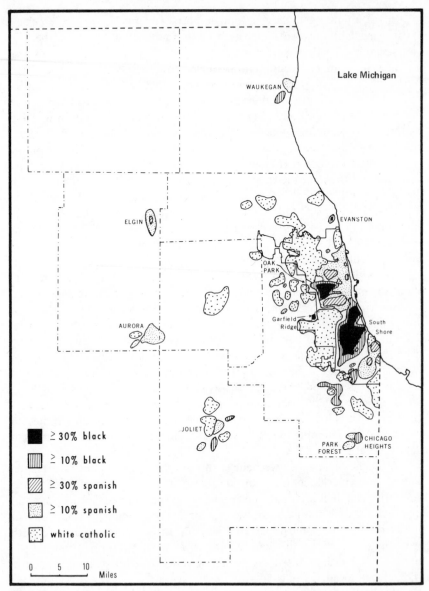

Figure 2.1
Black, white Catholic, and Spanish-speaking concentrations in the Chicago
metropolitan region in 1970.

organization recently joined a coalition of 19 northwest and southside groups
in charging that black public housing tenants threatened their communities
with "social pollution"; and South Shore, in which Molotch (1972) documented the failure of "managed integration." The other three cases are in

the suburbs of Evanston, Oak Park, and Park Forest, which tried by a variety of collective strategies to attain the sometimes conflicting objectives of containing black increase while maintaining viable open communities. Figure 2.1 maps the racial patterns of metropolitan Chicago in 1970 and locates the six case study areas.

West Englewood: a defensive network emerges

West Englewood sits astride the watershed between Chicago's southwest side blue-collar ethnic refuges and the growth of black residential areas westward from the main southside axis of ghetto expansion. The median family income in 1970 was $9,811, lower than the Chicago median of $10,242; housing is old, residential construction having virtually ceased since 1930. The median value of the community's homes was only $14,700 in 1970 and 47.6 per cent of all dwelling units were owner-occupied. For many years, a racial boundary line has been recognized along Ashland Avenue within the community, but this boundary came under increasing pressure as the black population in the area increased from 13 per cent to 50 per cent between 1960 and 1970. Finally, racial confrontation came to the fore as the Ashland line was breached north of Sixty-seventh Street, and blacks moved westward into neighborhoods heavily populated by Catholics of Polish, Irish, and Italian origins.

The public schools that serve West Englewood have been sources of major conflict and leading elements in community change. Overcrowding in black schools east of Ashland Avenue was "solved" by the Chicago Board of Education by changing attendance area boundaries and assigning black students to adjacent schools. Of the ten grammar schools serving the community, only two, in almost all-white census tracts, had low black enrollments in 1972: 8.7 per cent and 34.0 per cent. Black populations of the other eight schools ranged from 68.6 per cent to 99.9 per cent in that year. The three area high schools were, in 1972, 53.2 per cent, 85.9 per cent, and 96.1 per cent black. The result was growing racial sensitivity in the schools and along the Ashland Avenue line, compounded by block busting and panic peddling. Plans for location of dispersed public housing units in the area have been strongly opposed by West Englewood's whites, who fear it would bring low-income blacks into the area and precipitate even greater racial change. Fear of crime, declining city services, and the growth of "problem" businesses, such as taverns, have all played a part in the community's reactions.

Of the white community organizations that emerged in response to imminent racial change, three predominated. One, the South Lynne Community Council, had existed since 1957 as a liberal-oriented neighborhood improvement association. It expanded its activities in the 1960s to include a free rent-referral service and a center for rumor clearance. In 1963, the SLCC supported open housing, espousing "balanced integration" as preferable to inundation by blacks. SLCC's housing policy was that of encouraging white residents to remain along racial borderlines, aggressively recruiting new white families to move in, and assisting black families to locate in predominantly white areas west of the racial border. It vigorously fought panic peddlers, to little avail, and monitored property maintenance. SLCC battled for quality schools and pressured the Chicago Board of Education and the state for more financial support for education. The organization tried to welcome blacks into the community and prevent racial violence, but as one observer commented, it was "so successful that there are few whites left now to welcome blacks." SLCC is now largely black and concerned with post-transition problems.

The South West Associated Block Clubs, under the leadership of Father Francis X. Lawlor, was the most active local community organization operating in West Englewood. Father Lawlor organized the first block clubs in 1967 in defiance of the integration policies of the Catholic Archdioceses and, as a result, has been forbidden to perform his priestly functions. While SWABC was overtly a home improvement association, its implicit objective was to "hold the line at Ashland Avenue" by pursuing a community policy to exclude blacks from West Englewood until certain preconditions had been met. Primarily, these were the rectifying of what SWABC members believe to be inherent "socioeconomic, cultural" differences between blacks and whites. The black community was characterized, essentially, as riddled with crime, lacking respect for either persons or property, and completely "unprepared for home ownership." Until the black community solves "its own problems," Father Lawlor, who was elected to the political post of alderman in 1971, suggested buffer zones and black-imposed quotas for each block.

To organize white fringe neighborhoods, SWABC focused on three issues: crime, schools, and real estate practices. To offer protection against crime, the block clubs instituted nightly mobile patrols, and they regularly demanded increased police protection.

SWABC considered the Chicago Board of Education's integration policy its major enemy and has resorted to picketing to prevent its realization. "The Board of Education is the biggest block-buster in Chicago," Father Lawlor has been quoted as saying. Other enemies of stable housing according to SWABC are panic-peddling brokers, the federal government, Chicago politicians, and large developers. Hundreds of signs bearing the message "We are going TO STAY—down with FHA" could be seen in windows of homes west of Ashland Avenue in 1971, along with signs announcing "Jesus loves Father Lawlor." The city government is viewed by SWABC with even greater suspicion than the federal government. Says Father Lawlor:

> The powers-that-be have decided that the part of the city that counts is the lakefront, and they are determined to save it. The big realtors, the downtown politicians, the business community...are united in this situation. To them

the part of Chicago that matters is the high rises of Lake Shore Drive, the high income districts in and around the Loop, and the string of institutions on the south side running from Michael Reese Hospital to the University of Chicago. The rest of the city ? Let that go to the swiftest and the strongest, be he black or blue-collar white. Provided, of course, that their squabbles do not spill over into the enclave of civilization and high property values that the smart money boys have decided to save. (*Chicago Journalism Review*, February 1973)

One objective of SWABC has been to promote solidarity in the face of solicitations by real estate agents. As vacancies occur, SWABC has tried to fill them with "persons of the same cultural, social, and economic background as our own community," which has meant trying to prevent the sale or rental of properties to blacks. Thus the block clubs can be seen as little more than a continuation of the pre-1948 restrictive covenant policies.

The third organization operating in West Englewood is the Murray Park Civic Association (MPCA). Subsumed by SWABC, it includes some of Father Lawlor's strongest supporters and encompasses an area whose inhabitants are chiefly of Italian descent. Murray Park takes a reactionary approach to black encroachment. It has a history of racial violence: in 1971 four homes into which black families were moving were burned down. The year before, shots were fired and bricks thrown through the windows of a home owned by blacks in an all-white block. Earlier a storefront church was bombed on Easter Sunday, and there were three days of rioting around a house when its was rumored that blacks were planning to move in. Summed up the president of MPCA:

When you hear of an incident in this community of a garage being burned or a house being bombed, it's not because the people in this community are doing it for kicks. We're letting everyone know we mean business. (*Southtown Economist*, September 5, 1971)

Early black penetration of the Ashland Avenue dividing line was made into South Lynne, where less resistance was encountered than from Sixty-seventh to Eighty-seventh, the stronghold of SWABC and MPCA, although blacks have since moved across Ashland in this twenty-block strip. However, SWABC's success in "holding the line" since 1967 is largely attributable to the vigilance of the block clubs and the violent resistance to black move-ins in Murray Park.

In spite of these efforts West Englewood has experienced erosion of the remaining white community and block-by-block replacement by blacks. As the percentage of black students in the schools continues to rise, whites move away or temporarily enroll their children in parochial schools. A perceived lack of safety of the area is also a factor in precipitating flight. The critical question is when the situation is seen by the residents of any block as too serious to fight any longer. When that perception becomes widespread, the block not only undergoes racial transition, but undergoes it very rapidly— more rapidly than in other communities where turnover does not involve white flight, but is contained within the "normal" turnover sequence of the market.

Garfield Ridge: establishment of a defensive boundary

The establishment of a defensive boundary at Ashland Avenue in West Englewood may typify blue-collar response, for the same pattern has been repeated in many other central-city communities. One such is Garfield Ridge, a working-class community of single-family homes located on the western edge of the city of Chicago, north of Chicago Midway Airport and eight to ten miles southwest of the Loop. Small frame and brick homes are fronted by carefully tended lawns arranged along tree-lined streets, giving the community a suburban look. Garfield Ridge's 1970 population was 42,984, of whom 39,298 (91.4 per cent) were white and 3,686 (8.6 per cent) were black; the median family income was $12,603. Eighty-three per cent of the housing was in single-family units, and the mean value of owner-occupied homes was $22,800. The ethnic composition of the community is predominantly Polish, Italian, and other Eastern European: 90–95 per cent of the residents are Catholic.

The community differs from other areas experiencing racial change because it is a relatively new community located far from the central ghetto, and it has housed both black and white families for years. Blacks came to Garfield Ridge in 1951 with construction of the Chicago Housing Authority's LeClaire Courts public housing project. Although blacks were in the minority for the first five years after the project opened, half of the 615 families were black by 1956. The black percentage increased each year, and LeClaire has been more than 90 per cent black since 1964.

As has been the case with most public housing in white neighborhoods, there was substantial community opposition to construction of LeClaire Courts. The alderman at the time voted against the project in the City Council "because the people in the ward do not want it." The *Chicago Sun-Times*, editorializing in favor of public housing, asked: "What's your real reason for objecting to construction of a housing project in a sparsely populated area that private enterprise isn't the least bit interested in developing?" and the alderman later noted that "...I was able to get it cut down in size, but not to stop it" (Meyerson and Banfield, 1955).

In the tract within which the project is located, 98.5 per cent of all housing units in 1970 were built since 1950; the mean for the three adjacent tracts is 92.6 per cent, and the mean for the remainder of the tracts in Garfield Ridge is 54.1 per cent. In other words, at the time that LeClaire Courts was turning all-black, the surrounding residential area was being developed with homes which contained an all-white population. The extreme physical isolation of the Courts provides a partial explanation for this. To the north and west, access is terminated by the impassable barriers of railroad tracks, a six-lane expressway, the Sanitary and Ship Canal, and a vast industrial area. To the east, LeClaire is bounded by the six lanes of Cicero Avenue. Only to the south at Forty-fifth Street do the residents of the project have contact with the rest of Garfield Ridge.

In 1968, black families began moving out and buying homes in the blocks directly south of the Courts. As they did, the total black population of the community increased dramatically. Between 1960 and 1970, while the white population increased by 4.3 per cent, the black population increased by 32.9 per cent. Garfield Ridge thus was drawn into the confrontation between whites and blacks over housing, and today exhibits the full range of problems associated with racial change; the fabric of daily life is influenced by the presence of racial conflict. Whites now have established a barrier at Forty-seventh Street between white and black residential areas; most of the homes to the north of this dividing line have been purchased by black families, and the area has been abandoned by the white community. In 1970, the ward boundary was redrawn at Forty-seventh Street, removing the area north of Forty-seventh Street from the Twenty-third Ward, to which it is contiguous, and annexing it to the Twelfth Ward, from which it is separated by extensive railroad yards and industrial facilities. School attendance boundaries have been redrawn along Forty-seventh Street to further the de facto segregation.

Located adjacent to the black residential area north of the Forty-seventh Street barrier, residents of the area known as Vittum Park feel themselves to be the most threatened. "The community is like a tinder box...a bomb about to explode," said a resident. Meetings of the Vittum Park Civic League are closed to nonresidents and the press. Emotions at meetings run high, and several have ended in disorder when residents disagreed over policy or method. Homes that come up for sale in Vittum Park are listed with the Civic League and are not advertised through normal real estate channels. White respondents outside of Vittum Park estimate that up to a hundred homes there are available for sale in the course of normal turnover, but suggest that the Civic League is unable to find buyers.

Although white residents in the rest of the community feel less threatened than those in Vittum Park, racial tensions are high. Continued controversy over scattered site public housing has kept the issue in the forefront of attention:

> These liberals come on with these beautiful programs that are going to integrate us.... How are you going to integrate when they just inundate us.... Their kids got better bikes than we got. Why? They come across 47th Street and steal them. (*Chicago Sun-Times*, June 6, 1973)

> I want the privilege of living in an all-white neighborhood. (*Southwest News-Herald*, June 8, 1972)

> We don't want no other form of people to come and burn our houses and take away our families and tear them up.... I don't want no neighbor throwing no

garbage and barbecuing no beef out in the street. (*Chicago Daily Defender,* June 7, 1972)

The white residents of Garfield Ridge share close ties to the community and are vehemently concerned with protecting neighborhood status. They feel that they have "made it," that Garfield Ridge is a community that satisfies their status aspirations, at least for the present. The competition between races in Garfield Ridge is clearly defined in terms of territory. Because of the absolute refusal of whites to share residential space with blacks, white residents continue to leave Garfield Ridge in step with the continuing arrival of black families. Even a small infusion of black families in an all-white block results in the rapid white evacuation of the area. Thus, while the eventual transition of the community depends on the rate of growth of the black population, it is clear that only a small increase in the number of blacks is sufficient to place the entire area within the boundaries of the black housing market. The vehemence of white opposition ironically serves to speed transition in Garfield Ridge.

South Shore: the failure of managed integration

South Shore, a community of ninety thousand residents, lies ten miles south of Chicago's Loop, and has been undergoing racial change since the early 1960s. In 1960, South Shore was a solidly middle-class residential community inhabited by a relatively well-educated white population. There were indications, however, that South Shore was no longer a choice area for white middle-class residence. It possessed, according to Harvey Molotch (1972), many of the traits of an area likely to undergo racial change.

By the mid-1950s it was apparent to some South Shore residents that black in-migration was a distinct possibility in the near future, as the ghetto approached from the north and west. In 1954, a priest, a minister, and a rabbi decided to use the South Shore Ministerial Association as the basis for a more extensive community organization, from which the South Shore Commission emerged.

The organization's principal concern was how to contain ghetto expansion. Initially a split existed between the exclusionists, who wanted to prevent black entry and ensure that South Shore remain white, and the integrationists, who believed that the middle-class character of the community could be maintained by controlling the balance between white and black residents. By 1966, the integrationist viewpoint prevailed, but a growing proportion of South Shore's northwesternmost population was already black, as the tide of ghetto expansion moved in from adjacent Woodlawn.

Although the South Shore Commission described itself as a democratic, grass-roots organization, its middle-class character was always apparent. Low-income and blue-collar people were not represented in the leadership. Blacks were represented on the Commission Board and Executive Committee only after 1963, but remained underrepresented through 1967. They did not play a major role in Commission activities until after 1969.

The goal of the South Shore Commission throughout the 1960s was to make South Shore a community in which white middle-class individuals and families would choose to live. The strategy had two basic parts: (1) to create neighborhood and community conditions felt to be desirable to whites; and (2) to facilitate white move-ins by recruiting white, middle-class residents through actual intervention in the real estate market.

The first strategy involved tactics aimed at crime prevention and the improvement of education, housing, and other community amenities. The commission responded to a dramatic increase in serious crime between 1964 and 1966 by instituting citizen-manned radio patrols and pressuring police and courts for more arrests and convictions. The South Shore School-Community Plan was developed to give South Shore priority over other school districts in new educational facilities and programs, including authorization for a second high school (even though other southeast Chicago schools were significantly more crowded than South Shore High), a program for high school computer science training, creation of an evening junior college, and establishment of reading clinics and remedial classes. The South Shore Commission also served as a watchdog over building maintenance and code enforcement.

The most important phase of the housing market intervention strategy was the Tenant Referral Service. Using listings provided by local landlords and rental agents, the Commission took applications for apartments and referred applicants to vacancies. Referrals were made with the object of moving whites into integrated buildings; in other words, "reverse steering" was practiced. On the other hand the Commission did not want to refer blacks to a mixed building if white tenants could be found, and it was also reluctant to refer blacks to buildings that were all white. An attempt was made to increase the number of white applicants by advertising in local newspapers serving the nearby Hyde Park community, where the University of Chicago is located. Similar efforts to reach black prospects were not made. In fact, the Commission hoped to discourage black occupancy by cooperating with the Leadership Council for Metropolitan Open Communities, Chicago's major fair-housing group, in the operation of a neighborhood Fair Housing Center, which attempted to make black homeseekers aware of housing opportunities in all-white communities elsewhere in the metropolitan Chicago area. This center operated with little success from 1967 until it was terminated in 1969.

The Tenant Referral Service served a total of 750 applicants, 675 of whom did locate in South Shore. Two hundred of the 675 were black. The TRS became a major force in the local real estate industry. Landlords and agents were asked to hold units vacant until white renters could be found, and those who did not cooperate could be excluded from TRS or given the least desirable prospects. However, as more blacks came to the community, white demand sagged and black demand grew. TRS may have helped the market operate more efficiently by unifying listings, but it could not significantly affect the pattern of transition. By 1970, 57 per cent of South Shore's population was black, and by 1974, 80 per cent; the goal of the South Shore Commission had changed to the maintenance of South Shore as a stable middle-class *black* community.

Evanston: a traditional ghetto expands

Evanston, one of Chicago's oldest suburban towns, lies 13 miles north of the Loop on the luxurious North Shore. The community's early growth was linked to the commuter railroads and to Northwestern University, founded by the Methodist Episcopal Church in 1851. During the 1860s, Northwestern sold land at low prices to attract new residents. Purchasers were warned the land would revert back to the university if the owner permitted "intoxicating drink to be manufactured, sold or given away on the premises," or allowed "any gambling to be carried on." Evanston became the national headquarters of the Women's Christian Temperance Union, and, until 1972, no liquor was sold within the city limits. Many churches settled in the community and are still dominant in its landscape. Termed by its residents at various times, the "City of Gracious Living" and the "Athens of the Middle West," Evanston always has projected an aura of intense respectability.

The first blacks settled in Evanston in the 1800s and were trades- and craftsmen. After World War I, the opportunity for blacks in domestic service swelled the size of the black population. Evanston's total 1970 population was 79,808, of whom 12,861 (16 per cent) were black. Although some blacks live in predominantly white neighborhoods, and one area apparently has been stably and substantially mixed for over a decade, Evanston, on the whole, has been composed of two divided but coexisting communities. Throughout its history, white Evanston largely ignored black Evanston, reacting to it only in periods of real or fancied crisis.

Natural boundaries have maintained Evanston's social and ethnic distinctions on a geographic basis. Blacks, foreign born, and Jews have been concentrated in a roughly triangular area on the southeast side of Evanston, which covers approximately one half of the city's land area and is separated from the affluent lake front and northwest Evanston neighborhoods by the Chicago Sanitary District's North Shore Channel and the Chicago and Northwestern Railroad tracks. In 1900, the northern tip of this triangle formed the nucleus of Evanston's black ghetto, which has expanded southward since. More recently, the core of a second ghetto has developed along the eastern edge of the triangle.

Thus, in 1970, there were two vectors of black expansion in Evanston, south and southeast from the old black community and northwest from the newly forming ghetto. Both of these areas exhibited signs of the classic transition pattern, with blocks of high black occupancy rates flanked by blocks of rather abruptly decreasing percentages of blacks. This drop in black population in "fringe" blocks, however, was not as sharp as is typical in transitional neighborhoods in Chicago. Together with the presence of a few blacks scattered in white, nontransitional neighborhoods and the existence of one long-standing integrated neighborhood, this fact may account for the apparent perception on the part of some white residents that true integration rather than transition is taking place. This perception, however, is not shared by whites living in the neighborhoods close to the black residential areas.

In addition to the threat of continued ghetto expansion, Evanston has been confronted with declining white demand relative to black demand in the suburb as a whole. Between 1960 and 1970, the white population decreased as the black population grew. The aging of the housing stock and the increasing dependence of the city on residential property tax revenues, as nearby shopping centers cut into Evanston's share of the North Shore retail dollar, jeopardized the city's attractiveness to whites, whereas for blacks it still represented an opportunity for a desirable, middle-class suburban life style, with good schools and the other amenities which that implies.

Further aggravating the situation was evidence of racial steering on the part of North Shore real estate brokers. In July 1972, after nine months of investigation of real estate practices in Evanston and other North Shore suburbs, the Evanston Human Relations Commission released findings which showed discrimination taking much the same form as it had for a half century or more. One broker checked a buyer's choice of listings to make sure they were not located in a "Jewish, colored, or student area." Most salespeople assumed that the critical factor in the buyer's choice would be the racial or religious composition of the neighborhood. In general, whites were steered to houses in north and lakeshore Evanston or out of Evanston entirely, while blacks and Jews were shown houses in the triangle.

The threat perceived by white residents stemming from the expansion of Evanston's ghetto and the potential decline of white demand for housing in Evanston has been complicated over the past decade by increasingly vocal dissatisfaction and demands for social equality emanating from its black community. The black–white struggle has been played out most dramatically in the schools.

In 1966, Evanston's elementary schools were voluntarily desegregated through a combined strategy of redistricting and busing of black pupils. As controversial as school integration was, it appeared to have majority support of both white and black citizens. The real conflict arose around the new superintendent, Dr. Gregory Coffin, who had been hired specifically for his record on integration to implement the school integration plan. Friction developed between Dr. Coffin and the school board, which the board alleged was due to Coffin's failure to cooperate with and, in fact, deliberate deceit of the school board. However, to his supporters, which included the majority of Evanston's blacks, Dr. Coffin had become the symbol of a promise finally kept. When the school board voted in June 1969, to fire Dr. Coffin, Evanston was torn apart: the real meaning of the battle went far beyond the mere issue of schools, raising the issues of power and representation in Evanston's social and political life generally. The Coffin controversy blew the lid off the long-standing splits between liberal and conservative, black and white, "old Evanston" and "new Evanston."

In 1973, after six years of desegregated education, at least two serious problems remained. One was the old and unresolved issue of black representation and participation. The other involved the reciprocal relationship of school racial composition and neighborhood transition. In 1973, the school system was still substantially integrated, but the district was faced with the state-imposed necessity of further redistricting as the black enrollment of one school approached 50 per cent. Continuing neighborhood transition will mean that redistricting or other strategies for maintaining school integration must be an ongoing process; and continued redistricting, in turn, has the effect of speeding white flight. A May 1972 report by the U.S. Civil Rights Commission concluded that Evanston's school integration program had been a success, but said of the decline of white enrollment that

> some...may be attributable to such factors as birth rates, but several school officials view the decline as white flight from newly desegregated districts. (*Evanston Review*, February 1, 1973)

Racial conflict has become an increasingly open problem in the high school as well where, despite an extremely high black dropout rate, almost one fourth of the student population is black. Yet a relatively low level of organizational response to racial change is noticeable. In 1963, the West End Neighbors Organization, serving the area just south of the ghetto, formed "as a result of occupancy changes." The group held discussions among residents, pressured real estate brokers and the *Evanston Review* to refrain from listing housing on a racial basis, and proposed an ordinance for non-discriminatory advertising of housing. By 1970 its efforts had failed, for the neighborhood was then largely black. Other neighborhood organizations followed with similar tactics, but although there was a proliferation of groups involved with the issue of racial change on a local neighborhood basis, their efforts were fragmented and appear in most cases to have amounted to too little, too late. Nothing even resembling a concerted, community-wide attack on the problem of neighborhood transition existed until 1973, when some local activists united in an informal coalition to put pressure on the real estate industry to end racial steering. No program for stabilization has yet come from the city government, from Evanston's central institutions, or from the grass-roots level. Meanwhile, block-by-block expansion of the ghetto continues in the triangle.

Oak Park: reverse steering and quota systems

A more active response has come in another old Chicago suburb, Oak Park, which, despite its size (62,506 in 1970) and location (bordering Chicago, nine miles west of the Loop) has maintained the ambience of a respectable and quietly affluent suburban town. Solidly built up in the 1920s, its increasingly heterogeneous population and the rapid development of Chicago's western suburban fringe contributed to Oak Park's decline from the ranks of the elite. However, its residents remained assured of a high quality of life in the community, a quality sustained by such things as its symphony orchestra, its repertory theater, and an ample lecture schedule. Schools and public services are reputedly good, transportation to the Loop via rapid transit and an expressway is excellent, and the housing stock, despite its age, is sound.

From the viewpoint of its residents, Oak Park's main liability is that it is separated only by Austin Boulevard from Chicago's large expanding west-side black ghetto. By the summer of 1973, Oak Park had over a thousand black residents (up from 132 in 1970), who tended to be concentrated in neighborhoods closest to the Chicago ghetto. Only one block was over 50 per cent black, however, and nearly 40 per cent of the black population lived in areas further removed from the ghetto.

As in most white suburbs, blacks found it nearly impossible to buy or rent homes in Oak Park until the mid-1960s. A very few had obtained homes through white "straw buyers." However, a strong fair-housing group, the Oak Park–River Forest Citizens' Committee for Human Rights, had emerged in the early 1960s, and it succeeded in 1968 in persuading the village Board of Trustees to pass a fair-housing ordinance. The law, which was considered a strong one, seemed to be effective in controlling brokers' practices. Most apartment complexes, however, still excluded blacks. In the two years between the passage of the fair-housing law and the 1970 Census, the black population of Oak Park rose only negligibly, in spite of the considerable efforts by Citizens' Committee to attract black residents.

However, Oak Parkers generally did not welcome black neighbors, and the fair-housing law brought angry protests. Black families who did move into Oak Park suffered harassment and ostracism but no serious physical attacks. Overt violence was not Oak Park's style, and in this the village was sharply contrasted with its neighbor to the south, Cicero. Blue-collar ethnic Cicero also adjoins the Chicago ghetto, and some blacks who have unwittingly ventured across its corporate limits have been beaten or shot. The contrast between Oak Park and its neighbor, together with the fact that "liberals" dominated the village's political process, gave Oak Park a reputation for being "open."

Oak Park residents feel that the openness of their community makes it the most likely target for expansion of the ghetto, and the possibility that large sections of Oak Park will turn entirely black is perceived as a great threat. Furthermore, the liberal ideology of Oak Park's political leaders has posed the moral and practical dilemma of how to maintain an open, non-discriminatory housing market without white flight.

The solution was thought to be a strategy of "dispersal," that is, encouraging blacks to move into parts of Oak Park in a pattern that would minimize black concentration. Beginning in 1961, real estate brokers were asked to refer black clients to a "counseling program," the objective of which was to discourage blacks from buying in the southeast side of Oak Park abutting the Chicago ghetto, where blacks were tending to cluster. It was also intended as a means of monitoring real estate practices. The flaw in the plan, of course, was its completely voluntary nature.

The Oak Park Housing Center, a private no-fee housing referral service, opened in May 1972, with the goal of maintaining and promoting integration by attracting new white residents to the village. The Housing Center's advertising was directly aimed at a young, liberal population, and in its first year of operation about 85 per cent of its clients were white. The Housing Center engaged in "reverse steering." White clients were encouraged to locate in the southeast sector of Oak Park, while blacks were not shown listings in this "sensitive" area. The Housing Center had no formal connection with the village government, but the two worked in close cooperation, and the Housing Center activities were supported informally by village officials. The Housing Center thus represented the first instance since the passage of the fair-housing law in which the village establishment explicitly condoned some form of discrimination in the sale and rental of housing.

In April 1973, the Village Board passed a policy resolution calling for the maintenance of stable "dispersed integration." The Board had come to see serious limitations on the effectiveness of voluntary plans for dispersed black residency. Although the village authorities had condoned discrimination by the Housing Center, they did not officially sanction it until November 1973, when the "exempt location" clause of the fair-housing ordinance was applied for the first time. This provision, written into the 1968 fair-housing ordinance, allowed the village government to exempt areas from prosecution if it was determined that serious attempts at integration were being made and that their success might be jeopardized by strict enforcement of the fair-housing ordinance. The Board granted exemption to one block and one apartment building, both over 50 per cent black, when petitions signed by over half the white and black residents were presented. This action by the Board did not prevent black move-ins; it merely permitted racial discrimination in the two exempt locations. Soon after the Board's action, however, another step that would legally prevent black move-ins to certain areas was receiving serious consideration.

In December 1973, an amendment to the fair-housing ordinance, putting into operation a quota system, was proposed by a trustee and referred to the Community Relations Commission. If enacted into law, this quota provision would hold the black population of a still unspecified area of the village to 30 per cent. Public hearings turned up much opposition to the new quota plan, though it is probable that a majority of residents, most of whom do not attend such hearings, would have supported it. The quota was finally tabled, and apparently killed, primarily as a result of the extreme adverse publicity the plan received from the metropolitan news media. In May 1974, the village board began discussion of a multifaceted, but quota-less, integration maintenance program. However, selective discrimination has already replaced nondiscrimination as official policy.

Park Forest: a program of fair housing maintenance

Park Forest, the setting for William Whyte's *The Organization Man*, gained nationwide fame as one of the first "planned" communities in the Chicago area, and is now taking the most affirmative steps towards integration maintenance. Developed in the late 1940s, Park Forest won the 1953 "All American City" award given by the National Municipal League and *Look* magazine. The 4.5-square-mile village is located in the southernmost part of Cook County, thirty miles from Chicago's Loop. Anticipated industrial development in the village did not occur, and Park Forest has remained a dormitory suburb of Chicago. The 1970 Census showed that Park Forest had a total population of 30,638, of which 694 (2.3 per cent) were black; later surveys showed that by 1973 black population had risen to 7 per cent, scattered throughout the village.

Early integration efforts were begun by Park Forest residents acting independently of any organization. Whites bought houses in the village and transferred them to blacks contacted through fair-housing groups and black churches in Chicago. The sole criterion was economic: if an individual could afford housing in Park Forest, he could move into it. The first black move-in occurred in 1959, but "managed" integration proceeded slowly: in 1965 the village reported only 18 black families living in single-family homes and 19 in rental units. However, a great deal of effort was necessary to absorb peacefully even this limited number of blacks. As each move-in occurred, the village government and the Commission on Human Relations circulated memoranda to Park Forest clergy, rabbinate, and real estate brokers giving information about the family and the location of the property. It was hoped that brokers would concentrate on selling surrounding houses to whites to avoid clustering. (This was done until 1968 when federal civil rights laws made such tactics illegal.) When a difficult move-in was anticipated, members of the Human Relations Commission visited the neighbors involved. Nevertheless, there were black-white confrontations: epithets were hurled, a cross was burned near a home owned by a black, a fence facing a black family's house was painted black on one side and white on the other. One successful occupancy was described as "moved in today; no cakes and no problems."

The community remained stable during the 1960s and continued to attract white families. In 1969 the *Park Forest Star* noted that

> the village is now home to an estimated 160 Negro families, making it the most integrated of Chicago's predominantly white suburbs.... In January, 1968, the village adopted an open housing ordinance deemed to be among the strongest in the state.

As Park Forest tried to maintain an open-door policy, however, outside factors came into play that were to threaten its future as an integrated community. Adjacent to Park Forest and included within its School District 163 lies a section of suburban Chicago Heights whose proximity has had profound implications for Park Forest. Beacon Hill–Forest Heights is an area of over five hundred homes, 87 per cent of which are black-occupied. The land upon which Beacon Hill–Forest Heights is built was once owned by the developers of Park Forest, American Community Builders, and was intended for development as an industrial park. In 1952 when Park Forest consolidated its four school districts into one, this area was included in the annexation in the hope that future industry would augment the district's tax base. In 1959 when it became apparent that no industrial development was likely to occur, the land was transferred to Andover Development Corporation and rezoned and platted for about 550 houses. In 1960 Andover Development began building in Beacon Hill but sold out in 1962 to United States Steel Homes Division. By 1963, 270 homes in the $17,000–$18,000 price range had been constructed. Although all of the original purchasers were white, some of the houses were abandoned, placed on FHA foreclosure lists, and several were purchased by black families.

In 1970, Kaufman and Broad, the nation's largest homebuilding company, began to develop the land in Forest Heights to the east of Beacon Hill. By 1971 some 270 low-cost, poorly-constructed houses had been built (subsequently, the company's representatives were convicted of a variety of federal offenses in connection with this and other low-cost housing developments in the Chicago region). The FHA allotted the Forest Heights' development 255 Section 235 mortgages—about 95 per cent of the total development. Thus homes could be bought for as little as $200 down and $135 a month by persons eligible under HUD criteria. Although the development was advertised as integrated, the majority of the homes were sold to blacks. The marketing was done on a racial basis; the bulk of Kaufman and Broad's advertising was placed in the black media. By 1972, Forest

Heights was 99 per cent black, while adjacent Beacon Hill's black population rose to 87 per cent. In short, Beacon Hill–Forest Heights blossomed into a suburban mini-ghetto.

Between the end of the 1970 and the beginning of the 1971 school year the Beacon Hill grammar school, built originally for 350 students, changed from 25 per cent to 90 per cent black, and its enrollment soared to 700. Black enrollment in Blackhawk and Westwood Junior High Schools increased from almost zero in 1969 to 15 per cent in 1971, and these two schools were the first to experience racial conflict. More serious racial disturbances occurred at Rich Central High School in Olympia Fields where Beacon Hill–Forest Heights students are assigned after completing junior high in Park Forest. Two hundred and fifty Beacon Hill–Forest Heights students, many from families receiving welfare, were mixed in with classmates from a community with a median income of $30,000 a year.

The situation in the area's junior high schools and high school provoked a strong response from Park Foresters, but not as heated as that called forth by the decision to desegregate the District 163 elementary schools. In 1970 none of Park Forest's 11 elementary schools fulfilled the racial guidelines of 80 per cent white to 20 per cent black stipulated by the State Superintendent of Education. While Beacon Hill School was almost totally black, the other schools averaged 7 per cent black. In April 1972 the Board of Education adopted a voluntary magnet school plan whereby Beacon Hill School would have become the district's laboratory school, which students would have voluntarily requested to attend. Mandatory grade reorganization was accepted as the backup plan. The magnet school failed to attract sufficient volunteers, and the grade reorganization plan had to be implemented. In September of 1972 about 1,500 of 4,600 students in Park Forest were bused from their neighborhoods. Busing has been carried out in an orderly fashion, but the attitude of many Park Foresters has been ambivalent.

The greatest amount of criticism concerning busing has come from residents of the Eastgate area of Park Forest, the area closest to Beacon Hill Forest Heights. The majority of children assigned to Beacon Hill School are from this area, which has the least expensive housing in Park Forest. Real estate brokers had told residents the value of their property would decline because of the school assignments. Many whites wanted to sell their houses, real estate firms began steering whites away, and several black families moved into one section of Eastgate. At this point, the Eastgate Residents Association (ERA), organized in 1963 to protest a proposed highway and since restructured to encourage home maintenance and beautification of Eastgate, met with village officials and local real estate interests. A representative of ERA stated:

> In recent months there has been a new turnover of houses, and many new families have moved here...in this instance we do not feel that "clustering" is in our best interest. With the housing market so short of low-income housing,

and our homes being the lowest priced in Park Forest, we are particularly vulnerable at this time to become another racially dominated area such as Beacon Hill-Forest Heights. The fact that our village practices "open housing" and surround villages who apparently do not do their share makes us more vulnerable than ever. (*Park Forest Star*, May 11, 1972)

A study conducted in 1972 by ERA revealed that residents of Eastgate were disappointed with the school situation and busing program. According to the survey, if things do not change for the better, many may move from Park Forest.

Paralleling the Eastgate residents' concerns, other Park Foresters began to feel threatened by the influx of blacks. Since Beacon Hill–Forest Heights has no stores, its residents shop primarily in Park Forest. Blacks thus became highly visible in the community, and some whites began to shun Norwood Plaza shopping center, a small complex situated closest to Beacon Hill–Forest Heights, because they felt it had begun to take on a "ghetto-like" look. In cooperation with the Human Relations Commission, the village felt it necessary in 1971 to establish a rumor control telephone service to alleviate tense conditions. Some of the occupants in the inexpensive cooperative units south of the Eastgate area became apprehensive as move-outs occurred; they also felt very vulnerable to black inundation. 1970 Census figures show that ten courts were, at that time, from 4 per cent to 19 per cent black.

While the Beacon Hill–Forest Heights situation and the busing of school children created dilemmas for Park Forest, another, perhaps graver, threat to the community began to surface. Because Park Forest had been well publicized as an integrated community and many of the suburbs around it are known to be highly resistant to black immigration, increasing numbers of blacks began to seek homes within its confines. At an April 1973 meeting at the Village Hall, one real estate man remarked, "We are getting more and more blacks—almost more than whites—in the offices."

This particular meeting had been initiated by some residents of the Lincolnwood area of Park Forest, where the highest-priced homes are located, to protest the "clustering" of blacks on certain streets in their area. One resident said, "There is a difference between integrated housing and clustering. No one has objections to colored"; another remarked that

I once approved the rate of integration here, but it is getting reckless now.... Park Forest is a soft touch for colored.... The community has made a big mistake in publicizing integration so we have been attracting more than our share of what we are trying to get away from.

It was suggested that the village should provide by ordinance a definition for resegregation and make it illegal for real estate firms to sell property to blacks within a geographic distance as spelled out by village code; also that the village should supply real estate firms with minority residents' addresses. However, two days after the meeting the village attorney rejected both suggestions as violations of both the State and Federal Constitutions.

Park Forest confirmed its position as an equal opportunity community

by passing into law on November 26, 1973, a package of "integration maintenance" ordinances calling for

1. Prohibition of "steering" prospective home buyers into or out of
   neighborhoods on the basis of religious or racial composition.
2. The establishment of a Fair Housing Review Board, a quasijudicial body
   with the powers of subpoena to revoke brokers' licenses, and to refer
   complaints to court.
3. Reorganization of the Human Relations Commission establishing for it
   an advocacy role and granting it subpoena powers.
4. Authority of the Village Manager to take administrative action on housing
   discrimination complaints before referring them to any other bodies.
5. Prohibition of "redlining" by lending firms.

These ordinances back up the efforts of the village government to
eliminate discrimination. It has been characteristic of the village's official
response, since the first fair housing ordinance was passed in 1968, that the
emphasis has been on enforcing the constitutional right of minorities to
housing of their choice, but stopping short of attempts to "manage"
integration by direct intervention in the locational choices of individuals,
despite considerable external and internal pressures to do so. Suggestions
that there be legal restrictions on "clustering" or pressure on real estate
brokers actively to discourage it have been rejected. A proposed "exempt
location" provision in the new set of ordinances, which would have permitted
"steering" for the purposes of maintaining racial balance in certain areas,
created a furor among black residents and a controversy among whites, and
it was dropped from consideration. Nor have village officials been persuaded
by popular sentiment and the exhortations of real estate people to maintain
a lower profile with regard to open housing.

For Park Forest officials (as implied by their use of the term *integration
maintenance* for their new laws), the assumption seems to be that integration
can occur naturally in a housing market which is *free* from racial bias. If fair
housing brings with it some perils, the solution is more fair housing.

Contrasting attitudes: "No-CHA" vs.
the Regional Housing Coalition

The behavior represented by these case studies is summarized in two
broad-scale organizational responses to the region-wide pressure for racial
integration. The actions of a coalition of locality-based northwest- and southwest-side central-city white community organizations contrast with the
program of a regional coalition of suburban fair housing groups.

No-CHA. In the early years of the nation's public housing program,
the Chicago Housing Authority (CHA) sought to build public housing scattered throughout Chicago. The City Council, in the belief that most prospective public housing tenants were black, proposed, however, to restrict such

housing to ghetto areas. Since each alderman had a veto over site selection in his ward, the CHA had to go along with the City Council, if any public housing was to be constructed. Chicago's public housing was constructed and rented on a racially segregated basis.

As a result, the CHA administered regular (family), elderly, and Section 23 (leasing) public housing to keep blacks out of white areas. Regular public housing was located in ghetto neighborhoods to minimize the number of blacks displaced by slum clearance; on the other hand, the related programs were located in all-white neighborhoods. In the four housing projects in white neighborhoods, quotas kept the number of black tenants at zero or a minimal level. In projects for the elderly, tenant assignment policies ensured that white elderly occupied most of the housing in white neighborhoods. Only in racially changing areas were "integrated" projects—those with a second racial group of more than 10 per cent—to be found, and these projects and areas ultimately became all black. Leased housing was treated in the same fashion. Tenant selection was delegated to landlords by the CHA, giving them the right to refuse tenants because of "undesirability." Landlords were allowed to select tenants who were not on CHA's almost all-black waiting list. In sum, the white elderly were placed in public housing in white areas, while blacks were located in ghetto projects (Lazin, 1973).

Throughout the 1960s federal civil rights laws and regulations were valueless in producing any changes. While HUD was aware that the CHA was violating HUD regulations covering racial discrimination, it opted to serve rather than to regulate the local constituency. Finally, in August 1966, fair-housing interests filed suit in U.S. District Court, accusing the CHA of discrimination against blacks (*Gautreaux* vs. *CHA*). Judge Richard B. Austin, in February 1969, found the CHA guilty as charged and issued an order designed to promote integration by construction of public housing in all-white areas. As a result of the CHA's refusal to abide by the court's directive and HUD's refusal to enforce compliance, all public housing construction in Chicago ceased during the 1969–1974 period.

It was only under continuing pressure from Judge Austin that the CHA finally announced plans in 1973 for a program of scattered-site public housing in the white northwest and southwest sides of the city. Residents of the white neighborhoods were aghast: another element in their web of territorial defense and socioeconomic integrity was threatened. The attitudes that lay behind the ward politicians' earlier vetoes of public housing surfaced in an attempt by a consortium of 19 northwest- and southwest-side white community organizations (including two from Garfield Ridge) to use the National Environmental Policy Act of 1969 to prevent the Chicago Housing Authority from acting in accordance with Judge Austin's ruling. The argument of the Nucleus of Chicago Homeowners Association (No-CHA) was

> As a statistical whole, low-income families of the kind that reside in housing provided by the Chicago Housing Authority possess certain social class

characteristics which will and have been inimical and harmful to the legitimate interests of the plaintiffs.

Regardless of the cause, be it family conditioning, genetics, or environmental conditions beyond their control, members of low-income families of the kind that reside in housing provided by the Chicago Housing Authority possess, as a statistical whole, the following characteristics:

(a) As compared to the social class characteristics of the plaintiffs, such low-income family members possess a higher propensity toward criminal behavior and acts of physical violence than do the social classes of the plaintiffs.

(b) As compared to the social class characteristics of the plaintiffs, such low-income family members possess a disregard for physical and aesthetic maintenance of real and personal property which is in direct contrast to the high level of care with which the plaintiffs social classes treat their property.

(c) As compared to the social class characteristics of the plaintiffs, such low-income family members possess a lower commitment to hard work for future-oriented goals with little or no immediate reward than do the social classes of the plaintiffs'.

. . . .

By placing low-rent housing populated by persons with the social characteristics of low-income families described above in residential areas populated by persons with social class characteristics of the plaintiffs, defendant CHA will increase the hazards of criminal acts, physical violence and aesthetic and economic decline in the neighborhoods in the immediate vicinity of the sites. The increase in these hazards resulting from CHA's siting actions will have a direct adverse impact upon the physical safety of those plaintiffs residing in close proximity to the sites, as well as a direct adverse effect upon the aesthetic and economic quality of their lives.

U.S. District Court Judge Julius J. Hoffman ruled on November 26, 1973: "It must be noted that although human beings may be polluters, they are not themselves pollution." Assistant U.S. Attorney Michael H. Berman added, "Public housing residents are not untouchables and the judge rightly accepts this view." However, the suit indicated clearly enough that Chicago's northwest- and southwest-side communities do not accept it.

The regional housing coalition. Meanwhile, in the suburbs and in metropolitan-wide civil rights organizations two sets of interests have come together in a Regional Housing Coalition designed to promote "fair shares" of "low-income" (=minority) housing throughout the six-county area. On the one hand, there is the general desire to ensure civil rights, equity of treatment, and access to housing under conditions of accelerated decentralization of employment from the central city. On the other hand, there is the feeling that unless a regional solution to low-income minority housing is found, the more liberal communities like Park Forest and Oak Park will be beggared by their recalcitrant neighbors.

The Regional Housing Coalition (RHC)—a partnership consisting of the leadership Council for Metropolitan Open Communities;* the Northeastern Illinois Planning Commission; a steering committee of suburban mayors and village presidents; and business, religious, and civic organizations—announced its "Interim Plan for Balanced Distribution of Housing Opportunities for Northeastern Illinois" on October 1, 1973. The plan points up the immediate need for 167,600 housing units for people of low- and moderate-income throughout the six-county metropolitan area.

While housing for minorities is nowhere mentioned in the RHC proposal, its intention was to provide housing for blacks and other minorities in suburbs other than those few to which most suburban-bound blacks were migrating. It represented the convergence of interests of the Leadership Council, which was committed to fair housing in the metropolitan area, and residents of such suburbs as Oak Park and Park Forest, who also had committed themselves to fair housing and active integration efforts that now seemed destined to fail if the surrounding areas would not accept their "fair share" of the responsibility of providing housing for minority groups.

RHC appealed to those whose ultimate goal was increasing the number and proportion of blacks living in the suburbs, i.e., the Leadership Council and local fair-housing groups; at the same time it appealed to those whose concern was limiting the numbers of blacks moving into their own area, that is, Oak Park and Park Forest.

Clearly, if the RHC plan were expressly formulated as a plan for integrating the suburban area, the opposition would be far greater than the support. Therefore, it also was designed to appeal to yet a third set of interests: those who needed to provide a labor supply for the growing industrial and commercial base of the suburban area. Yet even phrased as a plan to encourage economic as opposed to racial integration, the RHC plan faced formidable resistance from suburban whites, as evidenced in their failure to date to support low-income housing. In 1972, only 3,000 out of the metropolitan area's 44,000 public-housing units were located in the suburbs, and many of these units were housing for the (white) elderly. Whatever the motivations of its backers, the RHC strategy of dispersal and "fair share,"

* The Leadership Council was formed in the wake of open-housing demonstrations by a civil rights coalition led by the late Dr. Martin Luther King in 1966. Leaders of Chicago's government, business, real estate, trade unions, religious organizations, and civil rights groups met to explore new ways to break down racial barriers in the real estate market throughout metropolitan Chicago. Currently the Council is engaged in a broad spectrum of programs to this end. During the past four years its legal action program, using the Civil Rights Act of 1968, filed more than 250 cases, with an 85 per cent success record. Through its affiliate, the Metropolitan Housing Development Corporation, it had developed, marketed, and made plans to initiate two low- and moderate-income suburban developments, and made plans to initiate two more. Its third strategy is one of affirmative marketing to overcome the past system of discrimination and to reach minority homeseekers with a positive invitation to consider the total city and suburban housing market.

developed in response to continued racial and economic segregation, stands in sharp contrast to the containment strategy pursued by No-CHA.

Summary and implications

To recap the situation in the Chicago metropolitan area, it can be seen that, since 1960, most changes in black residence have occurred in ever-expanding outward movements of the south- and west-side ghettos in Chicago proper. While this has provided more and better central-city housing for blacks on the peripheries, minimal integration has occurred because of white flight. The bulk of black move-ins to suburban areas has been in and around existing mini-ghettos. Even though numerically few blacks have moved into white suburbs, this "integration" may be the forerunner of suburban resegregation.

The Chicago local community case studies illustrate the variations in white responses to racial invasion. These responses ranged from violence in West Englewood to the erection of defensive barriers in Garfield Ridge, minimal organizational response in Evanston, reverse steering and "managed integration" in South Shore and Oak Park, and an "avowed commitment" to unrestricted open housing in Park Forest. *Across this range of white responses is the general conclusion that regardless of socioeconomic characteristics, a concentration of black families is perceived negatively by whites.*

This conclusion provides interesting insights into the National Academy of Sciences report to the U.S. Department of Housing and Urban Development (HUD), entitled *Freedom of Choice in Housing.* The report offered a framework for evaluating prospects for racial integration in housing and formulating national housing policy. The Academy concluded:

> For both blacks and whites, the quality and convenience of housing and neighborhood services take precedence over racial prejudice in housing decisions.

However, the studies summarized here contradict this. Whites are, in fact, abandoning "convenient" neighborhoods with good-quality housing, as in South Shore and Garfield Ridge.

The Academy report also concluded:

> There is no ratio of blacks to whites that is known to ensure success in racial mixing.

It would seem that predictions and perceptions are more important than numbers: South Shore changed racial composition gradually; transition in Evanston, with a 16 per cent black population, proceeded very slowly; but Garfield Ridge and other communities have turned over rapidly. Oak Parkers became panicky at the thought that large sections of their village might become all black when its black population was only 2 per cent. More important than the black-white residential ratio are the schools. When overcrowding in black schools necessitates changing boundaries of attendance areas and assigning black students to adjacent schools, white tensions rise. As total enrollments and black student populations increase, white families will transfer their children to parochial schools or move (South Shore, West Englewood). Busing to achieve racial guidelines of 80 per cent white to 20 per cent black (Evanston, Park Forest) also causes some white outmigration, while attracting only those white families actively seeking an integrated setting for the education of their children.

Another NAS conclusion was that

> To be successful, a marketing strategy should emphasize the positive racial attitudes that do exist and should take into account the variations in these attitudes.... Attitude changes are effected where the physical and social conditions encourage and support behavioral changes.

However, this is just the point: the variations exist, but the physical and social conditions, the larger structure, prevent successful integration even in areas where residents are most favorable. The responses at the No-CHA end of the scale patently obviates the success of responses emanating from the Park Forest end of the scale. Thus, the Chicago area studies confirm the NAS finding that "there is no evidence that socioeconomic mixing is feasible. The trend in the movements of urban population is toward increasing separation of socioeconomic categories. The tendency is manifested among blacks as well as among whites," thus casting serious doubts on the Regional Housing Coalition's ability to scatter "fair share" low-income housing throughout the six-county area.

As to racial mixing, the Academy concluded:

> It is unlikely that a policy of racial mixing can be consistently applied in metropolitan areas in which there are two or more autonomous governments. For any one locality to act in the total social interest is for it to put itself in a position to be beggared by others who do not accept similar responsibility voluntarily.

Thus, communities such as Oak Park and Park Forest which espouse "integration maintenance" policies will find themselves alone in the struggle as their reluctant neighbors maintain the status quo closed-door policy.

Summed up, the Academy's report described sets of attitudinal and institutional constraints which perpetuate housing segregation and limit the efficacy of strategies aimed at integration. In this chapter we have used the experience of the Chicago metropolitan area to explicate such patterns and constraints. The conclusion that must be drawn is that substantial residential integration by race is unlikely to emerge either in central city or in suburb in years to come; segregation will remain a fundamental feature of the American urban scene.

Chapter 3
The question of community attachment: linear development and systemic models of the local community compared

A basic issue to emerge in the previous chapter relates to the nature of the communal bond. Social scientists have long been concerned with the impact of urbanization and industrialization on local community bonds (Reissman, 1964; Short, 1971; Warren, 1972; Berry, 1973). One problem that has generated a substantial amount of scholarship is the influence of such ecological variables as population size and density on patterns of social participation and community attachment (Hauser, 1965b; Morris, 1968; Fischer, 1972b). In this chapter we shall explore some of these basic issues of local community organization, especially those factors which account for strong or weak community attachments.

Our theoretical point of departure is two models of the local community which have come to dominate the thinking and work of contemporary social scientists: the *linear development model* and the *systemic model*. These models explicate the work of early social philosophers and competing theorists of the Chicago school of urban sociology, who sought to specify the factors that affect local social bonds and community sentiments. Their divergent perspectives on local community organization generated quite different interpretations of social behavior in the urban setting.

The first model we call the linear development model because linear increases in the population size and density of human communities are assumed to be the primary exogenous factors influencing patterns of social behavior. This model has its intellectual roots in the philosophical writings of Sir Henry Maine (1861) and Ferdinand Toennies (1887). Maine saw the processes of urbanization and industrialization resulting in the gradual dissolution of family dependency and the growth of individual obligation. This involved the replacement of primary relationships based upon status and position within the family by impersonal secondary relationships based upon contract and limited liability. The change from status to contract was paralleled by the way in which property (particularly land) was owned. In rural areas, land was held in common by kinship groups; but land in cities became just another exchangeable commodity, so that the individual was no longer bound to it or to his kin-group.

Following Maine, Toennies argued that two periods could be distinguished in the evolution of all great cultural systems. The transition from the first to the second period was fostered by urbanization and the concomitant development of the modern state, science, and large-scale trade. In the initial period, which Toennies called *Gemeinschaft*, the basic unit of local organization was the family or kin-group, within which roles and responsibilities were defined by traditional authority and social relations were instinctive and habitual. Cooperation was guided by custom. The second period Toennies called *Gesellschaft*. Here, social and economic relationships were based upon rationality, efficiency, and contractual obligations among individuals whose roles became specialized. Returns to the individual were no longer derived from kinship rights but from competitive bidding for labor in the market place. Employers and professional peers replaced family and friends as the major groups influencing behavior in the urban setting.

Emile Durkheim (1893) proceeded directly from the foundations laid by Toennies. Durkheim saw increasing division of labor as a historical-biological process involving the evolution of human communities from segmental (mechanical) to differentiated (organic). The segmental community was based on blood relations, comprising a succession of similar self-sufficient kin-groups. Population growth generated increased physical density (person per unit of space) as well as increased social density (contacts per unit of time). Problems of scarcity created by population growth were resolved by functional differentiation limiting the sphere of competition to those in the same occupational group and by intracommunity and intercommunity exchange. One outcome was a division of labor in society, with individuals organized by the nature of the specialized functions they performed. Another was the aggregation of perviously isolated communities into interdependent territorial systems. Durkheim was explicit in pointing out that the movement from a segmental to a differentiated community required greater social density in addition to greater physical density, but both were assumed to be predicated on increased population size.

Georg Simmel (1902) also distinguished two types of communal life and examined their social-psychological correlates. He saw a fundamental contrast between rural and small-town relationships, on the one hand, and the large metropolis, on the other. Whereas in rural areas and small towns the individual is completely immersed in his immediate primary group, in the metropolis the individual becomes socially detached and assumes specialized part-time roles defined by mass society. The steady rhythm of habitual behavior that characterized small-town life is replaced in the metropolis by numerous external stimuli requiring constant conscious response. Simmel maintained that the metropolis actually sustains the individual's autonomous personality. The individual becomes more free, but there is also a threat: that impersonal accountability in external relations tends to displace primary attachments to friends and kin.

These contrasts were restated by William Graham Sumner (1906). Sumner differentiated *folkways*, instinctive or unconscious ways of satisfying human needs, and *stateways*, the contractual relations mediated by state institutions. The essential contrast he saw was between unconscious traditionalism and conscious innovation. In the former, mores are imposed by society; in the latter, they are mediated by the state and experienced in cities. Almost as an echo came the remarks of Delos F. Wilcox (1907):

> The city is, indeed, the visible symbol of the annihilation of distance and the multiplication of interests. . . . Among the business and professional classes, a man's most intimate associates may be scattered over the whole city, while he scarcely knows his next door neighbor's name. . . . The city transforms men as if by magic and newcomers are absorbed and changed into city men.

It remained for the Chicago sociologist Louis Wirth to synthesize these ideas of earlier social scientists into the single most widely accepted theory of the effects of urbanization on human behavior. In "Urbanism as a Way of Life" (1938), Wirth saw the essential character of urban society resulting from an increased number of inhabitants, high density of settlement, and heterogeneity of inhabitants and group life. Accepting the Toennies thesis, Wirth postulated the impact of these factors on local community organization to include (a) the substitution of impersonal secondary relationships for primary contacts, (b) the weakening of the bonds of kinship, and (c) a declining social significance of the neighborhood.

But other members of the Chicago school of urban sociology resisted this view, in particular W. I. Thomas (1967), Robert E. Park and Ernest W. Burgess (1921, 1925). They sought to account for the range of "social worlds" and social solidarities that emerged in the urban metropolis. For them, the local community was a social construction that had its own life cycle and reflected ecological, institutional, and normative variables. Their research and the writings of subsequent social scientists supply the basis for an alternative model of community attachment that we call the systemic model.

The systemic model is, in part, based on historical and anthropological materials that call into question the existence of a *Gemeinschaft* in preindustrial societies because of their internal discontinuities and complexity, and especially because of the dependence of these societies on some variant of bureaucratic or associated institutions. In good measure, the theoretical formulations of Toennies and other early social philosophers are incomplete and tautological at essential points. Much of the Toennies tradition is a normative value treatment of modern life, which reflects a reasoned moral position, but is not the basis for empirical research. The fundamental problem, however, with the *Gemeinschaft-Gesellschaft* approach is that it fails to explain the varying extent and forms of community organization found in modern society. Research on the social structure of urbanized Western societies is rich in those empirical observations which cannot be accounted for by this sociological tradition.

In the systemic model, community organization is treated as an essential aspect of mass society. The local community is viewed as a complex system of friendship and kinship networks and formal and informal associational ties rooted in family life and personal socialization requirements. At the same time it is fashioned by the large-scale institution of mass society. Indeed, it is a generic structure of mass society, whose form, content, and effectiveness vary over a wide range and whose defects and disarticulations are inherent aspects of the social problems of the contemporary period.

Community—that is the geographically based community—manifests varying and diffuse boundaries and reflects different intensity and scope of participation, depending, among other factors, on a person's position in the social structure and his locus in his life cycle. One can identify the social organization of communities in systemic terms by focusing on social networks (relationships) within local communities and analytically abstracting out those that are directly linked to the occupational system. The remaining social relations, to the extent that they have a geographical base, are manifestations of the social fabric of human communities, be they neighborhoods, local communities, or metropolitan areas.

Issues and method

To explore adequately hypotheses about the social fabric of communities derived from the linear development and systemic models, more than case-study data is required; sample surveys become essential. However, there are often limitations inherent in survey research data for examining local social networks. There is obviously an element of arbitrary delimitation both in the questions asked and in the sampling procedures utilized in survey research. Social networks require investigation by careful and detailed participant observation. Moreover, frequency of community contact and participation must take into consideration available opportunities. However, survey research strategies have been developed to recognize these complex issues and to produce relatively meaningful measures of social participation and local community attachment.

We should also reiterate that there are various cultural and normative definitions in different types of communities as well as the alternative meanings persons will attribute to the questions in a standardized survey research interview. In particular, community participation and attachment in an advanced industrial society has its own and various cultural meanings. In the context of contemporary mass society, the notion of community of "limited liability" seems appropriate (Janowitz, 1951, 1967). This notion emphasizes that community attachment in a highly mobile society can be very tenuous. That is, people's involvement in their local community is such that when it fails to serve their immediate needs or aspirations they will display a lack of participation and be prepared to leave the community for alternative opportunities.

A body of research literature has emerged on community participation— informal and formal—and on community attachments since the publication of Janowitz's *The Community Press in an Urban Setting* (1951), which roughly supports the systemic model (Axelrod, 1956; Babchuck and Edwards, 1965; Edwards and Booth, 1973; Laumann, 1973; Litwak, 1961; Zimmer, 1955; Zimmer and Hawley, 1959). However, the measures which have been employed are often so incomplete, and the sample design and size so limited, as to prevent comprehensive statistical analysis. A large-scale and detailed survey from the Royal Commission on Local Government in England makes it possible to explore more comprehensively than has been the case in the past research literature the impact of increased population size, density, and residential mobility on local community organization. This survey was conducted by Research Services, Ltd., in March 1967 and was published under the title *Community Attitudes Survey: England* (London: HMSO, 1969). The fact that the sample was drawn from Great Britain has certain advantages for the purposes at hand, especially in that it reflects a relatively high degree of cultural homogeneity; therefore we are able to highlight particular variables in which we are interested. The population is living under a single unified system of central and local government, although there is administrative variation according to size. The regional differences to be found in Wales and Scotland are excluded, and England at the time of the sample had a relative absence of racial and ethnic enclaves, although the concentration of foreign born was increasing. We expected that the findings would converge with those already encountered in the United States, but because of these factors, the relations would be more clear-cut and pronounced.

The sample survey was designed to assist the Royal Commission on Local Government in making recommendations to restructure the size and format of local government units. In this survey 2,199 adults were interviewed. A stratified random sample was drawn from a hundred local authority areas throughout England (excluding London) in numbers "correctly proportionate to the population which is contained within three main types of local authority of different population sizes within the Register General's Standard Regions." The focus of the survey was on information collected from individuals about their social position, their attitudes, and their social behavior both within and outside their local community. Data were also gathered on the demographic characteristics of the local jurisdiction in which they resided.

As stated, the purpose of our reanalysis of this body of survey data is to examine empirically some of the social and ecological factors which influence the character of local community participation and attachment. However, the main thrust of the reanalysis is not based on a direct search for multivariate findings of the highest aggregate explanatory value. Instead, the strategy is to explore a series of interrelated propositions which seek to examine the implications of the two competing models of the local community.

Our initial expectation, derived from the rejection of the linear development model, is that population size and population density will *not* be associated with important and significant differences in community participation and attachments. Under the Toennies-Wirth approach, the larger the population size and greater the density of an area, the more attenuated would be community participation and attachments. Of course, some differences would be expected, especially between the very largest and very smallest population concentrations, but the overall relevance of size and density as explanatory variables should be limited.

By contrast, the systemic model focuses on length of residence as the key exogenous factor influencing community behavior and attitudes. The major intervening variables are friendship and kinship bonds and formal and informal associational ties within the local community. The local community is viewed as an ongoing system of social networks into which new generations and new residents are assimilated, while the community itself passes through its own life cycle. Assimilation of newcomers into the social fabric of local communities is necessarily a temporal process. Residential mobility operates thereby as a barrier to the development of extensive friendship and kinship bonds and widespread local associational ties. Once established, however, these local social bonds tend to foster strong community sentiments.

This is not to suggest that length of residence is the only independent variable affecting community behavior and attitudes. The available literature leads us to investigate whether a person's social position and his stage in his life cycle likewise influence his friendship, kinship, and associational ties within the community (Axelrod, 1956; Bell and Boat, 1957; Gans, 1962b; Lazerwitz, 1962; Wilensky, 1961). Moreover, the influence of population size and density on community participation and attachments must also be determined and controlled if we are to compare the relative merits of the linear development and the systemic models. We have therefore sought to construct and analyze a general model of community organization which will examine the impact of five basic independent variables (population size, density, length of residence, social class, and stage-in-life cycle) on friendship, kinship, and associational bonds within the community, and the influence of all these eight factors on local community attitudes and sentiments.

Variables and measures

Community attitudes and sentiments

The Royal Commission survey contains three items that may be used as measures of community attitudes and sentiments:

1. Is there an area around here where you are now living which you would say you belong to, and where you feel "at home"?
2. How interested are you to know what goes on in [home area]? (N.B. In the actual interview questionnaire, the phrase "home area" was replaced, in all questions using it, with the *name* of the given local community.)
3. Supposing that for some reason you had to move away from [home area], how sorry or pleased would you be to leave?

Item 1, dealing with the respondent's answer as to whether there was a local community to which he belonged or felt at home, was given as a yes-no response. Responses to Item 2 on community interest were scored into those who expressed strong interest in the affairs of the local community and those who expressed little or no interest in the affairs of the local community. Responses to Item 3 on sentiments about leaving the local community were divided into those who responded that they would be very sorry or quite sorry to leave the local community and those who claimed that they would not be sorry to leave their local community.

Local social bonds (networks)

To measure friendship and kinship bonds within the local community, an additional series of questions was asked:

1. How many people would you say you know who live in [home area]?
2. How many adult friends do you have who live within ten minutes walk of your home?
3. How many adult relatives and in-laws do you have who live within ten minutes walk from your home?
4. Taking all your adult friends that you have now, what proportion of them would you say live in [home area]?
5. Taking all your adult relatives and in-laws, except the very distant ones, what proportion of them live in [home area]?

Responses to Item 1, relative number of local acquaintances, were scored as those who answered that they knew few or less people in their community and those who answered that they knew many or very many people in their community. Item 2, number of friends living nearby, was categorized into five or fewer friends and more than five friends; Item 3, number of relatives, into two or fewer relatives and more than two (non-nuclear) relatives living nearby; Items 4 and 5, proportions of all friends and relatives living in the local community, were categorized as half or less and more than one half of all friends or relatives residing in the local community.[1]

To measure the degree of the respondent's participation in formal associations within the local community, an extensive question was asked regarding participation in the following types of organizations.

1. Organizations connected with the respondent's work, such as trade unions, business clubs, and professional associations.
2. Public bodies or committees concerned with community affairs.
3. Organizations connected with politics.
4. Organizations connected with education and training.
5. Associations connected with churches or other religious groups.
6. Charitable organizations.
7. Civic or community groups such as a rentpayers association or parent-teacher association.
8. Formal social clubs such as a sports team, dance club, automobile club, hobby club, or fraternal organization.
9. Any other formal association not described above.

For each organizational membership listed by the respondent, it was noted whether the locus of participation was inside or outside the designated local community. Respondents were categorized into those who participated in fewer than two local community organizations and those who participated in two or more local community organizations.

Likewise, a question was asked regarding informal participation in local social activities. Included under this item were visits to cinemas, live theater, concerts, recitals, football, rugby or cricket matches, race tracks, bingo sessions, tenpin bowling, public dances, swimming pools, golf, tennis, public parks or gardens, or drives into the countryside. Again it was noted whether or not participation in such informal social activities occurred within or outside the local community, with the categoric breakdown being the same as for formal organizational membership.[2]

[1] In the survey, five response categories were utilized for the two questions regarding proportion of adult friends and relatives residing in the local community. The first four categories included: none of them; half or less of them; and all of them. An additional category was added to each respective question for those who claimed they did not have any friends and/or relatives, or only had one or two relatives. These respondents were not included in the first four categories because the relative proportions would not be meaningful and, therefore, were scored as missing data for those questions. A number of people also failed to respond to other items on the questionnaire, which reduced slightly the sample size for those particular items.

[2] Since we are utilizing a secondary source of data we recognize the limitations both in the variables included and in their operationalizations. For example, we recognize that in advanced industrial societies, forms of social participation and attachment are conditioned by the crucial role of the telephone, a dimension which continues to be overlooked in contemporary research. Likewise, the automobile broadens patterns of geographic participation and identification and renders less important local residences of friends and kin. But it is assumed that the interactions generated by the telephone and the automobile will influence the final patterns of community participation and attachment.

Independent variables

Population size was measured in terms of the size of the respondent's local authority. Five size categories in terms of a rural-urban continuum were employed. The continuum consisted of (1) rural districts; (2) municipal boroughs and urban districts up to 30,000; (3) municipal boroughs and urban districts of 30,000 to 60,000; (4) municipal boroughs and urban districts of 60,000 to 250,000; and (5) municipal boroughs and urban districts over 250,000, including conurbations.

Population density was measured in terms of persons per acre in the local ward or parish where the respondent resided. Six density categories were obtained, ranging from under one person per acre to over twenty persons per acre. Length of residence in the local community was scored into six categories ranging from less than one year to over twenty years/born there. Socioeconomic status (SES) was also scored into six categories ranging from unskilled to professionals. Nonworking housewives and retired persons were classified as to last job held. Finally, life cycle contained five stages: 21–29, 30–39, 40–49, 50–64, and 65 and older.

To analyze these data and examine the relative merits of the linear development and systemic models empirically, the Goodman modified multiple regression technique was used (Goodman, 1970, 1971, 1972a, 1972b, 1973). This technique is designed specifically for multivariate causal analysis of survey data which do not meet assumptions for conventional regression analysis. In short, the Goodman technique analyzes relationships among cell frequencies of multilevel cross-tabular tables and provides effect parameters analogous to partial slopes in regression analysis and measures of association analogous to the squares of zero-order, partial, multiple, and multiple-partial correlation coefficients. The Goodman technique also provides estimates of the magnitude of all interaction effects and their levels of statistical significance.

Since effect parameters computed by the Goodman technique are easiest to interpret when variables are dichotomized, and because our statistical analysis requires multilevel tables encompassing five or more levels of cross-classification, it is helpful to dichotomize variables where statistical properties warrant. However, if curvilinear relationships exist between some variables, the curvilinearity must be taken into consideration by including polytomous variables in the cross-classifications.

Examinations of bivariate frequency distributions among our variables showed that community size (i.e., the rural-urban continuum) exhibited curvilinear relationships with a number of indicators of community involvement and sentiment. Further examination of the bivariate distributions between the five independent variables and each local social bond and each community sentiment suggested the following breakdowns of independent variables for multiple cross-classifications. The percentage of persons in each category is given in parentheses.

1. Community size (rural-urban continuum). (a) Rural communities (24 per cent); (b) urban communities up to 60,000 (32 per cent); (c) urban communities over 60,000 (43 per cent);
2. Population density . (a) Up to ten persons per acre (55 per cent); (b) ten or more persons per acre (45 per cent);
3. Length of residence . (a) Less than one generation (twenty years) (57 per cent); (b) more than one generation (including born here) (43 per cent);
4. Social class. (a) Blue-collar occupations (53 per cent); (b) white-collar occupations (including farm owners) (47 per cent);
5. Life cycle (age). (a) Between 21 and 49 years old (54 per cent); (b) 50 years of age and older (46 per cent).

In employing the Goodman technique, our initial procedure was to construct ten six-way cross-classified contingency tables with each cross-classification encompassing the five basic independent variables along with either a social network or attitudinal variable. Each contingency table thus contained $3 \times 2 \times 2 \times 2 \times 2 \times 2 = 96$ cells. Next, modified multiple regression effect parameters of all saturated models (which include all possible interactions) were computed to estimate the main and interaction effects of size, density, length of residence, social class, and life cycle stage on each social bond (network) and each community sentiment. Results revealed that a multitude of small interaction effects exists, but that none was statistically significant at even the .05 probability level. Under such circumstances, matters are simplified, and we can recompute effect parameters from unsaturated models (which eliminate nonsignificant interactions) to compare the main (direct) effects of the five independent variables on each social bond and community sentiment.

Results

In Table 3.1 are presented the direct effects (Goodman's Betas) and standardized effects (Beta*) of community size, density, length of residence, social class, and life-cycle stage on the seven indicators of local social bonds.[3]

The parameters indicate that the effects of population size and density on local friendship, kinship, and associational bonds are mixed and not very significant. Noteworthy is the finding that persons residing in large urban areas tend to have more extensive social ties than those residing in rural communities. The differential effects of residing in large urban communities

[3] The standardized effect parameters [Beta*] are computed by dividing each Beta by its standard deviation. With large samples, the Betas are normally distributed with a mean of zero and a unit variance. Therefore, to determine statistical significance one consults a table of areas under a normal curve. For example, a Beta* [B* abbreviated] of greater absolute value than 1.96 is statistically significant at the .05 probability level. Likewise, a B* of greater absolute value than 2.58 is statistically significant at the .01 probability level.

Table 3.1
Modified multiple regression effect parameters (B) and standardized effects (B*) of independent variables on local social bonds

Independent variables	Number of friends		Number of relatives		Formal organization memberships		Participation in informal social activities		Number of acquaintances		Percentage of friends in community		Percentage of relatives in community	
	B	B*	B	B*	B	B*	B	B*	B	B*	B	B*	B	B*
Length of residence	.266	2.53	.664	5.40	.310	2.31	.214	1.53	.722	5.50	.578	5.30	.894	5.39
Social class	.014	.13	-.228	-1.85	.128	.95	.102	.74	.040	.30	-.152	-1.39	-.158	-.94
Life cycle (age)	-.034	-.32	-.332	-2.70	-.200	-1.49	-.336	-2.40	-.170	-1.30	.030	.27	-.318	-1.91
Density	-.142	-1.36	.044	.35	-.062	-.47	.064	.46	-.052	-.40	.042	.38	-.014	-.09
Size														
Rural	.054	.27	.030	.13	-.144	-.57	.196	.74	-.076	-.31	-.038	-.19	.186	.59
Small urban	-.130	-1.11	-.174	-1.28	-.062	-.42	-.350	-2.31	-.116	-.80	-.072	-.60	-.354	-1.96
Large urban	.076	.65	.144	1.03	.206	1.35	.154	.99	.194	1.34	.112	.91	.168	.89

Note: B* greater than 1.96, effect is statistically significant at .05 probability level.

as opposed to rural communities are especially pronounced with respect to relative number of acquaintances and memberships in formal organizations.

On the other hand, length of residence has positive and highly significant direct effects on all local social bonds, with the exception of participation in informal social activities. The effects of length of residence are particularly strong on relative number of acquaintances in the community, number of relatives living nearby, and proportions of all friends and relatives residing in the local community.

Social class and life cycle (age) have specific and limited effects on local social bonds. Higher-status individuals tend to have smaller proportions of their friends and relatives residing within their own community and fewer relatives living nearby. They also tend to belong to more formal organizations in the community. Both of these linkages reflect the greater mobility of higher-status individuals and their more extensive reliance on formal or secondary social networks.

As was to be anticipated, involvement in the social fabric of communities declines with advanced life stage. Most affected by older age are participation in informal social activities, local kinship ties, and membership in formal organizations. However, for comparative purposes, two main points emerge. First, social class and stage in life cycle are less powerful or consistent factors affecting local social bonds than is length of residence. Second, neither large population size nor high density leads to a significant weakening of local friendship and kinship bonds or of formal and informal social ties.

Let us now consider the relative impact of the five independent variables on local community sentiments. That is, we shall examine the direct effects of population size, density, length of residence, social class, and life cycle on (1) whether or not a person feels a sense of belonging to his local com-

Table 3.2

Modified multiple regression effect parameters (B) and standardized effects (B*) of independent variables on local community sentiments

	Sense of community		Interest in community		Sorry to leave	
Independent variables	B	B*	B	B*	B	B*
Length of residence	.604	4.60	.222	2.04	.270	2.48
Social class	−.070	−.53	.256	2.35	−.022	−.20
Life cycle (age)	.032	.24	−.112	−1.03	.124	1.14
Density	−.174	−1.33	−.024	−.22	−.232	−2.13
Size						
Rural	.244	1.01	−.146	−.71	.040	.20
Small urban	−.174	−1.17	−.048	−.40	−.114	−.94
Large urban	−.070	−.47	.194	1.61	.074	.60

Note: B* greater than 1.96, effect is statistically significant at .05 probability level.

munity, (2) whether or not he is interested in what goes on in his community, and (3) whether or not he would be sorry to leave his community if he had to do so. The effect parameters and their standardized values are presented in Table 3.2.

Again, we observe that population size and density have relatively weak and, for the most part, insignificant effects on local community sentiments. Conversely, length of residence has positive and statistically significant effects on all three community sentiments. Examining the three community sentiments separately, we see that whether or not a person feels a sense of community is clearly a function of length of residence. Although low population density and small community size (for example, a rural community) tend to have a positive influence on sense of community, their impact is relatively small when compared with length of residence.

Interest in the affairs of the local community is apparently influenced most by a person's position in the social structure. Higher-status persons have the skills and orientations which serve to articulate their interest in community affairs. Furthermore, their social position implies a greater stake (or vested interest) in the community, which usually generates a concomitant concern with community affairs. The effect parameters also show that community interest increases with length of residence and tends to decline with older age. However, contrary to the linear development model, interest in the affairs of the community tends to increase with community size. Rural communities have a weak negative effect on community interest, whereas large urban areas have a moderate positive influence on interest in local community affairs.

Whether a person would feel sorry to leave his local community is affected most by length of residence, followed closely by population density and then by advanced life-cycle stage. Since population density was shown to have essentially no effect on involvement in the social fabric of communities (Table 3.1), it may be inferred that in high-density areas, housing conditions or other social or physical features are such as to have a weakening effect on community attractiveness. This supports the notion of the community of "limited liability," namely, that a person may participate in the social fabric of the community but still be prepared to leave because of specific undesirable conditions.

Our analysis thus far of the direct effects of the five independent variables on local social bonds and community sentiments suggests that the length of residence is the key exogenous factor influencing local community attachment. It will be recalled that the systemic model also stipulates that the formation of extensive friendship, kinship, and associational ties further act to foster stronger community sentiments. Modified multiple regression analysis of basic cross-classifications of each social network variable with each community sentiment revealed that this was indeed the case. All 21 effect parameters (Betas) were positive and highly significant.

Table 3.3

Modified multiple regression effect parameters (B) and standardized effects (B★) of community ties on local community sentiments

Local social bonds	Sense of community		Interest in community		Sorry to leave	
	B	B★	B	B★	B	B★
Number of friends	.370	3.51	.320	4.01	.366	4.74
Number of relatives	.434	4.12	.022	.27	.208	2.7 2
Organization memberships	.050	.48	.500	6.25	.114	1. 48
Informal social activities	.226	2.14	.268	3.36	.044	.57

Note: B★ greater than 1.96, effect is statistically significant at .05 probability level.

The interesting research issue, though, is: What are the unique contributions of friendship bonds, kinship bonds, and formal and informal community ties to the formation of strong community sentiments? Table 3.3 gives us the answer.

In Table 3.3 are provided the effect parameters and standardized effect parameters from a modified multiple regression analysis of three five-way cross-classifications. Each multiple cross-classification included number of friends, number of relatives, memberships in formal organizations, and participation in informal social activities, along with a particular community sentiment.

The results indicate that number of friends is the overall most important type of social bond influencing community sentiments. Other types of local social bonds also exhibit some specific and significant partial effects. Number of relatives living nearby, for example, has a strong effect on a person's sense of community as well as a moderate effect on his desire to remain in the community. However, having greater numbers of relatives living nearby has essentially no impact on a person's interest in community affairs, once the degree of friendship and formal and informal ties are taken into account. Conversely, membership in local formal organizations has a strong direct effect on community interest but little independent influence on sense of community or desire to remain in the community. Finally, controlling for number of friends and relatives living nearby as well as for memberships in local formal organizations, participation in informal social activities has a moderate influence on sense of community and community interest but virtually no effect on desire to remain in the community.

In analyzing the effects of different social bonds we found that rather than replacing primary contacts, formal secondary ties fostered greater numbers of local primary contacts (cf., Axelrod, 1956; Bell and Boat, 1957). Controlling for the main effects of community size, density, and length of residence, memberships in formal organizations had a direct effect of +.33 on number of friends and a direct effect of $+.65$ on number of acquaintances. The corresponding coefficients of determination (r^2) between number of formal organizational memberships and numbers of friends and acquaintances were .50 and .79, respectively.

As a final step in comparing the relative merits of alternative models of community attachment, let us assess the relative amount of variation in each local social bond and each community sentiment that is accounted for by the main effects of population size, density, length of residence, social class, and stage in life cycle. This is done with the Goodman technique by first computing maximum-likelihood estimates of cell frequencies under different models (structural equations) employing an iterative procedure described by Goodman (1972a:1080–1085). The estimated cell frequencies generated by the different models are then compared with the observed cell frequencies by computing Chi-square, based on either the conventional goodness-of-fit statistic or the likelihood-ratio statistic.

In the Goodman system, the Chi-square values serve as measures of unexplained variation. Thus, the smaller the Chi-square value (that is, the better the estimated frequencies correspond to the observed frequencies) under a given model, the stronger is the explanatory power of the independent variable or variables included in the model. By computing the relative reduction in Chi-square values from various models containing different combinations of independent variables with the dependent variable, one obtains measures of association roughly analagous to coefficients of determination, multiple determination, and partial determination in conventional regression analysis.

Table 3.4 provides pertinent coefficients of determination, multiple determination and partial determination calculated using the likelihood-ratio statistics. The first five rows (coefficients of determination) indicate the per cent reduction in unexplained variation in each dependent variable accounted for by the main effects of each of the five independent variables. The next four rows (coefficients of multiple determination) indicate the relative reduction in unexplained variation in each dependent variable accounted for by the main effects of different combinations of independent variables. The last row (coefficient of partial determination) indicates the relative reduction in unexplained variation in each dependent variable accounted for by length of residence after the main effects of population size, density, social class, and stage-in-life cycle have been taken into account.

For the most part, the coefficients speak for themselves. Suffice it to point out that in six out of ten cases, length of residence alone explains a greater per cent of variation in local social bonds and community sentiments than do the combined effects of population size, density, social class, and stage in life cycle. Note also the large increments in the coefficients of multiple determination when the effects of length of residence are added to the effects of the other four independent variables, and the high coefficients of partial

Table 3.4

Goodman's coefficients of determination (r^2), multiple determination (R^2) and partial determination ($r^2_{12.3456}$) between independent variables and indicators of local community attachment

Independent variables	Number of friends	Number of relatives	Formal organization membership	Participation in informal social activities	Percentage of friends in community	Percentage of relatives in community	Number of acquaintances	Sense of community community	Interest in community community	Sorry to leave
Length of residence $(X_2)\, r^2_{12} =$	.33	.55	.20	.03	.79	.69	.74	.63	.13	.36
SES $(X_3)\, r^2_{13} =$	.00	.11	.06	.04	.09	.05	.00	.02	.32	.00
Age $(X_4)\, r^2_{14} =$	.00	.02	.05	.26	.08	.00	.00	.07	.01	.15
Density $(X_5)\, r^2_{15} =$	.14	.00	.12	.00	.00	.00	.03	.02	.10	.25
Size $(X_6)\, r^2_{16} =$	.13	.02	.14	.20	.01	.07	.05	.03	.17	.15
$R^2_{1.34} =$	.01	.14	.11	.29	.16	.05	.00	.09	.32	.15
$R^2_{1.56} =$	.18	.02	.15	.21	.01	.07	.06	.06	.16	.27
$R^2_{1.3456} =$	.18	.16	.25	.50	.17	.13	.06	.15	.45	.42
$R^2_{1.23456} =$	.52	.81	.56	.65	.84	.89	.85	.70	.66	.70
$r^2_{12.3456} =$	.39	.78	.42	.29	.81	.87	.84	.56	.39	.47

Dependent variables (X_1)

determination between length of residence and each social bond and community sentiment when the other four independent variables are held constant. These results, together with the effect parameters presented in Tables 3.1 and 3.2, clearly indicate that length of residence plays a far more important role in assimilation into the social fabric of local communities than do population size, density, social class, or stage in life cycle.

Summary and inferences

Using data from a large-scale community survey, we have shown that community sentiments are substantially influenced by participation in social networks. Whether or not a person experienced a sense of community, had a strong interest in the affairs of the community, or would be sorry to leave the community was found to be strongly influenced by his local friendship and kinship bonds and formal and informal associational ties.

Participation in these local social networks, in turn, was found to be influenced primarily by length of residence in the community. The longer the length of residence, the more extensive were friendship and kinship bonds and local associational ties. Even when population size, density, socioeconomic status, and life cycle were held constant, length of residence exhibited a significant direct effect on participation in local friendship, kinship, and associational networks.

On the other hand, neither population size nor density exhibited a significant relationship with participation in local friendship, kinship, or associational networks when length of residence, SES, and life cycle were held constant. The most general inference to be drawn, therefore, is that the systemic model of community organization based on length of residence is a more appropriate model than the linear development model (based on population size and density) for the study of community participation in urban society.

In drawing more specific inferences from these data about the inadequacy of Toennies' *Gemeinschaft-Gesellschaft* orientation and of Wirth's reformulations, it is necessary to underscore that we are dealing with a cross-sectional sample. But three conclusions continue to emerge and re-emerge from this study of the social fabric of the local communities. First, location in communities of increased size and density does not weaken networks of kinship and friendship. Instead, length of residence is a central and crucial factor in the development of these social networks. Second, location in communities of increased size and density does not result in a substitution of secondary for primary and informal contacts. Indeed, the results suggest that formal ties foster more extensive primary contacts within the local community. Third, increased population size and density do not lead to the weakening of community attachments. But there is support for the notion of "community of limited liability" in that community attachments are not incompatible with desire to avoid the negative and undesirable features of local community life.

This analysis emphasizes the importance of length of residence, supported by the impact of associational life, in accounting for community attachments. We are in this respect dealing with a process of "socialization," so to speak. In the contemporary setting length of residence requires careful analytical appraisal. Of course, length of residence is influenced by population growth. But we are making use of length of residence as an indicator of ecological, social-organizational, and normative processes. Length of residence is a reflection of the impact of large-scale organizations on the life chances and values of the residents of the massive metropolitan concentrations which advanced industrial society has created. The length of residence of a citizen is not to be thought of primarily or even essentially as an individualistic or voluntaristic act. It is in good measure the impact of the industrial order in the allocation and reallocation of employment and transportation opportunities. There is merit in speaking of this process as the social construction of community—a term offered by Gerald Suttles (1972b), in elaborating the older notion of the social world of the metropolis.

Yet a persistent question remains to be explained: why these empirical findings and those of other studies in the United States of the last two decades differ from the outlook developed by Louis Wirth. It can, of course, be argued that Wirth had insufficient empirical data—which was the case, especially data that could be statistically analyzed; that he was selective in his observation; and that his concern with social planning led him to accept Toennies' model as a metaphor for social criticism. But an alternative explanation, which delimits the scope of our findings, is in order. This alternative explanation reinforces the idea that Wirth was examining a slice of social reality and that we are also examining a slice of social reality—hopefully with the advantages that accrue from his efforts.

Wirth was writing at a time in which the expansion of cities in the United States—including the city of Chicago—had been fashioned by the massive waves of foreign-born immigrants. The patterns of migration and urban expansion drew these first-generation immigrants into the inner and densely settled portions of the urban metropolis. It was also a period of relative stability of the smaller communities of the United States, particularly the rural-based communities. Wirth no doubt was aware of the elements of intense social cohesion which developed within segments of the urban community. Everett Hughes, one of Wirth's close colleagues at Chicago, made the astute comment that "Louis used to say all those things about how the city is impersonal—while living with a whole clan of kin and friends on a very personal basis " (Short, 1971:xxix). But in the highly dense inner city of the metropolis, Wirth saw what he thought was a lack of integration or assimilation of immigrant population groups into the social fabric of the urban community. In retrospect, we see that he failed to give sufficient emphasis to mobility and the temporal sequence of assimilation, focusing instead on density and increased population size, variables which operate continuously and with a time lag. Had he "controlled" for length of residence, the differences in community attachment under conditions of increased urbanization would not have seemed as pronounced as he concluded.

Chapter 4
Systemic bases of
local community structure

Chapter 3 highlighted the importance of a systemic model of community attachment based upon length of residence, with additional effects attributable to social class and stage in life cycle. As yet, no body of theory built on these key variables has emerged to provide a systemic base that explains the nature of local community life in an urban society and that provides a viable alternative to the still widely used Wirthian construct. At its minimum, such a theory must comprehend the tension between increasing scale and mobility in a truly national society, on the one hand, and increasing insistence upon a mosaic of small and coherent communities with predictable life styles within a context of intensifying cultural pluralism, on the other. Such parts-within-wholes tensions are essential features of complex systems.

A beginning is to be found in the work of Janet Abu-Lughod (1968a), who points out that despite the fact that earlier terminologies are proving inadequate, it might be possible to develop the needed theory by substituting scale, interactional density, and internal differentiation for Wirth's causal trilogy of size, density, and heterogeneity. Let us look at these alternative variables in turn, spinning them into a web around the considerations of length of residence (mobility), social class, and stage in life cycle that were found to be of importance in the preceding chapters.

Scale and mobility

Scale differs from size in that it measures not the number of participants, but the extent of a given network of relationships among them—although as extent widens the number of persons directly or indirectly affected by system decisions naturally increases. What is missing from the concept of scale is a clear territorial referent. Whereas, in the urbanism described by Wirth, scale was territorially coterminous with size, in the new urbanism it is not, because what is involved is an increase in nationwide mobility.

Scott Greer (1962) has identified three aspects of increasing societal scale that are of major importance. The widening of the radii of interdependence means that, whether men know it or not, they become mutual means to individual ends; the intensity of interdependence increases. As a concomitant, increasing scale produces an increasing range and content of the communications flow. This results in a widening span of compliance and control within given social organizations, the salience of large-scale organizations, their nationwide span of control, and the similarity of their division of labor and rewards, tending to develop a stratification system cutting across widely varying geographical and cultural subregions of the country, and creating national citizens.

This nationwide quality, together with its accompaniment, increasing mobility, has been highlighted in a recent popular work by Vance Packard (1972). Packard observes that the average American moves 14 times in his lifetime, compared to 8 for a Briton, 6 for a Frenchman, and 5 for a Japanese. However, the rates are quite different for different social groups. The high mobiles are those with some college education and higher incomes, employees of large corporate or governmental organizations, in the mid-twenty to mid-forty age span. The low mobiles are the blue-collar employees and other working-class people, whose lives are built around kinship and ethnic ties within local neighborhoods. When the high mobiles move, it involves a shift from one urban region to another, but in moving they scarcely change their life style; there is a tendency to move between almost identical social environments, and indeed, their assessment of the quality of life in a community centers on the characteristics of its social environment. Professional real estate consultants aid them in their search for communities that offer the same environment in terms of schools and neighbors, income levels, education, family background, and clubs. Packard notes: "They will not be changing their environment, they will just be changing their address." The attachment to a type of environment that sustains a particular life style or social standing is the key to the way in which contemporary Americans have adjusted the need to retain a locally based sense of security and stability to the emergence of nationwide high-mobile society.

Packard equates mobility with rootlessness, and rootlessness with a proneness to malaise. Among the things he associates with the new migratory life style are low degrees of community involvement, small numbers of close friends, and high rates of alcoholism and infidelity. He also suggests a much greater tendency to flee the unpleasant, and consequently low tolerance for frustration, marked impulsiveness, and a growing tendency for various other social pathologies. Like the Progressives, Packard says there is a need to recover a sense of small-town community in America. Such perceptions are disputed by scholarly research, however. For example, Leo Srole (1972) notes that even if in the past mental-disorder frequencies were higher in cities, new evidence shows that this no longer is the case. No longer, he argues, should such place-bound variables as size or density be used to compare mental disorder rates, because mobility implies for many higher mobiles at least four successive life styles in significantly different physical settings: growing up in a smaller community; career building in the apartment neighborhoods of a city; child-raising in a suburb; and retirement by "empty nest" couples to better climates, to exurb country places, or to exclusive big-city apartment complexes. In this sense, then, the type of community occupied is not an antecedent causal variable, but a self-selected and interactive or circular variable related to life-style and life-cycle preferences. Further, during the rhythms of daily activity, the adult experiences many types of communities. The basic question from the viewpoint of social pathologies thus becomes: What kinds of people differentially stay in or are drawn to various kinds of milieus? And Srole concludes from his Midtown Manhattan Study (1962) that for *children* under special conditions of parental psychopathology, economic poverty, and family disorganization, both metropolitan and rural slums are more psychopathogenic than adjoining nonslum neighborhoods; and for *adults* seeking a change in environment, the metropolis under most conditions is a more therapeutic milieu than the smaller community, especially for nonconformists and many who are labeled "deviant."

Interactional density

Density also requires a redefinition to make it applicable to twentieth-century urbanism. Abu-Lughod has noted that although Durkheim made a conceptual distinction between material density (population concentration) and dynamic density (rate of interaction), he could still conclude that if the technology for increasing social contacts were taken into consideration, material density could be used as an index to dynamic density. This congruence, which Wirth's investigation of urbanism assumed, has been breaking apart. Interactional density facilitated by communication is far greater than physical density permits or requires, as Richard Meier reveals clearly enough (1962).

This new density is different in kind from the interactions that were measured indirectly through concentration. Increasingly, interaction on primary and secondary levels of involvement is being supplemented by an even more abstract form of interaction: *tertiary interactions* leading to tertiary relationships. If a primary relationship is one in which the individuals are known to each other in many role facets, and a secondary relationship implies a knowledge of the other individual only in a single formal role facet, then a tertiary relationship is one in which only the *roles* interact. Those performing the roles are interchangeable and, in fact, with the computerization of many interactions, are even dispensable, at least at the point of immediate contact. What are interacting are not individuals in one role capacity or another, but the functional roles themselves. Such tertiary relationships can be maintained only under conditions of *physical* isolation; once supplemented by physical contact, they tend to revert to the secondary. Thus, the isolation of different communities within urban regions promotes role and life-style stereotyping often created primarily by mass media imagery, particularly television.

Internal differentiation

In Wirth's view, heterogeneity arose primarily from external sources, and was continually reinforced and sustained by migration. The city, which brought into contact persons of diverse backgrounds, was conceived of as a fertile soil for cross-pollination; physical mobility was presumed to lead to cosmopolitanism and a questioning of inherited beliefs. The local community was conceived as being bound together primarily by ties of sentiment rather than by the instrumental usefulness of residents to one another, and these ties were felt to be weaker in the city than in small towns. Within cities, Robert Park's image of a "sorting-out process," into "little worlds that touch but do not interpenetrate," prevailed. The city was thought of as a mosaic of village-like units which stayed to themselves and closely controlled their members. Park, like other progressive thinkers, looked back to the days of the family, the tribe, and the clan with some sense of nostalgia: he looked to communications, education, and new forms of politics to reconstitute communities of sentiment and create a social order equivalent to that which grew up naturally in the simpler types of society, a positive outcome of heterogeneity and the melting pot. But Wirth and his contemporaries felt that the growth of large-scale organizations, centralization of organizational control, increasing functional division of labor, and widespread use of the automobile were all reducing the significance of the local community. Neither the predicted decline in heterogeneity nor increasing homogeneity through blending in the melting pot have occurred, however; rather, the coalescence of society has facilitated an elaborate internal subdivision.

First, the extent to which many of the immigrant groups have been assimilated into the larger society now appears to be quite limited. Milton M. Gordon (1964) sees the process of assimilation as involving several steps or subprocesses. Each step represents a "type" or "stage" in the assimilation process. He identifies seven variables by which one may gauge the degree to which members of a particular group are assimilated into the host society which surrounds them. The stages and subprocesses are:

Type or stage of assimilation	Sub-process or condition
1. Cultural or behavioral assimilation	Change in cultural patterns to those of host society.
2. Structural assimilation	Large-scale entrance into cliques, clubs, and institutions of host society on primary-group level.
3. Marital assimilation	Large-scale intermarriage.
4. Identificational assimilation	Development of a sense of peoplehood based exclusively on host society.
5. Attitude receptional assimilation	Absence of prejudice.
6. Behavioral receptional assimilation	Absence of discrimination.
7. Civil assimilation	Absence of value or power conflict.

Using this sequence, white Protestant Americans are the most assimilated; indeed, it is they who most frequently constitute the mainstream or host society (Anderson, 1970). Yet in primary group life, even they tend to clique. Much of the New England upper class has consisted, for example, of a group of self-conscious Yankee families clustered in their own exclusive social institutions.

At the other extreme, black Americans display minimal assimilation (Pinkney, 1969). They are by and large acculturated, but there is minimal structural, marital, and identificational assimilation. They continue to experience widespread prejudice and discrimination, and conflicts are increasing rather than decreasing as Black Power advocates achieve a broader constituency. Where there have been deliberate attempts to integrate, as with busing school children, racial frictions have escalated and the result has been greater polarization rather than increased tolerance (Armor, 1972).

Jewish Americans have become a thoroughly Americanized group, acculturated to the American middle-class way of life (Goldstein and Goldscheider, 1968). Yet, at the same time, there is an increasing emphasis on "being Jewish," including association with Jewish culture, religion, and organizational life. Third- and later-generation Jews, in particular, are seeking to temper assimilation with separate group identity.

As for other groups, Japanese Americans have experienced the pluralistic development of congruent Japanese culture within the larger American society (Kitano, 1969), whereas for both Indian and Mexican Americans there remains a bifurcation of Indian and white (Wax, 1971; Moore, 1970), increasing because of the militancy of the Red Power and Chicano movements.

In the case of other ethnic groups, especially the blue-collar eastern and southern European Catholics, expressions of cultural pluralism are increasing, too. Let us focus once again on Chicago as an example. Race and ethnicity now dominate the public life of Chicago. Chicago's residential patterns, neighborhood schools, shops, community newspapers, hospitals, old-age homes, cemeteries, savings and loan associations, and charitable, fraternal, and cultural organizations attest to the role of ethnicity in Chicago's culture and politics. Public decisions affecting home ownership, schools, public housing, police, merchandising, allocation of state and federal funds, and welfare are increasingly perceived in terms of nationality-group or racial-group attachments. Ethnicity defines interest groups in the city, is recognized in the public decision making of the city, and is rewarded and encouraged by politicians and established institutions. Ethnic and racial quotas have been informally adopted by public officials on a large scale. The formation in Chicago of associations of policemen and public schoolteachers along ethnic and racial lines, and the revitalization of ethnic and racial vocational and professional associations, confirm the trend to define interests in these terms.

When the heterogeneity of American cities was caused primarily by the influx of successive immigrant waves, the policy of encouraging assimilation was ideologically taken for granted. Consumers might demonstrate a wide range of behaviors and preferences, but this variety was viewed as being both temporary and expendable. A white, middle-class, "Americanized" standard could be imposed from the outside and justified in terms of the shared higher goal of assimilation. People behaved the way they did only because they had not yet *learned* the better way. The segregated local residential community was regarded as a passing entity that might be maintained only so long as temporary patterns of racial and socioeconomic segregation persisted, but ultimately the local community would decline as people found other, pre-ferable, nonterritorial bases for association. Territorial groups were, it was felt, coercive in character and far less attractive than voluntary forms of association. The latter would shortly replace local community ties, and these "interest communities" would result in a better response from government and big business. The local community would then decline as racial and socioeconomic segregation declined and interest communities replaced residential communities.

What is indicated in American urban regions today is, however, that a new type of heterogeneity exists and is intensifying. This heterogeneity results from internal differentiation and may be understood from another ideological position: that of cultural pluralism. In such a framework, the forms of community that emerge are in no way vestigial remnants of more fragmented localized society.

A major advance in understanding these new forms of community has been made by Gerald Suttles (1972b). Suttles argues that a useful point at which to begin is by retrieving the cognitive maps of childhood. For the child, awareness of the city radiates outward, with the density of information diminishing rapidly with the distance from home. The area of comfortable familiarity constitutes the experience of neighborhood.

Yet cities do not consist of an infinitely large number of neighborhoods, each centering on one of millions of inhabitants at only a slight spatial remove from his fellows. Rather, there is a small number of social labels applied to definable geographic areas. Because population characteristics of a city are continuously variable, with no clear demarcation between one side of the street and the other, society imposes categorical labels on specific geographic realms. Neighborhood categories are not simply found in nature, but are consensually imposed definitions.

A neighborhood label, once affixed, has real consequences, Suttles points out. For outsiders it reduces decision making to more manageable terms. Instead of dealing with the variegated reality of numerous city streets, the resident can form a set of attitudes about a limited number of social categories and act accordingly. For those who live within it, the neighborhood defines an area relatively free of intruders, identifies where potential friends are to be found or where they are to be cultivated, minimizes the prospects of status insult, and simplifies innumerable daily decisions dealing with spatial activities. Thus, the mental map of neighborhoods is not superfluous cognitive baggage, but performs important social functions.

The boundaries of neighborhoods are set by physical barriers, ethnic homogeneity, social class, and other factors that together contribute to the definition of homogeneous areas that are supportive of particular life styles. But if a neighborhood exists first as a creative social construction, it nonetheless possesses a number of important properties. First, it becomes a component of an individual's identity, a stable judgmental reference against which people are assessed. A neighborhood may derive its reputation from several sources: first, from the master identity of the area of which it is a part; second, through comparison and contrast with adjacent communities; and third, from historic claims. In this framework, the idea of a community as first and foremost a group of people bound together by common sentiments, a primordial solidarity, represents an overromanticized view of social life. Communities do lead to social control, they do segregate people to avoid danger, insult, and status claims; but whatever sentiments are engendered by neighborhoods are strictly tied to functional realities.

There are multiple levels of community organization in which the resident participates. The smallest of these units is the *face block*. For children it is the prescribed social world carved out by parents. It is here that face-to-face relations are most likely, and the resulting institutional form is the block association. Next, in Suttles' typology, is the *defended neighborhood* or *minimal named community*, which is the smallest segment of the city recognized by both residents and outsiders as possessing a particular character, and which possesses many of the facilities needed to carry out the daily routine of life. Third, the urban resident also participates in the *community of limited liability*, a larger realm possessing an institutionally secure name and boundaries. The concept, as noted earlier, emphasizes the intentional, voluntary, and especially the differential and tenuous involvement of residents in their local communities. Frequently an external agent, such as a community newspaper, is the most important guardian of such a community's sense of boundaries, purpose, and integrity. Finally, even larger segments of the city may also take shape in response to environmental pressures, creating an *expanded community of limited liability*. Thus an individual may find himself picketing to keep a highway not just out of his neighborhood, but out of the entire South Side. In this way, varied levels of community organization are created as responses to the larger social environment. The urban community mirrors the social differentiation of the total society.

Life styles in the mosaic culture

The communities in which Americans thus live vary in their racial, ethnic, and socioeconomic composition and hence in their available life styles; in their physical features, which can be used to create images and boundaries; and in their historic claims to a distinct reputation or identity. Members of a mobile society select among communities in terms of the life style they are perceived to offer. What, then, are some of the principal life-style differences that are to be found within American society today, setting aside the differentiation associated with cultural intensification based on race and ethnicity?

These differences of life style appear to arise from the experience by all Americans of two common developmental processes: (1) passage through stages of the life cycle, with especially sharp breaks associated with the transition from one state to another, as in marriage, family expansion, entry into the labor force, retirement, and so on; and (2) occupational career trajectories that may necessitate, preclude, or otherwise pattern geographic mobility alongside social mobility. These developmental processes are cross-cut by several different value systems: *familism*, in which a high value is placed upon family living and there is corresponding devotion of time and resources to family life; *careerism*, in which there is an orientation toward upward social mobility and a corresponding disposition to engage in career-related activities, to at least a partial neglect of family ties; *localism*, a parochial orientation implying interests confined to a neighborhood and reference to groups whose scope is local; and *cosmopolitanism*, an ecumenical orientation implying freedom from the binding ties to a locality and reference to groups whose scope is national rather than local, so that the cosmopolitan resides in a place but inhabits the nation.

From these bases, one can distinguish between *working-class communities*, *ghettos*, and *ethnic centers*, in which the broad pattern of interaction is one where informal meeting places, street-corner gangs, church groups, and precinct politics tend to dominate the collective life; *middle income familistic areas*, in which informal relations seem to be heavily shaped by the management of children, and formal organizations are much more extensively developed than in lower-income areas; the *affluent apartment complex* and the *exclusive suburb*, which generally have a privatized mode of interaction and organization such as social clubs, private schools, country clubs, and businessmen's associations; and *cosmopolitan centers*, which have long existed in some cities and which seem to be sprouting up in other cities as they grow and are able to provide a critical mass of local-grown talent and misfits, to create their own symbolic milieus of tolerance (Suttles, 1972b).

It is in the latter cases that there emerge the most extreme forms of *subcultural intensification*—the strengthening of the beliefs, values, and cohesion both of groups that previously existed as social entities outside the city, and of new groups emerging within expanding urban systems as new cultural values and norms are established (Fischer, 1972b, 1975). Two ingredients are involved: growth of "critical masses" such that subcultural institutions can develop (for example, political power and national churches for ethnic groups, hangouts for "bohemians," bookstores for intellectuals, museums for artists, new communities for the elderly, and "turfs" for each group) which strengthen the subculture and attract more of its members to the city; and contrasts with other subcultures that intensify people's identification with and adherence to their own. Whereas Wirth thought that the clash of different values in the city might negate all values, we observed in Chapter 2 that internal cohesion is strengthened by the conflict that arises on contact.

One important form of intensification is that which appears in criminal subcultures. Whereas the Wirthian model explains deviance by the breakdown of norms, the alternative is to see deviance as constituting *subcultural* deviance from center-society values. The group nature of much crime, especially urban crime, seems clear. The state of Illinois, for example, classifies its prison population into four subgroups: the sociopathetic, the immature, the neurotic, and the gang-related, with the latter the largest group. Greater criminality in cities might, therefore, be explained by an intensification process for criminals, such as the growth of an underground or street gang, as well as by an intensification process that would operate in regard to other deviant subcultures: homosexuals, prostitutes, "hippies," political dissidents, and so on.

What appears to have emerged and to be emerging in America as a result of these changes is a *mosaic culture*—a society with a number of parallel and distinctively different life styles. While one result is divisive tendencies for the society as a whole, at another level mutual harmony is produced by mutual withdrawal into homogeneous communities, and by exclusion and isolation from groups with different life styles and values. A mosaic of homogeneous communities maintains different life styles that are internally cohesive and exclusive but externally nonaggressive unless they are threatened by intrusion of those from other subcultures. Mobility within the mosaic leads to a high degree of expressed satisfaction by residents with their local communities, and the option for those who are dissatisfied to move to an alternative that is more in keeping with their life-cycle and life-style requirements.

Concluding remarks

Drawing the foregoing together, we propose that a systemic base for understanding the nature of local community life can be formed by combining the individual-level variables of mobility, life-cycle, and socioeconomic attainment with the aggregate-level concepts of scale, interactional density, and internal differentiation. We further propose that the essential driving force of local community life resides in social rather than economic dynamics, particularly the social dynamics of status achievement.

The drive for achievement (improved status) is a variable of utmost importance within the "mainstream" American culture (see McClelland, 1961). Children must "get ahead" and "improve themselves" through education. Workers must ascend the job hierarchy. Earnings must be spent on the best possible homes and material possessions in the best possible neighborhoods. Any increase in job or financial status must be matched by a move to a better neighborhood. "Downgrading" of the neighborhood through entry of those perceived to be of lower status must be fought, and if it cannot be contained one must flee to avoid the inevitable resulting loss of status. To do otherwise would be to abandon the aggressive pursuit and the outward display of "success."

When a family seeks a home they look for other things too, of course. The prime decision relates to the home—its price and type, determined by socioeconomic attainment and by the family's needs at the stage-in-life cycle during which the choice is made. Since a large number of homes qualify within these first bounds for all but the poor, neighborhood considerations then come into play. The scale of urban regions has brought complexity, and the rapidity of urban change produces uncertainty and insecurity. In the search for self-identity and security in a mass society, he seeks to minimize disorder by living in a neighborhood in which life is comprehensible and social relations predictable. Indeed, he moves out of "his" neighborhood when he can no longer predict the consequences of a particular pattern of behavior or patterns of behavior of his neighbors. He seeks an enclave of relative homogeneity: a territory free from status competition because his neighbors lead similar life-styles are at similar stage in the life-cycle; a safe area, free from status-challenging ethnic or racial minorities; a haven from the complexity of urban scale, to be protected and safeguarded by whatever means: legal or extra-legal institutional measures, and frequently illegal violence—each a symptom of defensive territoriality protecting that which has been achieved.

The resulting homogeneous niches are exquisitely reticulated in urban space. High-status neighborhoods typically are found in zones of superior residential amenity near water, trees, and higher ground, free from the risk of floods and away from smoke and factories, and increasingly in the furthest accessible peripheries. Middle-status neighborhoods press as close to the high-status ones as is feasible. To the low-status resident, least able to afford costs of commuting, are relinquished the least desirable areas adjacent to industrial zones radiating from the center of the city along railroads and rivers, the zones of highest pollution and the oldest, most deteriorated homes. In the cores of the ghettos, widespread abandonment of properties marks the extremes of neglect. The whole is more a Balkanization than a melting pot. What the staff of the White House *Commission on Crimes of Violence* predicted in 1969 has today become reality:

> we can expect further social fragmentation of the urban environment, forma-
> tion of excessively parochial communities, greater segregation of different
> racial groups and economic classes. . .and polarization of attitudes on a variety
> of issues. It is logical to expect the establishment of the 'defensive city,' the
> modern counterpart of the fortified medieval city, consisting of an economically
> declining central business district in the inner city protected by people shop-
> ping or working in buildings during daylight hours and 'sealed off' by
> police during nighttime hours. Highrise apartment buildings and residential
> 'compounds' will be fortified 'cells' for upper-, middle-, and high-income
> populations living at prime locations in the inner city. Suburban neighbor-
> hoods, geographically removed from the central city, will be 'safe areas,'
> protected mainly by racial and economic homogeneity and by distance from
> population groups with the highest propensities to commit crimes (Mulvihill,
> et al., 1969).

These polarization patterns, formerly varying substantially from one region of the country to another, are now nationwide in their appearance, another indication of the emergence of mass society.

One might say, then, to conclude this section of the book, that the most pervasive feature of the American urban scene today is segregation—of income groups, family types, ethnic and racial minorities, as well as of land uses and activity networks. And the action space within which the dynamics and interactions of opposing forces are being played out is the larger socio-spatial system of the metropolis to which we shall now direct our attention.

PART III
INTRAURBAN FORM
AND STRUCTURE

Chapter 5
Internal structure of the city: classical views

A question left unresolved in Chapter 4 is whether any unit of sociospatial organization exists between the local community and the nation as a whole. One does—the metropolis, comprising a set of local communities bound together into a common labor market and housing market, and by shopping, services, local media, and other interrelationships. It is to this unit of organization that we now turn. Exactly how does a mosaic culture find its expression in the spatial structure of the metropolis?

We return once again to our systemic view. This chapter is devoted to a very brief review of the external determinants of internal city structure, and to various components of that internal structure—population densities, social and economic patterns, and so on. Each of the components is then examined in much greater detail in the remaining chapters of this section of the book. The emphasis throughout is upon now-classical views of urban structure. Part IV then turns to questions of contemporary urban dynamics, and of the changes needed in these classical views.

External determinants of internal structure

It is critically important to realize that one cannot study the ecology of an urban area in isolation, because cities are the central elements in the spatial organization of regional, national, and supranational socioeconomies by virtue of the interregional organization in a total "ecological field" of the functions they perform (Friedmann and Alonso, 1964; Pappenfort, 1959). In a specialized society, economic activities are undertaken by design, or survive in the market place, at those locations which afford the greatest competitive advantage. Among these activities, those most efficiently performed in limited local concentrations provide the basic support for cities. The location theorist commonly classifies locally concentrated economic activities into those which are raw-material oriented, those located at points which are intermediate between raw materials and markets, and those which are market oriented (Isard, 1956). Raw-material orientation includes direct exploitation of resources and the processing of raw materials, and its character is that of the developed resource endowment of different places. Activities in intermediate locations are usually of a processing kind, involved in intermediate and final processing and transformation of raw materials, and most frequently locate at some favorable spot on the transport network, such as an assembly point, a gateway, a break-of-bulk point, or a port. Market-oriented activities may be secondary (for example, where there is a weight gain involved in the final processing of raw materials or intermediates prior to delivery), but are dominantly tertiary, concerned with the direct service to the consuming population through wholesale, retail, and service functions. The consuming population comprises the workers in the other specialized activities, of course, plus the local population supported by the tertiary trades. Thus, market orientation implies a location best suited to serve demands created by prior stages of the productive process. The three classic principles of urban location derive from the three types of locational orientation of economic activities: cities as the sites of specialized functions; cities as the expressions of the layout and character of transport networks; and cities as central places (Harris and Ullman, 1945). All three factors, or some combination of them, may operate in the case of any particular city. However, whereas all cities will have a central business district providing retail and service functions to the city and surrounding populations, the role of the other two factors will vary greatly from one city to another.

In the internal structure of cities these specialized functions have priority. The central business district has traditionally been a point of focus about which land uses and densities, the spatial patterning of the urban population, subsidiary retail and service locations, transportation and commuting patterns, and the like, have evolved. When other specialized activities are performed, they create supplementary or additional nodes. Thus, cities are supported by "basic" activities ("staples") whose locations are determined exogenously to the city by comparative advantage in larger regional, national, and international economic systems. These always include the central business district, the focus not only of the city itself but also of its tributary region, and may include other specialized activities. The "skeleton" of the city comprises the locations of these basic activities, plus the urban transport network. "Flesh" is provided by residential site selection of workers with respect to the skeleton, and "blood" comes from the daily ebb and flow of commuters. Further patterning is provided by the orientation of subsidiaries and business services to the basic activities, and by local shopping facilities to the workers. Shopping trips create another ebb and flow. Further "second-"

and "third-round" effects can be described, but these follow logically from the first.

The residential pattern: urban densities

Apparently, a simple expression summarizes the population density pattern of cities:

$$d_x = d_o e^{-bx} \tag{1}$$

where d_x is the population density d at distance x from the city center, d_o is density at the city center, e is the natural logarithmic base, and b is the density gradient. The "city center" is, of course, the central business district, so that when the natural logarithm of population density of small areas within the city is calculated, along with their distance from the central business district, and a scatter diagram is constructed with distance along the abscissa, the points in the diagram lie around a straight line with downward slope b, or:

$$ln \cdot d_x = ln \cdot d_o - bx \tag{2}$$

The gradient b may be considered an index of the "compactness" of the city, just as differences in central density d_o index the overall level of "crowding." Equation (2) has been shown to be applicable to cities regardless of time or place (Clark, 1951). Why should this be so? Muth (1969) has shown how this negative exponential decline of population densities with increasing distance from the city center is a condition of locational equilibrium that stems logically from the operation of a competitive housing market. Seidman (1964) has shown it to be a natural consequence of Alonso's locational theory of land use. The theoretical bases of the empirical regularity are thus readily available.

The density gradient b, like the densities it indexes, also shows consistent behavior. For example, it falls consistently with city size P as follows:

$$ln \cdot b_j = ln \cdot P_o - c \cdot (ln \cdot P_j) \tag{3}$$

so that cities have experienced progressive "decompaction" with increasing size. Further, since in the United States, recent expected growth of metropolitan areas between two time periods t and $t + l$ is a constant proportion of size P such that

$$ln \cdot P_{t+1} = k + ln \cdot P_t \tag{4}$$

which further implies exponential growth of population with time

$$P_t = P_o e^{kt} \tag{5}$$

then

$$b_t = b_o e^{-ckt} \tag{6}$$

which states that the density gradient diminishes through time in a negative exponential manner, which is the case. (This "law of proportionate effect" may be seen by plotting populations of U.S. cities in 1960 against their populations in 1970. The scatter of points is linear and homoscedastic with a slope of $+1.0$ on double logarithmic paper. Satisfaction of this assumption means that in steady-state the distribution of towns by size will be lognormal so that Zipf's rank size rule for city sizes holds: $P_r = P_1/r^q$ or $\text{Log} \cdot P_r = \text{Log} \cdot P_l - q \cdot \text{Log} \cdot r$. In these equations P_r is the population of the city of rank, r, P_l is thus the largest city, and q is an exponent.)

Newling (1965) has shown, additionally, that the two generalizations (that population density declines exponentially with increasing distance from the city center; and that the density gradient itself falls through time in a negative exponential manner) together lead to a third regularity, which he calls the "rule of intra-urban allometric growth." This is that the rate of growth of density is a positive exponential function of distance from the city center:

$$(l + r_x) = (l + r_o)e^{gx} \tag{7}$$

where r_x is the percentage rate of growth of density at distance x, r_o is percentage growth at the center, and g is the growth gradient, measuring the rate of change of the rate of growth as distance from the center of the city increases. He goes on to show that since both density and the rate of growth are functions of distance from the city center, the rate of growth may be expressed as a direct function of density:

$$(l + r_x) = mD^{-q} \tag{8}$$

where r_x is as above, m is a constant, D is initial density and q relates the rate of change of the rate of growth to the rate of change of density. As density increases, the rate of growth drops. Moreover, Newling argues for the existence of a "critical density" above which growth becomes negative, that is, population declines. In several cases he shows a convergence upon thirty thousand persons per square mile as this critical density, and in one study he concludes:

> The inverse relationship between population density and the rate of growth, the identification of a critical density, and the observation that negative growth, occurring as it does above the critical density, is not solely attributable to competition between commercial and residential use of land, all lead one to speculate that perhaps there is indeed some optimum urban population density to exceed which inevitably incurs social costs. We may speculate that certain events in the history of the city will cause this optimum to be exceeded (for example, heavy immigration without a commensurate expansion of the housing stock and supply of social overhead capital), with deleterious consequences for the areas concerned (such as blight, crime and delinquency, and

other social pathological conditions) and leading to an eventual decline in the population of the affected areas.

. . . .

If this is so, then consistent relationships are available between size of city and the pattern of population densities within cities, between growth of the urban population and change of densities within. Further, there is the strong suggestion that this chain provides direct links between an overall urbanization process and the occurrence of pathological social conditions in particular parts of particular cities.

Social and economic patterning of the residents

The generalizations in the preceding section are strong. Equally strong generalizations are now possible concerning the social patterning of the urban residents who live at the density patterns in the changing ways already described.

There has been a long tradition of research by sociologists, geographers, and economists dealing with the social and economic characteristics of urban neighborhoods. Among the earliest descriptive generalizations were those of Hurd (1903), who related neighborhood characteristics, especially income and rentals, to two simultaneous patterns of growth, which he called *central* and *axial* growth. Later, Burgess (1925) emphasized the importance of outward growth from the center which caused concentric zonations of neighborhoods. Change occurred by the outward movement of the wealthier to the periphery, and the continued expansion of inner zones upon the outer in a process of invasion and succession by the lower-status groups living closer to the city center. Hoyt (1939), on the other hand, emphasized the significance of axial growth when he developed his sector concept. According to this notion, status differences established around the city center are projected outwards along the same sector as the city grows, thus creating a wedge-shaped distribution of neighborhoods by type with the higher status groups following scenic amenities and higher ground. In addition, the literature of sociology has been replete with studies of the segregation of ethnic groups in particular localities, a segregation conforming neither to the concentric nor to the axial schemes.

Considerable debate has taken place about the relative merits of each of these models. A succession of large-scale factor analytic studies conducted since the end of the Second World War now makes it possible to state definitively that the three models are additive contributors to the total socioeconomic structuring of city neighborhoods. In each of the studies the answer is the same: there are three dominant dimensions of variation. These are (1) the axial variation of neighborhoods by socioeconomic rank; (2) the concentric variation of neighborhoods according to family structure; and (3) the localized segregation of particular ethnic groups.

Neighborhood characteristics involving educational levels, type of occupation, income, value of housing, and the like, are all highly correlated—as they should be, for they are also functionally related at the individual level in terms of status-achievement mechanisms. Each varies across the city in the same way: according to sectors. High-status sectors search and follow particular amenities desired for housing, such as view, higher ground, and so on. Lower-status sectors follow lower-lying, industrial-transportation arteries that radiate from the central business district and which, together with that district, form the exogenously determined skeleton of the city. This is consistent with the idea, also, that the lower the income the closer is home to work in the contemporary American city.

Conversely, the age structure of neighborhoods changes concentrically with increasing distance from the city center, along with age of housing, densities, existence of multiple-unit structures, incidence of ownership by residents, participation of women in the labor force, and the like. Thus, at the edge of the city are newer, owned, single-family homes, the residences of larger families with younger children than those nearer the city center; the wife more often stays at home. Conversely, the apartment complexes nearer the city center have smaller, older families, fewer children, and are more likely to be rentals; in addition, larger proportions of the women work. This "family structure" pattern is consistent with the ideas of Burgess, and has been called by sociologists the "urbanism-familism" scale.

Third, particular ethnic groups will be found to reside segregated in local subcommunities. The most obvious case of segregation today in the American city is that of blacks, although every new migrant group has also experienced this pattern of living. Along with segregation go such other variables as lack of household amenities, deterioration of housing, over-crowding, and the like.

If the concentric and axial schemes are overlaid on any city, the resulting weblike pattern of cells will contain neighborhoods remarkably uniform in their social and economic characteristics. Around any concentric band communities will vary in their income and other characteristics, but will have much the same density, ownership, and family patterns. Along each axis communities will have relatively uniform economic characteristics, and each axis will vary outwards in the same way according to family structure. Thus, a system of polar coordinates originating at the central business district is adequate to describe most of the socioeconomic characteristics of city neighborhoods. The exception is in patterns of segregation, which are geographically specific to the particular city, although segregation is a phenomenon which is found in them all. The three classic principles of internal structure of cities are thus independent, additive descriptions of the social and economic character of neighborhoods in relation to each other and to the whole.

Services for the residents: local business

The central business district provides a range of goods and services for the entire urban population and for the larger tributary region served by the city. In addition, a system of smaller business centers exists within the city to serve the city population with the commodities they require on a weekly or monthly basis. Such purely internal or endogenous business appears even in small towns with populations of less than one thousand. It does not begin to assume any identifiable structure, however, until it reaches the level of county seats, and a variety of internal forms is clearly distinguishable only in the cities which serve as centers for multicounty "functional economic areas." At this stage the structural differentiation of centers and ribbons is clear. Ribbons follow the major section and half-section streets and the radial highways, performing a variety of service functions (building materials and supplies, household requirements) and automobile-oriented activities (gas, repair, parts); they contain many large single-standing, space-consuming stores (discounters, furniture, appliances), in addition to being interspersed with convenience shops (food, drugs, cleaners) for adjacent neighborhoods. Certain stretches of ribbon are devoted to the activities of "specialized functional areas," such as Automobile Row. At the major and minor intersections of the street system are business centers, differentiated from the adjacent ribbons by the functions they perform, by the ways consumers shop in them, and by land values. The centers provide both convenience and such shopping goods as food, drugs, clothing, shoes, and luxuries. Consumers generally shop on foot from store to store, in contrast to their single-purpose trips to ribbon establishments. Land values within the city fall with increasing distance from the city center, but commercial values add extra texture. The ribbons create ridges that rise above the adjacent residential areas. Steeply rising cones at the intersection of ridges clearly indicate the location and extent of centers. Four levels of outlying centers have been identified beneath the central business district: neighborhood and community shopping centers at the convenience level, and larger shoppers' and regional centers providing specialized goods and services along with standard convenience items. The differentiation between these levels is made in terms of the number and variety of service functions performed and in the size of trade area served.

It is axiomatic that retail and service activities are consumer-oriented, because internal business has developed entirely to serve the population residing within the city. Consistent with the earlier sections of this chapter, it is also possible to place internal business provision within the same frame. Consider a city divided by the concentric-axial scheme described above, and let R indicate the total retail and service provision of any of the cells defined, with P representing total population of the cell, D the population density, F an index of its family structure, and S an index of its social rank. Then

$$R = sP^uD^vF^wS^z \tag{9}$$

which yields an extremely close fit in every city studied. Moreover, the provision of local business, and local business change, may clearly be related back to the socioeconomic pattern of the city. Similar expressions may be developed for ribbons and centers separately, although in the case of centers certain problems emerge concerning the use of arbitrary cells instead of the market areas of the centers as the units of observation, even though a properly drawn set of circles and radii will, by their intersections, locate the outlying business centers of many cities.

As in the case of socioeconomic structure, however, segregation creates problems for generalization. In Chicago, for example, retail systems assume not one but two equilibrium positions. In segregated nonwhite residential areas there is a two-level hierarchy of business centers, comprising the neighborhood convenience type and the smaller shoppers' goods type, whereas in the rest of the city there exists a four-level hierarchy of outlying centers. All retailing is experiencing changes because of increased scale of retailing, increasing consumer mobility, and rising real incomes. Yet as the nonwhite residential area expands outwards, still another element of retail change is added. Simply, neighborhood transition means loss of markets, since real income among the nonwhite population is approximately one-third lower than that of the population displaced. The effects are felt in several stages:

1. *Anticipation of neighborhood transition.*   In this phase the normal replacement of businesses that fail, or close because the businessman retires or dies, ceases. Vacancy rates begin to rise. Also, a "maintenance gap" appears because property owners, increasingly uncertain about prospective revenues, reduce normal maintenance expenditures. Dilapidation grows.

2. *During turnover.*   Demands drop precipitously, especially for higher-quality goods, and the specialty shops in the larger business centers fail. Vacancies in centers rise to levels as high as one third to one half of the stores.

3. *Stabilization phase.*   The neighborhood settles down into its lower-income character. Because incomes and revenues are lower, it is almost impossible to eliminate the effects of the earlier maintenance gap, and so a general run-down appearance persists. Rents in the business centers drop, and activities from the ribbons and new businesses directed at the changed market move in and fill up the centers once again. Vacancies mount in the abandoned ribbons, settling down in more than 20 per cent of the stores, but concentrated in the older buildings which, through lack of use, deteriorate more. Zones of segregated housing are thus crisscrossed with ribbons of unwanted and blighted commercial property. Much that is critical to an understanding of business within the city thus depends less upon the

structuring implicit in use of a model such as Equation (9) than upon the existence and nature of segregation in the housing market.

Although the skeleton of the city is determined by broader regional and supraregional forces, the flesh shows certain simple systematic regularities which are tightly knit into a locational system of simultaneous concentric and axial dimensions. Let us now begin to explore each of these regularities more thoroughly.

Chapter 6
Urban population densities: structure and change

As noted in Chapter 5, it was Colin Clark (1951) who was responsible for resurgence of interest in the mathematics of urban population densities, with his argument that regardless of time or place the spatial distribution of population densities within cities appears to conform to a single empirically derived expression

$$d_x = d_o e^{-bx} \tag{1}$$

where d_x is population density d at distance x from the city center, d_o is central density, as extrapolated, and b is the density gradient, indicating the rate of diminution of density with distance. Clark argued that Equation (1), which says that urban population densities decline in a negative exponential manner with increasing distance from the city center, "appears to be true for all times and all places studied, from 1801 to the present day, and from Los Angeles to Budapest." In fact, he provided 36 examples in which the natural logarithm of that equation

$$ln \cdot d_x = ln \cdot d_o - bx \tag{2}$$

appeared to be a good fit to the sample data at his disposal.

One thing he did not do, however, was to provide a theoretical rationale for his formula. And 36 cases hardly enable one to assert complete universality for Equation (1) regardless of time or place, especially when a sound theoretical base is lacking. Other empirical support is necessary, and is abundantly available.

In Chicago, the classic laboratory for urban analysis in the United States, Winsborough (1961) shows the pattern to hold for every census year from 1860 to 1950, with the weakest correlation between density and distance − 0.97. Similarly, Kramer shows that the fit is better for net residential densities in Chicago than for gross densities, and that the rate of decline varies by radial sector (1958). Muth (1969) and independently Weiss (1961) found that the pattern holds for all large United States cities studied in 1950. Sherratt fitted a similar model to data for Sydney, Australia; and Newling found that Equation (2) provided a satisfactory fit for Kingston, Jamaica (1960). However, the only evidence provided for Asia relates to Calcutta, in which Kar (1962) shows that a negative exponential pattern existed in 1881, 1901, 1921, and 1951. Kar's graphs are presented in Figure 6.9. We can now fill in this gap a little: Robert J. Tennant (1961) has found that Clark's model also applies in Colombo, Hyderabad, Manila, Rangoon, Singapore, Djakarta, and Tokyo (and, independently of Kar, in Calcutta). The relationship for Hyderabad is shown in Figure 6.1.

Almost a hundred cases are now available, with examples drawn from most parts of the world for the past 150 years, and no evidence has yet been advanced to counter Clark's assertion of the universal applicability of Equation (1). To be sure, the goodness of fit of the model varies from place to place, but in every place so far studied a statistically significant negative exponential relationship between density and distance appears to exist.

Theoretical rationale

The basic theory of urban land use is by now well known. The argument runs as follows. Sites within cities offer two goods—land and location (Alonso, 1960). Each urban activity derives utility from a site in accordance with the site's location. Utility may be translated into ability to pay for that site. The most desirable locational property of urban sites is centrality (or maximum accessibility in the urban area, as transport routes converge at the center); for any use, ability to pay is directly related to centrality. The less central the location, the greater are the transport inputs incurred and the lower the net returns. Bid-rent functions thus decline with distance from the city center. However, the intercept (utility derivable from maximum centrality) and the slope (rent-distance tradeoff) of this function differ for different activities; and in competitive locational equilibrium, with each site occupied by the use that pays most for the land, the resulting spatial structure of land use is one that is zoned according to relative accessibility. Land prices diminish outward; and as they do, regardless of other changes, land inputs will be substituted for other inputs, and intensity of land use will diminish. Thus declining residential densities should be expected (Garrison et al., 1959).

Most parts of a city are occupied by residential land uses of different kinds. Alonso (1960) has shown that bid-rent functions are steeper for the poorer of any pair of households with identical tastes. Hence, in equilibrium, one expects the poor to live near the center on expensive land, consuming little of it, and the rich at the periphery, consuming more of it. Since land consumed by each household increases with distance from the city center, population densities must drop, with due allowance for variation in size of household.

Muth (1969) goes further. Making assumptions about relative perfection of competition in the housing market, maximizing or minimizing behavior as appropriate, diminution of price of housing with distance from city center, and a demand function for housing linear in the logarithm of price at the center and population, he develops a model in which price per unit of housing, rent per unit of land, and output of housing per unit of land all decline, and per capita consumption of housing increases, with distance from the city center. Net density (output of housing per unit of land divided by per capita consumption of housing) must therefore also decline. Moreover, if the price-distance function is assumed to be negative exponential and the production function for housing logarithmically linear with constant returns to scale, then net population density must decline negative-exponentially with distance from the city center.

For the model to hold, a negative-exponential price-distance relationship must exist. Also, since Muth's model considers only residential competition, we should expect the negative exponential to be a better fit for net than for gross residential densities, since the gross includes all land regardless of use.

Figure 6.1
Density–distance relationship for Hyderabad.

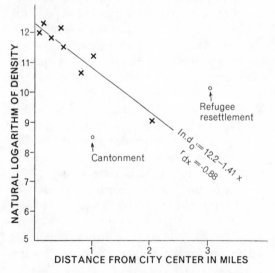

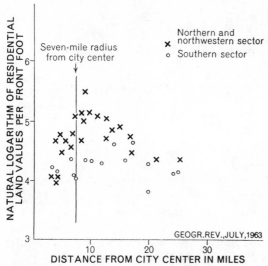

Figure 6.2
Residential land values, two sectors of the Chicago
metropolitan area, 1961. Source: Olcott's
Blue Book of Land Values for Chicago, 1961.

Evidence can be provided to verify both these points. Figure 6.2 shows that front-foot values for residential land diminish negative-exponentially in all parts of Chicago where undeveloped lots are available for sale. However, the more central parts of the city are largely filled, since the older the development, the less land is available, the less active the land market, and the lower the price. The earliest developments are the most central and are occupied by the lowest socioeconomic groups. Indeed, within a six- or seven-mile radius of the center of Chicago the only market for residential lots is related to renewal activity or intended for conversion to new highways, at condemnation prices (except within the Loop or where private investment is committed to residential renewal). Hence it should be no surprise that the negative-exponential price-distance relationship does not begin to take form until newer areas with an active land market for residential use are reached. Notice also in Figure 6.2 how the rate of decline varies by sector.

Since a negative-exponential price-distance relationship holds, we should now find a negative-exponential decline of densities. Figure 6.3 shows this to be the case, and the fit for net densities is better than that for gross. Equation (1) is therefore a logical outcome of urban-land-use theory. The theory of urban land use, which originated with Hurd and Haig, and which owes its recent improvements to Ratcliff, Alonso, and Muth, provides the needed rationale for the appearance of Clark's empirical regularity.

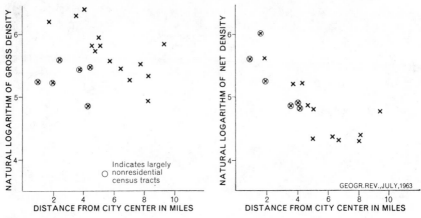

Figure 6.3
Gross and net density gradients, Chicago, 1960. Note how use of gross densities is complicated by presence of substantial nonresidential areas in certain census tracts.

Implications of the model

From (1) it follows that the population residing within a distance r of the city center is

$$\int_0^r d_o e^{-bz}(2\pi x)dx = P_r \tag{3}$$

which equals

$$2d_o\pi b^{-2}[1 - e^{-br}(1 + br)] = P_r \tag{4}$$

assuming, of course, that a full 360° is concerned. When $r = \infty$, this becomes

$$2d_o\pi b^{-2} = P_\infty \tag{5}$$

If population is held constant in (5), we thus expect

$$d_o = (2\pi)^{-1}b^2 \tag{6}$$

and this indeed appears to be true.

In Figure 6.4, d_o is graphed against b for United States cities in 1950. Isolines of city size have been interpolated to hold P constant. Cities in any size class trend upward with the appropriate slope.

However, another expression exists for central density, d_o. Weiss (1961) found that for the United States in 1950 the density-distance gradient b could be calculated for any city by using the expression

$$b = (10^5/P_m)^{1/3} \text{ in mi}^{-1}, \tag{7}$$

where P_m is the population of the metropolitan areas. From this

$$P_m = 10^6 b^{-3} \tag{8}$$

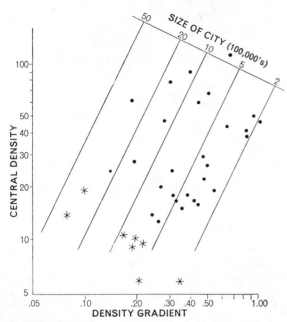

Figure 6.4
Central densities and density gradients related to city
size, United States cities, 1950.

Muth's data, for central cities, P_c, indicate the exponent of b to be -2.65, which is of the right relative order of magnitude.

Substitution of (8) in (5) results in the expression

$$d_o = (10^5/2\pi)b^{-1} \tag{9}$$

which obviously differs from (6), in which population is held constant.

Figure 6.5
Relationship between city size and density
gradient, United States cities, 1950.

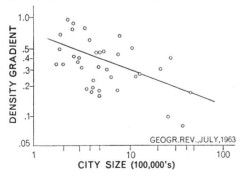

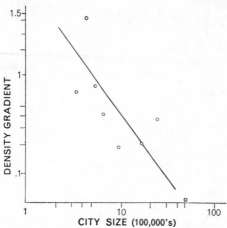

Figure 6.6
Relationship between city size and density
gradient, selected Asian cities, postwar.

Apparently (9) applies to different subsets of cities (for example, the subset
starred in Figure 6.4, which has the correct slope of -1) defined in terms of
factors influencing central density, d_o, regardless of city size. Equation (6)
applies to subsets of cities of the same size, and Equation (9) to subsets of
cities with similar histories of development.

Equation (8) is of further interest. It says that as the population of a
city increases, the density gradient diminishes, or that small cities are more
"compact" than larger cities by virtue of their steeper density gradients.
Figure 6.5 shows how the relationship holds for 36 United States cities in
1950 (Muth, 1961). Asian cities are graphed in Figure 6.6 (Tennant, 1961).
It is obvious that Asian cities, with a greater intercept and a steeper slope, are
far more compact than their United States counterparts of equivalent size,
yet the generalization that compactness diminishes with size again holds.

Factors influencing central density and density gradient

Enough has been said in connection with Equations (6) through (9) to
indicate that a more penetrating analysis is required of factors influencing
central density and the density gradient in any cross section of cities. For
purposes of analysis, Muth's data for United States cities will be used, as he
has already provided the parameters d_o and b for each in 1950.

The most obvious influence on population densities near the city center
is the age and mode of development and building. Older cities built with small
lots and subdivisions will have higher densities than cities built at other times

with other modes of subdivision. Winsborough (1961) follows Boulding (1956b) in arguing for the controlling influence of timing of development on subsequent form. "At any moment the form of any object, organism, or organization is the result of its laws of growth up to that moment," and "Growth creates form, but form limits growth." (This is Boulding's "first principle of structural growth." Winsborough says that "the timing of growth affects the density patterns .... Different influences on the pattern of development at different times and the resultant structure of the city sets limits on its subsequent growth....")

But age of cities is useful only in defining similar subsets of cities for which Equation (9) holds regardless of size. If size is held constant, central density is related to the density gradient, which in its turn may be influenced by a variety of additional factors. Thus central density is a function both of age and, as a composite surrogate for these other factors, of density gradient. A regression equation computed to quantify this functional relationship yielded the expression

$$d_o = 0.5302 - 0.6362 \text{ age} - 3.495 \, b^{-1}. \tag{10}$$

Both age (years since the city reached 50,000) and density gradient were significant at the 0.01 significance level, and 61 per cent of the variance of d_o was accounted for (Simmons, 1962). This is pleasing, since Muth's estimates of d_o were subject to error, but it also indicates that additional factors should be investigated in connection with the definition of subsets of cities in Equation (9). A better measure of age should be found, probably one indicative of the nature of the growth process, accounting in particular for differences in local transportation technology.

Equations (3), (4), and (5) assume a circular city, with integration of Equation (1) proceeding over the full 360° of the circle and the city center located at the center of the circle. Yet cities that conform to these assumptions are hard to find. Asymmetry and lopsidedness are common, elongations and crenulations many. Theoretically, at least, one would expect the density gradient to diminish as shape distortions increase, because areas that would normally be occupied by certain densities are now no longer available, and uses that prefer these densities must move outward to the nearest available sites (though with some inevitable changes because of substitution effects). We must find out whether this is so, and if shape distortion, size of city, and so on interact to create the overall density gradient.

Muth (1961), after a detailed multiple-regression analysis, rejected no fewer than nine different variables hypothesized to have some influence in determining density gradient: density of local transit systems, quantity of local transit trackage, area of the standard metropolitan area (SMA) in 1950, proportion of SMA growth between 1920 and 1950, median income, proportion of SMA sales in the central business district, proportion of substandard central-city dwelling units, proportion of urbanized-area male employment in manufacturing, and average density of central city. Only size of SMA and proportion of manufacturing outside the central city clearly appeared to bear significant relationships to b, though per capita car registrations showed a significant partial correlation, and the signs of other items, such as median income, indicated behavior in the right direction. Muth argued that size of city was significant only because other variables existed that were significant, and that could be approximated in sum by such a surrogate.

This leads us to postulate that density gradient is a function of size of city, shape distortion, and proportion of manufacturing outside the central city.

The index of shape distortion, A, was constructed as the ratio between the sum of distances of points arranged in a regular network within the boundaries of the city's urbanized area from the city center, and the sum of distances of points in the same regular network from the center of a circle of the same area as the city. A is 1.0 for a perfectly circular city and increases with distortion in shape. The index is highly sensitive to elongation or lopsidedness, but not to crenulations, and only slightly to a starfish pattern created by radiating transportation routes. It appears to be fairly highly correlated with physically created distortions, especially the presence of water bodies; thus A is also an index of the influences of city sites on population density patterns. The correlation between A and a site index S was 0.584. S was defined as WT where $W = 1 - $ (water area/total area) in a circle of the same area as the city centered on the CBD, and $T = 1 - $ tan (average slope). $S = WT = 1.0$ for a circular city on a level plain. Of the correlation between A and S, the greatest proportion of the covariance was accounted for by the W component, and indication that major shape distortions are largely a function of location alongside water bodies. A regression equation of the form

$$\log b = 3.08 - 0.311 \log P - 1.0 \log A + 0.407 \log M \qquad (11)$$

resulted for the sample of 46 United States cities. Only size of city (P) was significant at the 0.05 level, and scarcely 40 per cent of the variance of b was explained. However, there is reason to believe that at such conventionally high significance levels the risk of making errors of the other kind (rejecting true hypotheses) is somewhat too large for comfort, and there is thus reasonable doubt whether shape distortion (A) and spatial pattern of employment in manufacturing (M) should be rejected, the more so since Muth did find M to be significant. A final decision cannot be made at this time, and further work is required. The only positive conclusion is to reiterate the relationship already found by Weiss and Muth, and remarked briefly in passing by Clark, that b diminishes as size of city increases, so that smaller cities are more compact than larger.

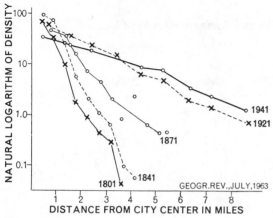

Figure 6.7
Density–distance gradients for London, 1801–1941
(after Clark).

Changes through time

The size-compactness relationship so far derived is cross-sectional. It applies to different cities in a region at the same period of time. One can argue from this cross-sectional pattern that as cities grow they should experience diminishing density gradients and degrees of compactness.

All the evidence indicates this to be true in cities of Europe, North America, and Australia. Clark found diminishing gradients through time for London, Paris, New York, Chicago, Berlin, and Brisbane, for example. The curves for London are reproduced in Figure 6.7. Table 6.1 lists central densities for Chicago from 1860 to 1950 by decades. A progressive decline

Figure 6.8
Chicago's diminishing density gradient,
1860–1950, compared with the 1950 United
States cross section.

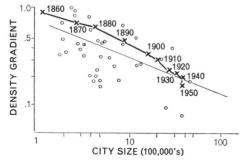

Table 6.1

Central densities and density gradients in Chicago,* 1860–1950

Decennial census	Central density	Density gradient
1860	30.0	0.91
1870	70.8	0.87
1880	96.6	0.79
1890	86.3	0.50
1900	100.0	0.40
1910	100.0	0.36
1920	73.0	0.25
1930	72.8	0.21
1940	71.1	0.20
1950	63.7	0.18

* Urbanized area.

in density gradient is evident, together with first an increase and later a decrease of central density. This phenomenon reappears in Clark's results, and Winsborough associated it with a shift in local transport technology. For example, in 1890, when mass transit was the most rapid and flexible transportation system available, there was a positive correlation between population concentration and intensive use of mass transit in Chicago; but by 1950, when mass transit was competing with automobiles, the direction of the correlation was reversed. Figure 6.8 shows the relation of Chicago's density-gradient time path to the 1950 cross-sectional picture for the United States.

Contrasting changes in western and non-western cities

As Western cities grow through time they experience steady decreases in density gradient, and therefore in degree of compactness, whereas central densities first increase and later decrease. But the same changes do *not* occur in non-Western cities. Figure 6.9 shows density gradients and central densities for Calcutta from 1881 to 1951. Central density increased steadily, but although the urbanized area did expand, the density gradient remained constant. This tendency toward increased overcrowding, with maintenance of a constant degree of compactness, appears to be characteristic not only for the rest of the Indian urban scene but also more generally in the non-Western world.

Figure 6.10 summarizes the cross-sectional and temporal patterns that may therefore be identified. At any point in time the empirical regularities

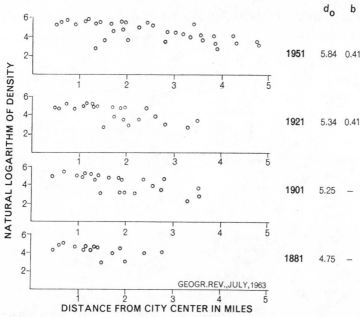

Figure 6.9
Density gradients for Calcutta, 1881–1951.

to be observed are the same for both Western and non-Western cities. But through time the patterns differ. In the West central densities rise, then fall; in non-Western cities they register a continual increase. In the West density gradients fall as cities grow; in non-Western cities they remain constant. Hence, whereas both degree of compactness and crowding diminish in Western cities through time, non-Western cities experience increasing over-crowding, constant compactness, and a lower degree of expansion at the periphery than in the West.

Colin Clark (1951) observed that there are "two possibilities for develop-ment, if the population is increasing. Either transport costs are reduced, enabling the city to spread out; or they cannot be reduced, in which case density has to increase at all points." In this, however, he identified the permissive factor for the accelerated *sprawl* of Western as opposed to non-Western cities (on the supply side) rather than the real reason for accom-panying differentials in density gradient (which is on the demand side).

Alonso showed that the rich, in Western cities, live at the periphery on cheap land and consume more land at lower densities than do the poor who live at the center. The Western world has also experienced a revolution in levels of living such that the richer, more mobile groups have increased not only numerically but also proportionally. Hence accelerated sprawl facilitated by improved transportation systems has been stimulated by greater

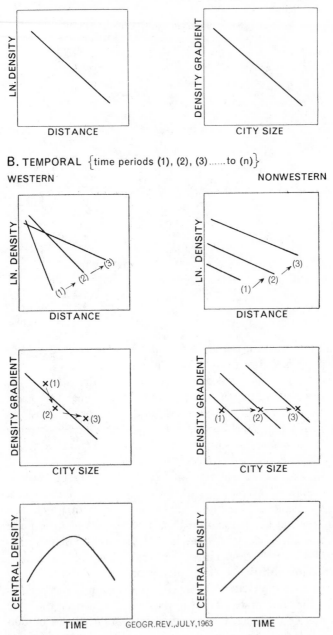

Figure 6.10
Cross-sectional and temporal comparisons, Western and
non-Western cities.

demands for peripheral lower-density land, with attendant reductions of the density gradient. The Western world has experienced significant changes in the nature of demand for residential land. Changed transport systems have merely ensured an adequate supply to meet the demands.

However, the socioeconomic pattern of non-Western cities is markedly different. In Western cities the poor live at the center and the more mobile rich at the periphery; in non-Western cities the reverse is true. The least mobile groups occupy the periphery. Any income improvements lead to greater demands for central locations and to increased overcrowding. Sprawl reflects projection of the overall surface outward as densities increase throughout, in a periphery of degrading and depressing slums. The degree of compactness of the non-Western city remains, therefore, relatively unchanged, with the least mobile groups located at the periphery. In spite of reductions of transport costs in non-Western cities, the groups located where the possibilities of saving are greatest are the groups least able to take advantage of the possibilities. Changes on the supply side occasioned by transport improvements are of little utility. Differences in movements of central densities and density gradients through time are a function of the inverted locational patterns of socioeconomic groups within Western and non-Western cities, and attendant contrasts in demands for residential land. These inverted locational patterns and other remarkable differences in the internal structure of Western and "preindustrial" non-Western cities will be examined in more detail in the following chapter.

Chapter 7
The social areas of the city: from classical to factorial ecology

The nature of the contrasting social and spatial patterns of Western and non-Western cities and the processes that give rise to these differing patterns should now be probed more thoroughly.

Relevant insights are many. For example, in his mid-nineteenth-century classic, Kohl (1841) devoted an entire chapter to the internal structure of cities. Urban structure, he said, might be viewed vertically as a series of layers, as in Figure 7.1: (1) the ground floor contains the establishments and living quarters of the businessmen; (2) the first floor is the "area of wealth and pleasure," the seat of the nobility; whereas (3–4) "upwards" people with lower income reside; the same (a and b) applies to the subterranean levels. These layers, he continued, translate into "arches" of homogeneous social structure, so that the economic and social importance of the population declines as you move out from the center of the city. However, some outer rings might contain a wealthy population because of "a growing sympathy for the countryside."

Parallels with this nineteenth-century European view can be found in studies of Indian cities. For example, Chatterjee (1960) concluded, in looking at the ecology of Calcutta's neighbor, Howrah:

> The influence of the caste system is reflected in the usual concentration of the higher castes in the central areas of good residential localities, while the lower caste groups usually occupy the fringe.... The people still attach more importance to these centrally-situated residential areas.... Thus, in spite of the modern development of road transport, the residential decentralization or movement towards the fringe outside the old residential areas is not very marked.

Sjoberg (1960) generalized the pattern to all "preindustrial" cities and contrasted it with "industrial" urbanization:

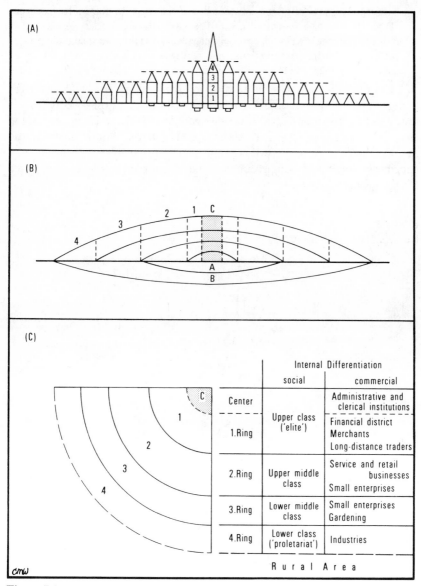

Figure 7.1
The ideal-type city: Its internal socioeconomic and
morphological differentiation (after Kohl).

The feudal city's land use configuration is in many ways the reverse of that in
the highly industrialized communities. The latter's advanced technology
fosters, and is in turn furthered by, a high degree of social and spatial mobility
that is inimical to any rigid social structure assigning persons, socially and
ecologically, to special niches.... [There are] three patterns of land use

wherein the pre-industrial city contrasts sharply with the industrial type: 1) the pre-eminence of the "central" area over the periphery, especially as portrayed in the distribution of social classes, 2) certain finer spatial differences according to ethnic, occupational, and family ties, and 3) the low incidence of functional differentiation in other land use patterns.

On the other hand, according to the classical papers of the Chicago School, the ecology of the industrial city—Sjoberg's norm—stems from a process of competition between various population or interest groups in the city, the dominance of particular groups within the natural areas of the metropolitan community, with change taking place by the invasion of a natural area by a competing group, and the ultimate succession to dominance of that group in the area.

Figure 7.2
Burgess' zones and ethnic communities.

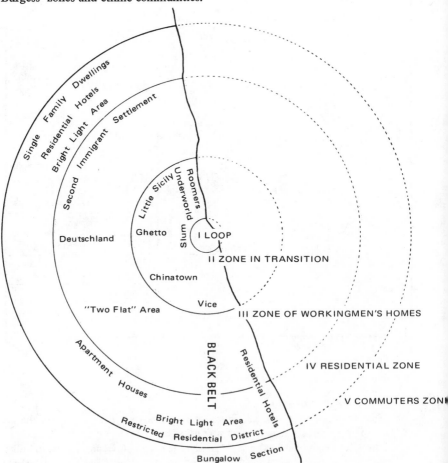

The process, the argument continues, derives energy from city growth, such that older population groups move from their original home to the periphery as their economic status improves, to be replaced at the city center by new arrivals. It results in Burgess's concentric zones of successively *increasing* status, subdivided into small natural areas on the basis of race ("the Black Belt"), ethnicity ("Little Sicily"), or type of structure ("residential hotels"), as depicted in Figure 7.2. It is this zonal result that Sjoberg contrasts with the preindustrial form.

There have, of course, been alternative views of the spatial patterns or the determinants of the internal structure of the twentieth-century industrial city. Hoyt (1939), for example, argued that high-rent and low-rent neighborhoods occupied distinct subareas of the city, and that these were not aligned concentrically about the city center but were distributed in a sectoral fashion. The spatial pattern of the city's rental areas was determined in Hoyt's view by the choice of those who could afford the highest rents. They pre-empt land along "the best existing transportation lines," "high ground—free from the risk of floods," and "land along lake, bay, river, and ocean fronts where such water fronts are used for industry." Others have suggested the significance of persistent cultural values attached to areas (Firey, 1947) and of "multiple nuclei" (Harris and Ullman, 1945). Still others have tried to identify and characterize new "postindustrial" ecologies characterizing contemporary metropolitan growth (Friedmann and Miller, 1965). How may all of these distinctive views be rationalized into a single comparative scheme? That is the question addressed in this chapter, using as a mechanism an investigation of Calcutta based upon the method of factorial ecology. But first, we review a more traditional mapping approach to Calcutta's social ecology. This is followed by an explication of the methods of factorial ecology, application of those methods to Calcutta, and then by a return to the questions of comparative ecology.

Bose's analysis of Calcutta's ethnic diversity

Bose's ecological study of Calcutta (1965, 1968), one of the most important available to date, was phrased largely in terms of ethnic variability and cultural differences. Bose depicted Calcutta as "the scene of a major confrontation between the enduring institutions of old India—her caste communities and diversity of ethnic heritages—and the pressures and values arising from the process of urbanization that presages India's industrial revolution."

Traditionally, Indian society was based upon caste: Brahman priests and teachers; Kshatriya warriors and rulers; Vaisya landowners, merchants, and money-lenders; Sudra cultivators, workers, and artisans; and the "untouchables" (today's "scheduled" castes)—these were the classical ideals

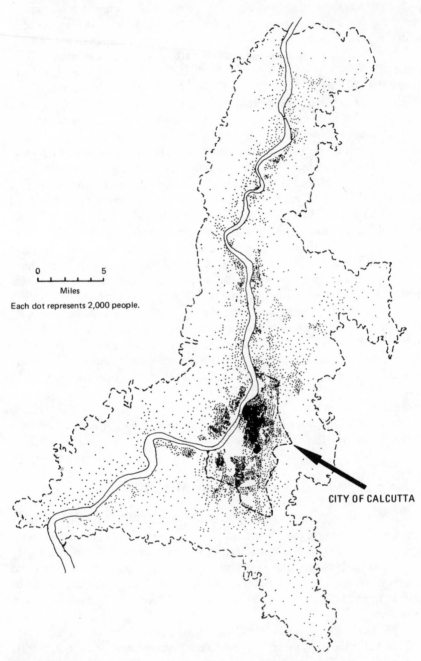

0 |___|___|___| 5
      Miles

Each dot represents 2,000 people.

CITY OF CALCUTTA

Figure 7.3
Population distribution in the Calcutta metropolitan district, 1961.

in a complex system of more than 12,000 caste communities. Each was identified with occupation and maintained by intramarriage and spatial monopoly of particular trades in particular locales. Ranking of occupations

Figure 7.4
Dominant land use.

BUSTEES
INDUSTRIAL
INSTITUTIONAL
COMMERCIAL

into a hierarchy that generally coincided with ritual practices and residence of different castes in separate quarters maintained distinctiveness in the system and provided social content for the urban ecology.

It is this pattern, according to Bose, that is being modified in Calcutta, India's largest city. The city of Calcutta is a legal unit of more than three million people set within a metropolitan area of seven million (Figure 7.3). Created by the British during the eighteenth century, and once the seat of the British raj, it is today a leading industrial center, a major port, a national metropolis, and the focus for eastern India. One quarter of the city's population arrived after 1947 as refugees from East Pakistan. Seventy-five per cent of the population lives in tenements or bustees (Figure 7.4). More than half the multimember families have only one room to live in.

Figure 7.5
Bengali-speaking Hindus of upper, middle, and artisan classes (after Bose).

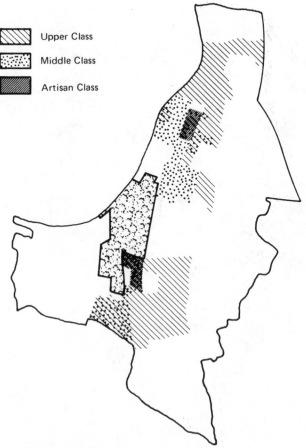

Upper Class

Middle Class

Artisan Class

There are three males to every two females in the city, and half the "households" consist of single men helping to support families remaining in villages elsewhere in India. Within the bustees there is but one water tap for every 25 persons and no means for disposing of sewage. Cholera is endemic. Life is often brief and brutal.

Bose pointed out that Calcutta still houses many traditional communities (Figs 7.5–7.10), for, although the city was a British creation, it was occupied by Indians who brought to it traditional Indian institutions. Among the Bengali-speaking Hindus, the upper castes (Brahmans, Kayastha scribes, Vaidya physicians) originally lived in the northern half of the old "native quarter" at the heart of the city (Wards 6, 8, 9, 10, 19, 20, 21; see Figure 7.11), but today they have also moved southward into former European and Anglo-Indian residential areas. The middle commercial and artisan castes

Figure 7.6
Bengali poor (after Bose).

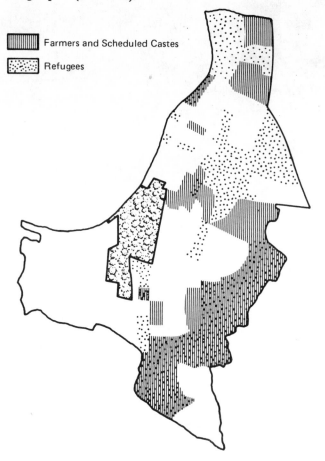

Farmers and Scheduled Castes

Refugees

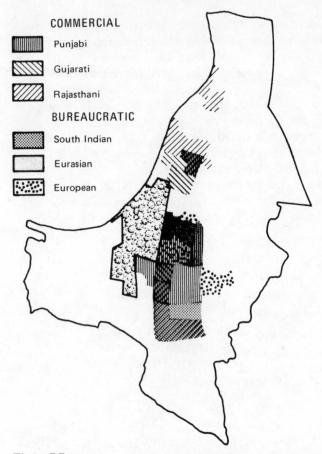

Figure 7.7
Non-Bengalis of the commercial and bureaucratic classes
(after Bose).

(Subarnabanik bankers, Gandhabanik spice merchants, Kansari brass-workers, Tantubanik weavers) were concentrated in the commercial segment of the native quarter to the north and east of Dalhousie Square (Wards 21–27, 41–42). The lower scheduled castes (Jaliva fishermen, Hari and Dom bamboo workers and agricultural laborers, Dhopa washermen) lived around the outer eastern and northeastern fringes of the city. These fringes have, since partition, also become the residential areas of many castes of Bengali-speaking Hindu refugees from East Pakistan. The refugees also moved into many formerly Muslim wards at the northern and southern extremities of the city.

Bose also described the wide variety of non-Bengalis in the city. Oriya craftsmen can be found scattered in bustees throughout the Bengali wards.

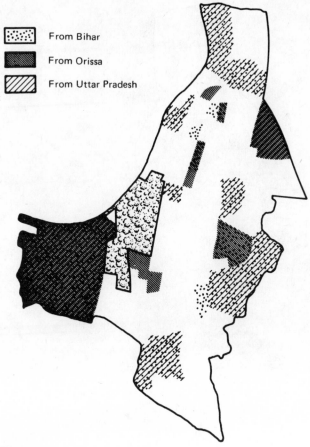

From Bihar

From Orissa

From Uttar Pradesh

Figure 7.8
Non-Bengali poor (after Bose).

Hindi speakers from Uttar Pradesh supply porters for the carrying trade in the commercial wards (19, 22, and so on) and form the bulk of the industrial laborers elsewhere. Kalwar traders live alongside the Bengali commercial castes. The Rajasthanis and Marwaris traditionally occupy the commercial core (Wards 24–26, 37, 39). South Indian immigrants are concentrated in Wards 62–64.

The former European section was in the contiguous wards immediately east and south of the Maidan (a large park including old Fort William, stretching along the bank of the Hooghly River). The Europeans formerly occupied the uppermost levels of the city's social hierarchy. They lived in palatial houses with large gardens and open spaces, which they owned or rented from Bengali landlords. Their residences are now being bought by Rajasthanis, Punjabis, and other prosperous non-Bengalis. Anglo-Indians

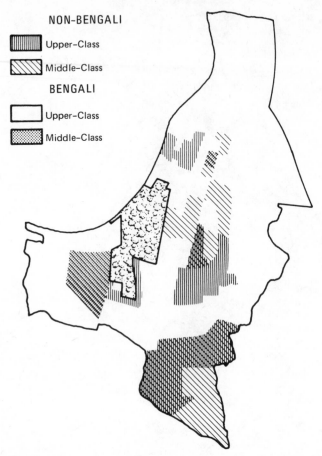

Figure 7.9
Upper- and middle-class Muslims (after Bose).

and Indian Christians used to be employed, under European patronage, in the railways, docks, and commercial establishments. Today the Indian Christians are occupationally indistinguishable from, say, other Bengalis if they are Bengali-speaking. The Anglo-Indians have been migrating away from the city and even from India. The Muslim population, although it is not fractionated by caste, is quite explicitly stratified by class. Two large Muslim quarters surround the places of residence in the southwest and south of the city that were furnished by the East India Company to the Nawab of Oudh and the descendants of the Tippoo Sultan of Mysore. The Muslim commercial middle class lives in wards near the central business district; the lower-class Muslims live in the tenement and slum districts of the east and northeast and in large tracts of the city surrounding the old centers of the Muslim aristocracy in the south and southwest. Many lower-class

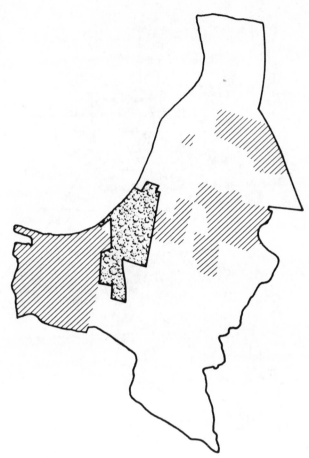

Figure 7.10
Lower-class Muslims (after Bose).

Muslims used to be employed in the soap and leather industries, regarded among Hindus as "polluting" occupations reserved for "low" caste people.

These differences can be seen in myriad and minute ways in the daily life of the city. For example, Dutt shows, in a study of Calcutta's local markets (1966), that

the nature of commodities sold in a market depends largely on the community composition of the locality. For example, in Burrabazar Market, we find a large number of fruit stalls. Mechua Bazar Market has no fish or meat stall. This can be explained by the fact that the community composition of that particular area is of non-Bengali vegetarians. In the Muslim- and Christian-dominated localities there is a large concentration of beefstalls in the markets. The examples are Rajabazar, Mullik Bazar, Taltola, Park Circus and New

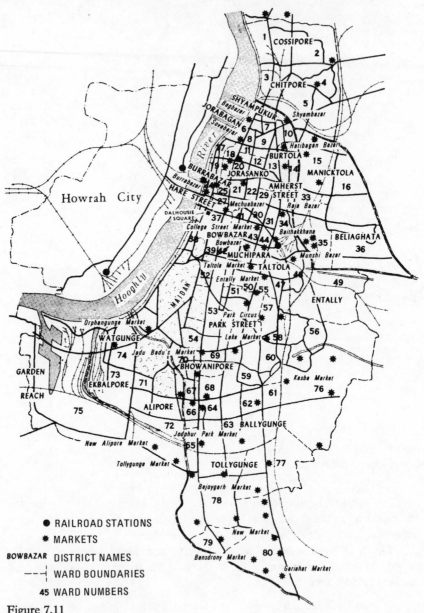

Figure 7.11
Wards, districts, and markets.

Market. In Bengali Hindu-dominated localities, fish stalls predominate. The examples are Shyambazar, Hatibagan, Sovabazar, Gariahat, Kasba, Lake, Bagbazar, Minshi Bazar markets. In the "mixed" localities which consist of

Muslims, Christians, and Bengali Hindus, markets have stalls of vegetables and fruits on one side, and fish, meat and beef on the other. Such markets are New Market, College Street, Entally and Park Circus markets.

Bose concludes:

> The map of Calcutta thus shows a highly differentiated texture. Ethnic groups tend to cluster together in their own quarters. They are distinguished from one another not only by language and culture but also by broad differences in the way they make their living. Naturally there is a considerable amount of overlap, but this does not obscure the fact that each ethnic group tends to pursue a particular range of occupations.

> . . . .

> It can be said, therefore, that the diverse ethnic groups in the population of the city have come to bear the same relation to one another as do the castes in India as a whole. They do not enjoy monopoly of occupation, as under caste, nor are they tied to one another by tradition in reciprocal exchange of goods and services. There is also no ritual grading of occupation into high and low. But preference for or avoidance of some kinds of work are expressed in class differences among occupations, as can be observed elsewhere in the world. The social order of Calcutta might therefore seem to be evolving through a transitional stage, in which caste is being replaced by an increasingly distinct class system.

> . . . .

> Actually, the superstructure that coheres the castes under the old order seems instead to be reestablishing itself in a new form. Calcutta today is far from being a melting pot on the model of cities in the U.S. There the Irish, Italian and eastern European immigrants have merged their identities within a few generations. The communal isolation of the first generations was quickly reduced by occupational mobility in the expanding American economy and by the uniform system of public education that Americanized their children.

> . . . .

> In Calcutta, the economy is an economy of scarcity. Because there are not enough jobs to go around everyone clings as closely as possible to the occupation with which his ethnic group is identified and relies for economic support on those who speak his language, on his coreligionists, on members of his own caste and on fellow immigrants from the village or district from which he has come. By a backwash, reliance on earlier modes of group identification reinforces and perpetuates difference between ethnic groups.

An approach via factorial ecology

If Calcutta is, as Bose argues, the scene of confrontation between traditional Indian institutions and industrial urbanization, then one should be able to see in its ecology not simply the ethnic diversity of region, caste, and culture that he describes, but also the two further ecological themes of the preindustrial city categorized by Sjoberg (pre-eminence of center over

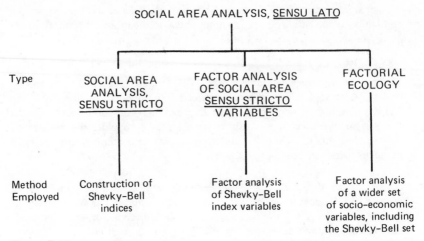

Figure 7.12
Typology of social area analysis.

periphery, low incidence of functional differentiation in land use) and an increasing admixture of the ecological patterns of the industrial city. One way to determine whether this is the case is by assuming the analytic stance of factorial ecology.

Factorial ecology is an outgrowth of social area analysis. Since there exists semantic confusion about this term, some clarification is required (Fig. 7.12).

Social area analysis, sensu stricto

The term *social area analysis, sensu stricto,* applies only to that mode of analysis originally outlined by Shevky, Williams, and Bell in their studies of Los Angeles and San Francisco (1949, 1953, 1955). From a number of postulates they derived three basic constructs which, they considered, describe the way in which urban populations are differentiated in industrial society. The three constructs were called social rank (economic status), urbanization (family status), and segregation (ethnic status). They then proposed three indexes, one per construct, each made up of from one to three census variables, designed to measure the position of census-tract populations on scales of economic, family, and ethnic status. The analysis also made possible the classification of census tracts into social areas based upon their scores on the indexes.

Social area analysis, *sensu stricto,* has been criticized both on theoretical grounds (the theory underlying the constructs) and for empirical reasons (the method of dimensioning the constructs) (Hawley and Duncan, 1957;

Duncan, 1955). Bell (1955) attempted to meet the empirical objections that the social area analysts selected measures on the assumption that the constructs were correct but failed to provide a test of their validity. He used factor analysis to show that, in the cases of Los Angeles and San Francisco, the census measures selected to construct the indexes in fact formed a structure consistent with Shevky's formulations. Van Arsdol, Camilleri, and Schmid (1958) went on to extend Bell's test of the Shevky model to some ten cities, six of which confirmed the validity of the Shevky indexes as measures of his constructs. The fact that four cities did not conform to Shevky's constructs suggested, however, that the existence of the constructs should be left as an empirical question to be determined by the patterns in the variables, rather than one to be assumed correct a priori. The logical extension of this argument is that many more variables detailing the way in which census-tract populations vary according to socioeconomic characteristics should be included in any study and that factor analysis should be used to isolate the fundamental patterns of variation in the data, be they Shevky-Bell construct patterns or otherwise. Whether this viewpoint is valid may be debatable, but nontheless it constitutes the premise on which the balance of this chapter is founded.

Factorial ecology

Factorial ecology is the term now used to characterize studies involving this application of factor analysis to ecological study. A data matrix is analyzed containing measurements on m variables for each of n units of observation (census tracts, wards...), with the intent of (1) identifying and summarizing the common patterns of variability of the m variables in a smaller number of independent dimensions, r, that additively reproduce this common variance; and (2) examining the patterns of scores of each of the n observational units on each of the r dimensions. The dimensions isolated are an objective outcome of the analysis. *Interpretation of the dimensions (factors) depends on the nature of the variables used in the analysis and the body of concept or theory that is brought to bear.* Theory provides the investigator with a set of expectations regarding the factor structure which can be compared to the actual set of factors produced. This comparison was made formally by Van Arsdol, Camilleri, and Schmid but somewhat less formally by later writers. What is impressive, however, is that studies of American cities have, by and large, succeeded in isolating the three social area dimensions originally proposed by Shevky: socioeconomic status, family status, and ethnic status.

Moreover, it becomes increasingly evident that each of these dimensions captures the essential features of one of the classic spatial models: socioeconomic status (Hoyt); family status (Burgess); ethnic status and "segregation" studies (Firey). Thus, in Western metropolises born in the industrial age and populated by a variety of races or national groups, "the models are

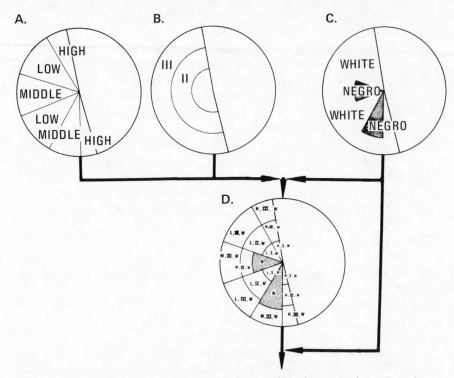

independent, additive contributors to the total socioeconomic structuring of city neighborhoods."

Those indexes that measure the socioeconomic status of individuals or groups vary principally by sector; those variables that measure the family and age characteristics of the population vary principally by concentric zone; those measures that isolate a minority group within the population show a tendency for that group to cluster in a particular part of the metropolis. Such is the conclusion of Anderson and Egeland (1961) in their analysis of the spatial variance of a number of socioeconomic measures within four medium-sized American cities (Akron, Dayton, Indianapolis, and Syracuse). Those variables that were thought to measure social status were combined into a "social rank" index, and those variables that were considered indicative of family status and degree of urban acculturation were combined into an index of "urbanization" in the manner originally developed by the social area analysts. An analysis of the variance of these two indexes by sector and by concentric zone revealed that social rank varied principally by sector, and urbanization by concentric ring. A third dimension corresponding to their "segregation index" is also present and describes the concentration of particular minority groups in limited neighborhoods of the city.

This basic triad of spatially arranged social dimensions can be super-

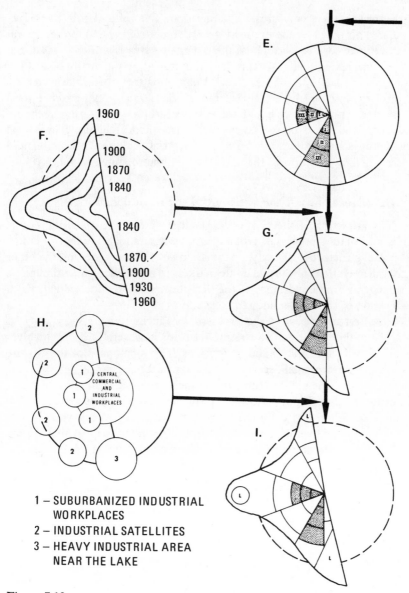

Figure 7.13
Integrated spatial model of the metropolis.

imposed to form, at the intersections of sectors, zones, and segregated areas, *communities* of similar social, family, and ethnic status. Figure 7.13 provides an idealized picture into which distortions can be successively introduced to approximate reality more closely.

However, there is a further complication. The zones within the segregated area occupied by the minority group do not correspond to the general life cycle zones of the metropolis; the segregated area is a microcosm of the whole, compressed spatially, reproducing in miniature the metropolitanwide pattern (Figure 7.13, E). This modified pattern is then further distorted by city growth (Figure 7.13, F). "Tear faults" develop as zones cross sectoral boundaries, with displacement of zones outward in the early-growth sectors. Finally, introduction of secondary work-place nodes—a heavy industrial area in the southern part of the city, industrial satellites in a crescent— further changes the form of the model for the metropolitan region.

Behavioral basis: the residential location decision

The attraction of the classical ecological view of the industrial city is that it postulates a process of group competition and mobility that produces the spatial structure of the city's population. In Hoyt's model, on the other hand, the operations of the real estate market are not spelled out, although he does recognize the importance of the decisions of prestigious individuals in the location of the high-grade rental sector.

An alternative view of the processes producing the structure may be proposed in the case of the North American city, however. This view involves the behavior of individuals and institutions. The inhabitants of the city are faced with a fundamental decision: where to live. The principal determinants of such a housing choice are three in number: the price of the dwelling unit (either in rental or in purchase-value terms); its type; and its location, both within a neighborhood environment and relative to place of work. These determinants have parallels in the attributes of the individual making the housing choice: the amount he is prepared to pay for housing, which depends on his income; his housing needs, which depend on his marital status and family size, that is, his stage-in-life cycle; his life-style preferences, which will affect the type of neighbor he will want; and, finally, the location of his job. When the values of the two sets of characteristics match, a decision to purchase housing will be made (see Table 7.1).

The most important determinant is undoubtedly income. A large family

Table 7.1

Determinants of housing choice

Individual characteristics		Housing characteristics
Income	↔	Price
Stage-in-life cycle	↔	Type of home
Life-style preferences	↔	Neighbors, type of community institutions
Attitude toward journey to work	↔	Location with respect to the job

with many children may need a large single-family home, but unless family income is above that minimum threshold required by the mortgage company, the purchase of this type of home will not be possible. Similarly, low wages may make it essential for a worker to live near his place of employment in order to minimize the costs of the journey to work. He may therefore be unable to satisfy his preferences as to type of neighborhood.

It may be objected that other factors such as the nature of the school system in the community or the location of the prospective home relative to a golf course play an important role in the decision, but these influences become significant only if the individual's income is large enough for him to be able to ignore the more basic determinants. Gans, in his participant-observer study of the inhabitants of a new suburban town in New Jersey (1967), asked a sample of Levittowners what their principal reason was for moving to Levittown. Over 80 per cent gave house-related reasons for moving.

Gans's study also contains many valuable illustrations of the way in which some of the residential choice determinants outlined above operate. In response to a perceived demand for middle-income homes, Levitt built houses during the 1950s with prices ranging from $11,500 to $14,500. To ensure that he got paid for the houses he built, Levitt told his salesmen to apply an informal minimum-income limit of $5,500 a year for the would-be house purchaser. Anyone with less income than that was not to be sold a house. In fact, a small percentage of house buyers did have lower incomes, but the majority had higher. The house price acted as a selection mechanism: the middle-range prices attracted predominantly lower-middle-class families with small minorities of working-class and upper-middle-class families.

The houses were free-standing single family residences with three or four bedrooms, and, not unnaturally, young families with (or intending to have) children became their occupants. They were spacious and comfortable homes for a four- or five-member household, and the surrounding yard provided play space for the young children, over whom the mother could keep an eye from the kitchen window. Since Levittown was an entirely new settlement, there were no neighbors in the community when the first inhabitants arrived, but the builder was careful to project a prestigious image in nearby Philadelphia (in order to counteract the unfavorable impression created by his other venture in the urban region, which had experienced some trouble over racial integration and subsequent status decline), and, for the first two years, until legal action forced Levitt to integrate, discrimination was practiced against prospective black buyers.

The residential location decision was thus clearly a product of the interaction of the home purchaser and the home builder—not an unexpected conclusion. Their choices either effect (in the case of a new community) or affect (in the case of an established community) the urban ecology, creating relatively homogeneous areas containing similar sorts of people (areas called "natural" in the earlier ecological literature and generally "social"

A. SOCIAL SPACE Units: Individual or Families

B. HOUSING SPACE Units: Dwellings

High Rise Apartment   Low Rise Apartment   Town House   Duplex   Single Family House

Value and Quality

Hi

Type

High Price or Rent

High-Medium Price or Rent

Average Price or Rent

Low-Medium Price or Rent

Low-Price or Rent

No individual lot or private yard High Density   Smaller Lot and yard Medium Density   Small Lot and Yard Medium Density   Large Lot and Yard Low Density

Hi = individual i's home in housing space

Single Small Family Childless Couples   Large Family Married with Children

Socioeconomic Status

Si

Stage in Life Cycle

Professional
College Education Managerial
Sales
High School Clerical Education
Craftsman
Operative
Grade School Education Laborers
Underemployed
Little Schooling Unemployed

High Income
High Middle Income
Average Income
Low Middle Income
Low Income

Luxurious
Most Spacious
Spacious Comfortable
Average Average
Crowded Could be improved
Overcrowded Blighted

Post Child-bearing Ages and/or Early Adulthood   Pre-School, School and Child-bearing Ages

Si = individual i's position in social space

128

C. COMMUNITY SPACE Units: Tracts or Larger Sub-areas   D. LOCATIONAL OR PHYSICAL SPACE
Units: Tracts or Larger Sub-areas

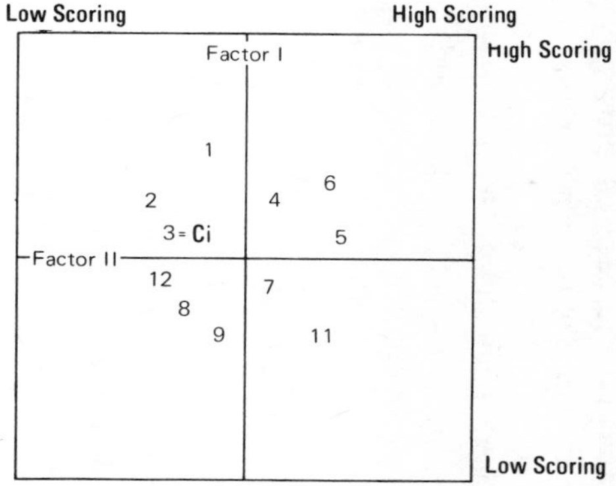

C_i = the community in which i's home is located

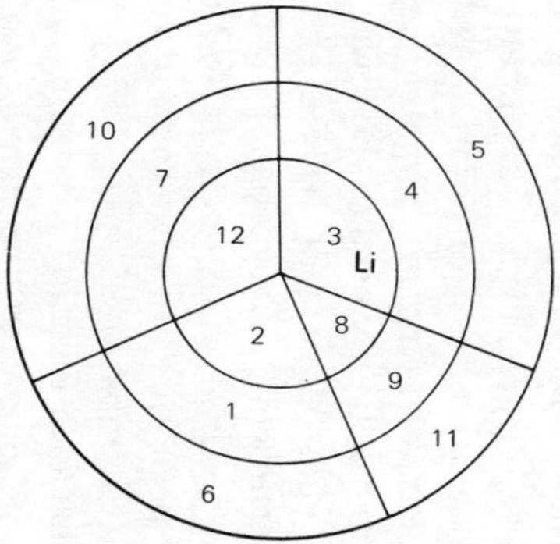

L_i = the zone in the community in which i's home is located

Figure 7.14
Residential location decision process.

today). The character of established social areas, in turn, influences subsequent housing choices of later neighborhood occupants, because people seek compatible neighbors who share the same views on life (particularly on such important questions as the way in which children should be brought up), and who find this compatibility by selecting their residence in the appropriate social area.

The choice of house and community type is only one part of the complete residential location decision, however. The housing consumer is also faced with the problem of *locating* his residence. This involves attitudes toward the journey to work; the time and cost of commuting have to be traded off against the relative benefits of living in alternative communities that meet, within budgetary constraints, family needs. The lower the family or individual income, the more constrained will be the choice. Thus, lower-status people live closer to their work than people of higher status.

Therefore, to summarize more generally with reference to Figure 7.14: the individual or family occupies a position, s_i, in *social space*, determined by economic status and family status (Chombart de Lauwe, 1952). The household matches this position with that of a dwelling located in an analogous position, h_i, in *housing space*, and of housing in a similar location, c_i, in an equivalent *community space*, whose axes comprise socioeconomic status on the ordinate and familial characteristics on the abscissa. From a range of possible communities found in the same zone of community space, one dwelling in one community is selected on the basis of proximity to job location (if this is a constraint) or on the basis of other important neighborhood characteristics, thus fixing the choices in *physical space*.

An orderly social ecology results through like individuals making like choices, through regularities in the operation of the land and housing markets, and through the collaboration of similar individuals in excluding those of dissimilar characteristics from their neighborhood or in restricting certain minority groups to particular areas. The autonomous suburb is the prime example of the process of exclusion, and the ghetto the most glaring illustration of the process of restriction. Sectoral patterning of such attributes of the neighborhood residents as education, occupation, and income, and of neighborhood structural characteristics such as rent or value and quality of housing, is a product of the differing abilities of various income groups to bear the costs of the journey to work. Lower-income workers, because of their restricted budgets, must live close to their work (concentrated in the inner city around the central business district and along the rail and water routes radiating outward from it). The higher incomes of upper-status workers give them the freedom to locate their homes in areas of higher residential amenity—away from their places of work, away from the smoke and dirt of industry, and close to amenity features such as lakeshore and open space. The age structure of the population, average family size, and female labor force participation change as distance from the city center increases: young families locate farther from the center than do older families. This pattern is a response to the change in house age and type as distance from the center becomes greater—the houses are newer, and single-family homes predominate as the city center is left behind. It is the lower land values toward the urban periphery that make possible this land-voracious construction, and the increasing real income of home buyers makes possible the purchase of such newer houses. Finally, minority groups find themselves segregated from the rest of the population to a greater or lesser degree as a result of recent arrival in the city, discrimination in the housing market, or through choice of home in congenial communities.

Comparative evidence: several possible factor structures

From these bases of residential choice, the ecological pattern of American cities is dimensioned cumulatively by socioeconomic status, family status, and the constraints of race and ethnicity. What of non-American cities? Do they exhibit the same patterns of variation in census tract populations?

Pedersen's study of Copenhagen (1967) provides one of the most comprehensive urban ecological analyses to date. From a matrix of 14 socioeconomic variables (related to age distribution, employment status, distribution by industry, household size, sex ratio, and female employment of the population) and 76 zones of the city, three basic factors emerged in both 1950 and 1960: (1) an urbanization or family status factor; (2) a socioeconomic status factor; and (3) a population growth and mobility factor. The first factor, when mapped and graphed, displayed the classic Burgess pattern of concentric rings; the second factor exhibited sectoral distribution, with the exception that the central zones of the city were of uniformly low status; the third factor exhibited highest values in the zone of new suburban development and in the low-status areas of the inner city. Pedersen devotes a separate chapter to detailed consideration of each of the three principal factors. Age structures of the communities of the city are considered as functions of the population growth and migration process. Income differentials in the suburban portion of the metropolitan region are shown to have narrowed over time as the area has been converted from farmland into residences for city dwellers, whereas the gap between inner-city and suburban incomes has widened. Absent from Pedersen's findings is a differentiation of Copenhagen's population on linguistic or racial lines—such differentiation does not exist in Denmark.

Confirmation of such homogeneity of population in Scandinavian cities is found in Sweetser's comparison of Helsinki and Boston (1965). In addition to the basic social area dimensions, the Boston study yielded a number of ethnic dimensions (Irish middle class; Italian blue collar). The Swedish-speaking population of Helsinki, however, failed to load on a separate factor but rather loaded on the socioeconomic status factor, since they were of generally higher status than the Finnish-speaking population.

Studies of non-Western cities highlight the cultural context of the American and Scandinavian factor structures and have led recent investigators to try to isolate those basic conditions in the urban system, social and spatial, which are necessary in order to produce the observed factor structures. A principal conclusion of Janet Abu-Lughod's study of Cairo (1968b) was that "no factorial separation between indicators of social rank and the in-

Table 7.2

Factor conditions	Types of variables used	Necessary conditions based on Abu-Lughod (1968b)
Socioeconomic status factor	Education, occupation, income	(1) That the effective ranking system in a city be related to the operational definition of social status; (2) that the ranking system in a city be manifested in residential segregation of persons of different rank at a scale capable of being identified by the areal units of observation used in the analysis
Family status factor	Family size, portions of the age pyramid, fertility	(1) That family types vary, either due to "natural" causes such as those associated with sequential stages in the family cycle, or to "social" causes such as those associated with other divisions in society, whether ethnic, socioeconomic, or other; (2) that subareas within the city are differentiated in their attractiveness to families of differing types at a scale capable of being identified by the areal units of observation used in the analysis
Disassociation between socioeconomic status and family status dimensions		Either (1) that there exists little or no association between social class and family type; or (2) if there is some association between social class and family type, that (a) there is a clear distinction between stages in the family cycle, each stage being associated with a change of residence; and that (b) "subareas within the city offer, at all economic levels, highly specialized housing accomodations especially suitable to families at particular points in their natural cycle of growth and decline" at a scale capable of being identified by the areal units of observation used in the analysis, and (c) "cultural values permitting and favoring mobility to maximize housing efficiency, unencumbered by the 'unnatural' frictions of sentiment, local attachments, or restrictive regulations"

Table 7.2—*continued*

Factor conditions	Types of variables used	Necessary conditions based on Abu-Lughod (1968b)
Separate minority group	Proportion of the subarea population in the minority group, measures of the relative concentration of minority groups	(1) That the characteristic(s) differentiating the minority group from the rest of the population be of perceived significance in the social system, i.e., that the urban population be truly heterogeneous; (2) that the minority group be residentially segregated from the rest of the population, at the scale of observation used in the analysis
Disassociation between minority group and socioeconomic status dimensions	Measures used above for minority group, measures of education, occupation, and income (and the distribution of resources)	(1) That the minority groups be residentially segregated from the rest of the population, at the scale of observation used in the analysis; and either (2) that there exists little or no association between minority group status and socioeconomic status; or (3) that there is some association between social class and minority group status, that (a) the minority group still spans most of the social status range, though it may be concentrated at the lower end of the range; (b) a fairly full range of housing accommodation quality be available for families within the residentially segregated area
Disassociation between the minority group and stage-in-life-cycle dimensions	Measures used above for the minority group, measures of age distribution, family size, and fertility	(1) That the minority group be residentially segregated from the rest of the population at the scale of observation used in the analysis; and either (2) that there exists little or no association between minority group status and family status; or (3) that if there is some association between family status and minority group status that (a) minority groups still span most of the family status range, though it may be concentrated at one end of the range; (b) a fairly full range of housing accomodation types be available for families within the residentially segregated area

dicators of family cycle stage could be obtained." This contrasts with the normal separation of these two sets of indicators in factor analyses of American city data matrixes. As a result, Abu-Lughod was led to outline, in an extremely effective way, the conditions that were necessary and sufficient to produce the dimensions of socioeconomic status and family status that are found to have independent existence in almost all American cities, conditions that were not fulfilled in the case of Cairo, as shown in Table 7.2.

These conditions shown in Table 7.2 in fact give rise to several alternative factor structures representing permutations of the three basic sets of variables (SES: socioeconomic status set; LC: stage-in-life cycle of family status set; MG: minority-group set [only one minority group is discussed here, although clearly the argument could be extended to several]):

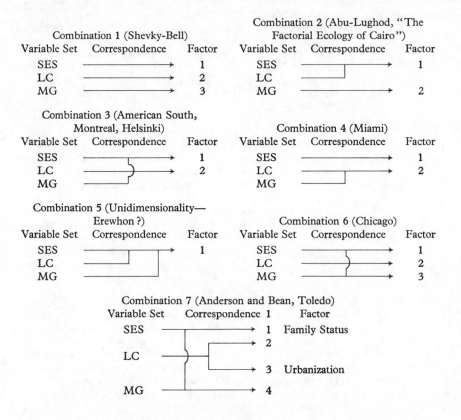

Combination 1 (Shevky-Bell)

Variable Set	Correspondence	Factor
SES		1
LC		2
MG		3

Combination 2 (Abu-Lughod, "The Factorial Ecology of Cairo")

Variable Set	Correspondence	Factor
SES		1
LC		
MG		2

Combination 3 (American South, Montreal, Helsinki)

Variable Set	Correspondence	Factor
SES		1
LC		2
MG		

Combination 4 (Miami)

Variable Set	Correspondence	Factor
SES		1
LC		2
MG		

Combination 5 (Unidimensionality— Erewhon?)

Variable Set	Correspondence	Factor
SES		1
LC		
MG		

Combination 6 (Chicago)

Variable Set	Correspondence	Factor
SES		1
LC		2
MG		3

Combination 7 (Anderson and Bean, Toledo)

Variable Set	Correspondence 1	Factor
SES		1 Family Status
		2
LC		
		3 Urbanization
MG		4

These combinations constitute the frame within which a factorial ecology of Calcutta can be evaluated.

Calcutta's ecology: dimensions of variation, 1961

Principal axis factor analysis with normal varimax rotation of all factors with eigenvalues exceeding unity was used to analyze an 80-ward × 37-variable data matrix for Calcutta. The variables came from the 1961 census and represent all the quantitative materials available at the time we performed the analysis. They were obtained as special hand tabulations (as of the time of writing, the census tables had yet to be published for Calcutta), and they relate to family structure, literacy, type of employment, housing characteristics, and land use.

The units of observation were the city's eighty municipal wards—administrative units which, as Bopegamage (1966) points out, possess social, economic, and demographic homogeneity only by accident. This lack of homogeneity is regrettable, but perforce to be tolerated. Because, too, the wards vary in their total population size, this variable was left in the analysis as a "screen" for the possible influence of size differences. All other variables were converted to ratios (population density, average size of household) or to percentages (percentage of workers employed in manufacturing, and so forth).

The factor analysis routine utilized the University of Chicago's computation system and produced as output matrixes of descriptive statistics for each variable, simple correlations among variables, factor loadings for ten factors (correlations of variables with factors), eigenvalues indicating that the ten factors reproduced 79 per cent of the total variance or 95 per cent of the common variance, and orthonormal factor scores for each ward with respect to each factor.

Not all of these materials can be reproduced here; instead, only the principal results of the analysis will be reviewed. However, the findings do reveal that alongside the rich ethnic variability described by Bose, Calcutta is also characterized by a broadly concentric pattern of familism; an axial arrangement of areas according to degree of literacy; and both substantial and increasing geographic specialization of areas in business and residential land uses, gradually replacing the former mixture of businesses and residences that were instead separated into occupational quarters. This mixture of preindustrial and industrial ecologies thus lends support to the idea that the city is in some transitional developmental stage.

Factor 1: a land use and familism gradient

According to Factor 1, the larger peripheral wards of the greatest population size (factor loading 0.762) are also those with the greatest percentages of females over fifteen (loading 0.817) and married women (0.749), the greatest non-Bengali component (0.591), proportion of buildings solely dwellings (0.892), with a higher proportion of the houses mud-walled (0.545). Conversely, the smaller wards at the heart of the city have greater proportions of their buildings occupied by businesses and offices (-0.825) and by commercial establishments and restaurants (-0.518), greater proportions of the population employed (-0.796), a greater mixture of the "houseless" male population (-0.727), and a greater percentage of the population residing in institutions (-0.813).

A sixth-degree trend surface fitted to the factor scores reproduced 78 per cent of the variance in the scores in terms of spatial regularity and shows a clear concentric pattern around the city center (Fig. 7.15).

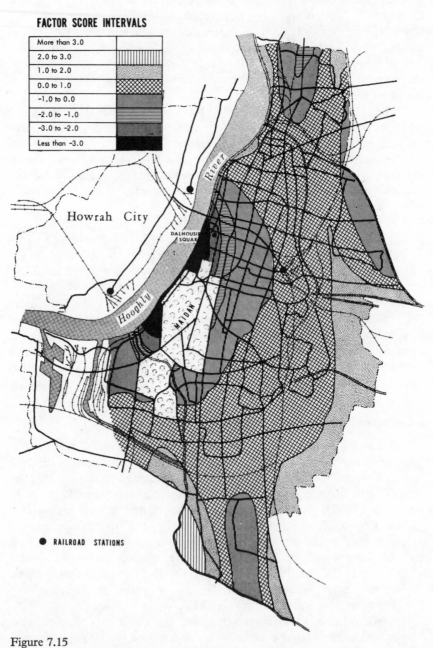

Figure 7.15
Factor 1: land use and familism.

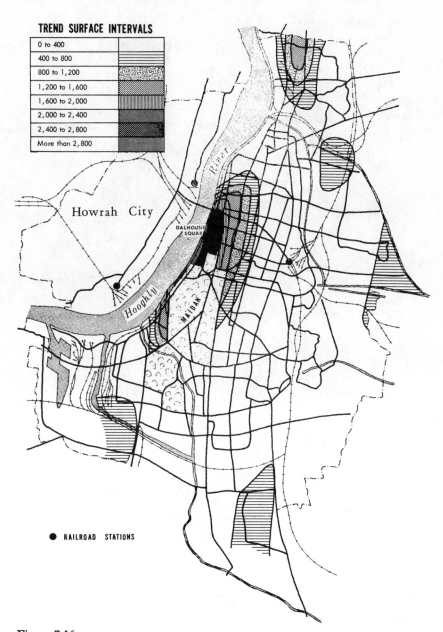

Figure 7.16
Trend surface of business and office land use.

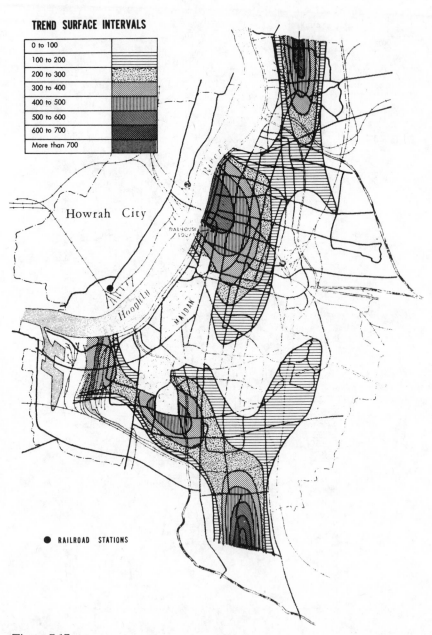

TREND SURFACE INTERVALS

0 to 100	
100 to 200	
200 to 300	
300 to 400	
400 to 500	
500 to 600	
600 to 700	
More than 700	

Howrah City

● RAILROAD STATIONS

Figure 7.17
Houseless population.

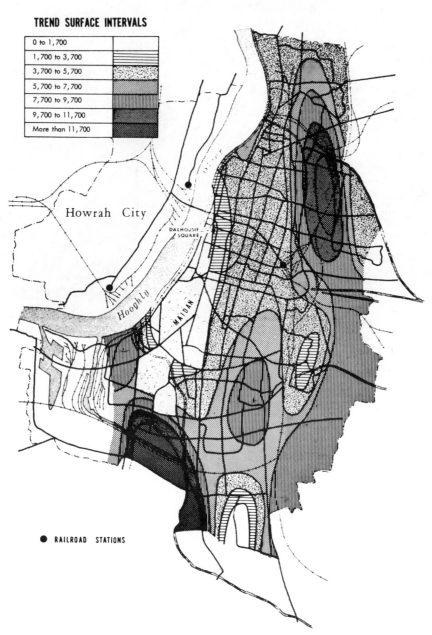

TREND SURFACE INTERVALS

0 to 1,700	
1,700 to 3,700	
3,700 to 5,700	
5,700 to 7,700	
7,700 to 9,700	
9,700 to 11,700	
More than 11,700	

Howrah City

DALHOUSIE SQUARE

Hooghly

MAIDAN

● RAILROAD STATIONS

Figure 7.18
Distribution of married females.

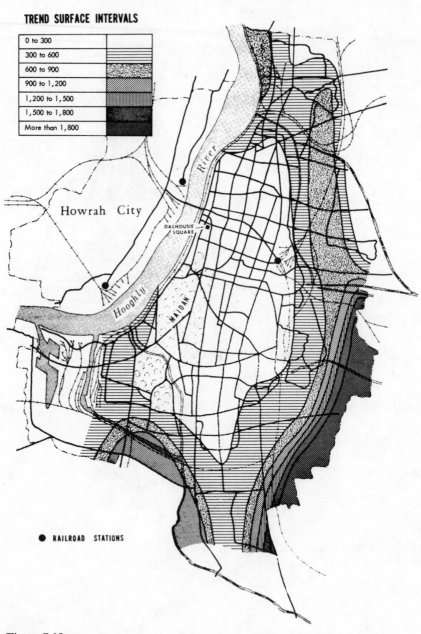

TREND SURFACE INTERVALS

0 to 300	
300 to 600	
600 to 900	
900 to 1,200	
1,200 to 1,500	
1,500 to 1,800	
More than 1,800	

Howrah City

DALHOUSIE SQUARE

River

Hooghly

MAIDAN

● RAILROAD STATIONS

Figure 7.19
Houses with mud walls.

(The equation is

$$F = \alpha_0 + \alpha_1 X + \alpha_2 Y + \alpha_3 X^2 + \alpha_4 Y^2 + \alpha_5 XY + \alpha_6 X^3 + \alpha_7 Y^3$$
$$+ \alpha_8 X^2 Y + \alpha_9 Y^2 X + \alpha_{10} X^4 + \alpha_{11} Y^4 + \alpha_{12} X^3 Y$$
$$+ \alpha_{13} Y^3 X + \alpha_{14} X^2 Y^2 + \cdots + \alpha X^3 Y^3 + \cdots$$

where X and Y are latitude and longitude coordinates for [approximate] geometric centers of the wards.)

There is thus a clear and strong land use and familism gradient in Calcutta from commercial core to residential periphery, modified only by the interjection of peripheral nodes at the cardinal extremities of the city. The commercial core (Figure 7.16) houses most of the "sidewalk sleepers" (Figure 7.17), largely young males, whereas the dominantly residential areas of the outer zones of the city have more balanced sex ratios (Figure 7.18) and concentrations of mudwalled dwellings (Figure 7.19). Comparison of trend surfaces for some of the variables (as depicted in Figures 7.16 to 7.19) with Figure 7.15 reveals how their commonalities are captured by this factor.

Factors 5 and 6: traditional commercial communities and the peripheral ring

Surrounding the city core, two traditional ethnic commercial communities–cum–functional areas were identified by the factor analysis. Principal loadings on Factor 6 were: percentage of workers employed in trade, −0.560; percentage of buildings in commerical use, −0.422; and percentage of population in scheduled castes, 0.541. In the factor score map, areas of former Bengali commercial caste concentration in the southern half of the old native quarter stand out (Figure 7.20; compare with Figure 7.5), in stark contrast to the traditional peripheral ring of the lower-status schedule castes (Figure 7.21).

Variables loading on Factor 5 were the following: percentage of buildings in mixed residential-commercial use, 0.825; percentage of buildings in business and offices, −0.298; percentage of population Hindu, −0.678; and percentage of population non-Bengali, 0.303. This cluster of variables indexes the relative concentration of the non-Bengali commercial classes in the wards immediately east of the Maidan (Figure 7.22; compare with Figure 7.7).

The zones identified by Factors 5 and 6 are not simply traditional ethnic community areas, however. They occupy the principal zones of commercial growth between 1911 and 1961 (Figure 7.23), so that if their ethnic bases are weakening, as Bose suggests, their functional role in the city, as areas of specialized land use, is becoming more marked.

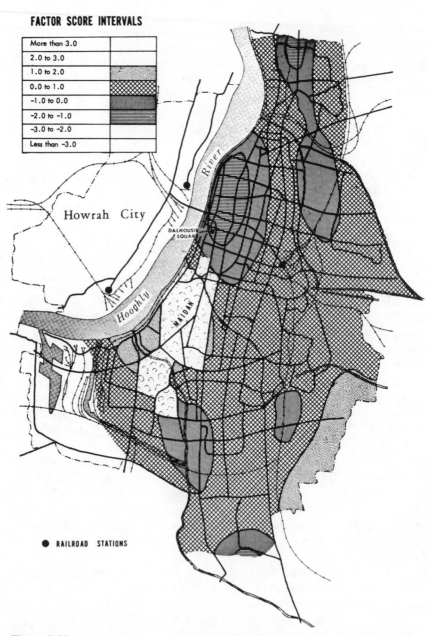

FACTOR SCORE INTERVALS

More than 3.0	
2.0 to 3.0	
1.0 to 2.0	
0.0 to 1.0	
-1.0 to 0.0	
-2.0 to -1.0	
-3.0 to -2.0	
Less than -3.0	

● RAILROAD STATIONS

Figure 7.20
Factor 6: traditional Bengali commercial community.

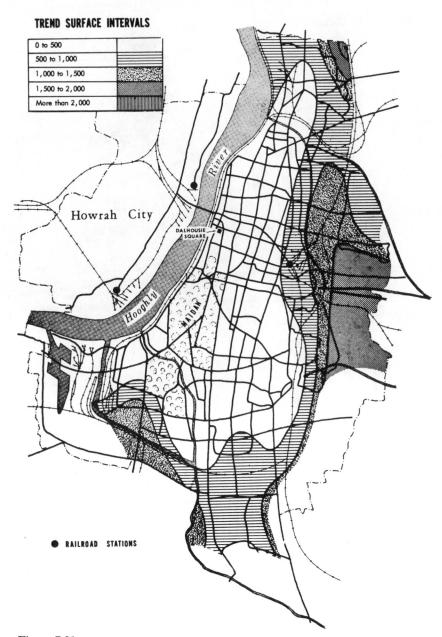

Figure 7.21
Females in scheduled castes.

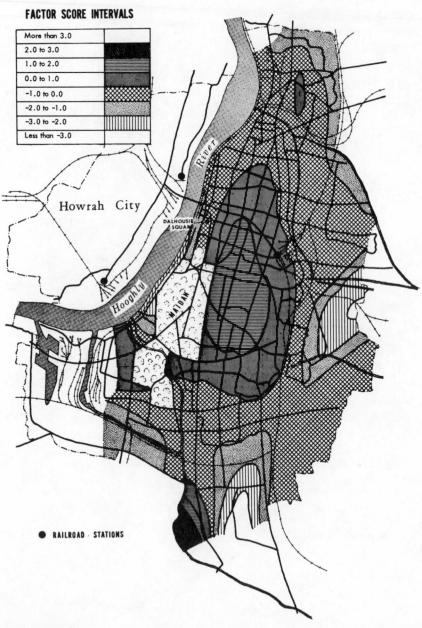

Figure 7.22
Non-Bengali commercial wards.

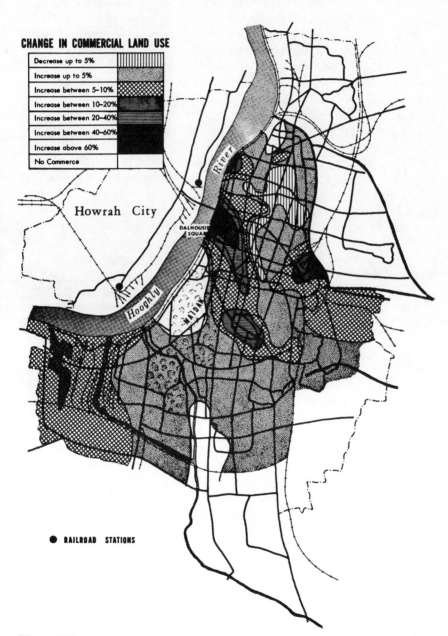

Figure 7.23
Changes in commercial functions, 1911–1961 (after Bose).

Factor 4: the substantial residential areas

The variable configuration and factor scores isolated by the fourth dimension identify the most substantial residential areas of Calcutta. Factor loadings are as follows: percentage of houses with more than one room, -0.784; percentage of children under five, 0.778; percentage of houses with walls and roof, 0.556; percentage of houses with mud walls, 0.361; percentage of population literate, -0.494; percentage of population in scheduled castes, 0.357; percentage of workers employed in manufacturing, 0.465; percentage of workers employed in services, -0.357; and percentage of buildings that are schools, -0.392. Scores are mapped in Figure 7.24.

Two zones are identified: the Bengali upper-caste northern half of the old native quarter, and the Europe residential zones east of the Maidan, with the latter rating somewhat higher than the former. Over the years, as their ethnic base has weakened, those areas have increased their residential function (Figure 7.25).

Somewhat higher literacy rates than elsewhere and lower proportions of children under five (an inverse relationship noted in transitional communities elsewhere) are among the features that distinguish the residents of these two areas of larger, sounder homes from more peripheral residential wards in which there are greater proportions of mud-walled single-roomed bustees housing higher proportions of manufacturing workers (compare with Figure 7.4). Thus, the factor apparently comes close to indexing status.

Factor 3: axiality in literacy

However, a substantial variation in percentage of literate males (loading 0.808) and females (0.917) overlaps the city's mixed and dominantly residential areas, especially in the areas of houses with more than one room (0.399), where the proportion of buildings that are places of education is greatest (0.566). The high-literacy wards stretch along a southward axis from Park Street through Ballygunge to Tollygunge and include sections of Alipore (Figure 7.26). Characteristically, wards with the greatest concentrations of scheduled castes have the lowest literacy levels (-0.345); these are the wards with the greatest admixture of Oriya and Bihari in-migrants. This separation of Factor 3 from Factor 4 indicates that the links between status and literacy remain far from absolute in Calcutta.

Factor 2: higher-status Muslim concentrations

Loadings on this factor are: percentage of workers employed in manufacturing, 0.696; percentage of workers employed in construction, 0.807; percentage of workers employed in trade, 0.722; percentage of workers employed in transportation, 0.776; percentage of workers employed in

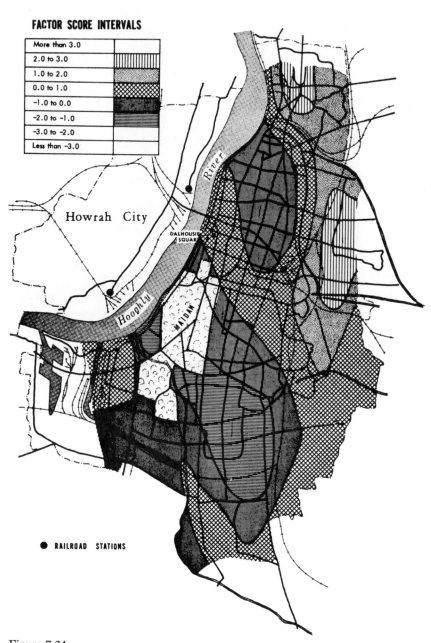

Figure 7.24
Factor 4: substantial residential areas.

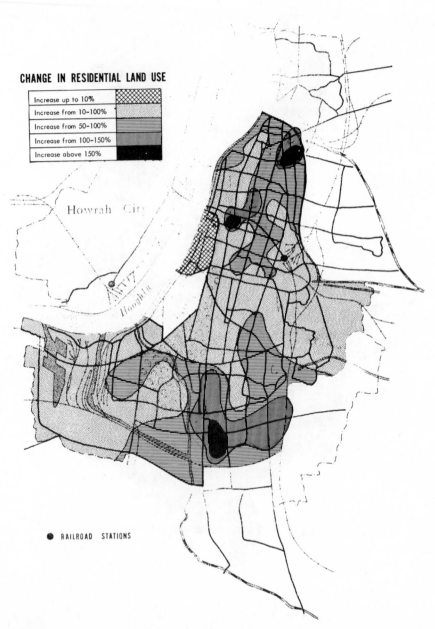

CHANGE IN RESIDENTIAL LAND USE

Increase up to 10%	
Increase from 10-100%	
Increase from 50-100%	
Increase from 100-150%	
Increase above 150%	

Howrah City

Hooghly

● RAILROAD STATIONS

Figure 7.25
Changes in residential functions, 1911–1961 (after Bose).

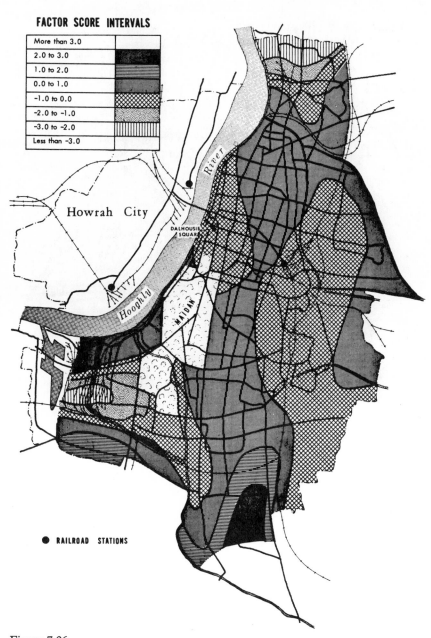

FACTOR SCORE INTERVALS

More than 3.0	
2.0 to 3.0	
1.0 to 2.0	
0.0 to 1.0	
−1.0 to 0.0	
−2.0 to −1.0	
−3.0 to −2.0	
Less than −3.0	

Howrah City

River

Hooghly

DALHOUSIE SQUARE

MAIDAN

● RAILROAD STATIONS

Figure 7.26
Factor 3: literacy in the population.

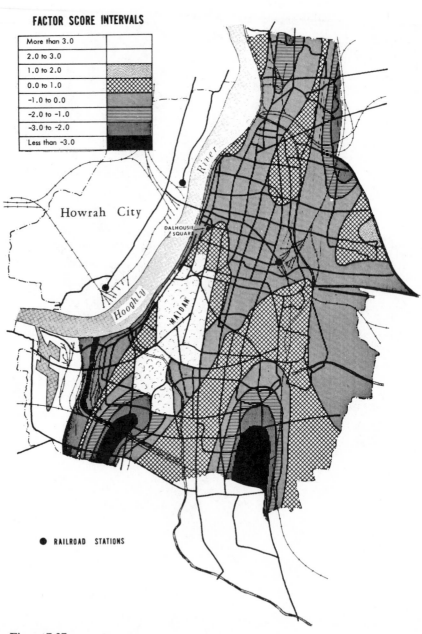

Figure 7.27
Factor 2: Muslim concentrations.

services, 0.851; percentage of females employed, 0.879; and percentage of houses with more than one household, − 0.633. Interpretation of this configuration of variables is not particularly transparent, until the map of factor scores (Figure 7.27) is studied simultaneously. Broad areas of the city have uniform levels of positive scores, but certain extreme-negative concentrations stand out, identifying communities having many houses containing more than one household, low percentages of females employed, and so forth. These communities (compare with Figure 7.6) are the principal areas of upper- and middle-class Muslim concentration.

Factors 7 through 10: special land use configurations

The remaining four factors with eigenvalues exceeding unity produced in the analysis identified special features of Calcutta's land use pattern:

1. Peripheral, essentially rural, areas with substantially higher sizes of households than the rest of the city (0.838).
2. Hotel-restaurant-entertainment districts (percentage of buildings commercial and restaurants [0.661]; percentage of places of entertainment [0.869]).
3. The medical center (percentage of buildings medical institutions [0.576]).
4. The factory zones (percentage of buildings factories [0.691]).

These round out the meaningful variations extractable from the original 80 × 37 data matrix, and they further indicate the substantial functional differentiation of land use achieved in Calcutta by 1961.

Calcutta's social areas

One valuable outcome of the analysis was an 80 × 10 matrix of ortho-normal factor scores. The relative similarity of each ward to every other could thus be measured as the distance separating the wards when plotted as points in the Euclidean space of the ten dimensions (Gower, 1966). An 80 × 80 distance matrix portrays these similarities. Application of a grouping algorithm to this matrix permits the wards to be allocated to relatively homogeneous subsets, consistent with the notion of social areas.

In the present case, a stepwise grouping algorithm that successively minimizes within-group variance was utilized. The city was found to subdivide into a set of subregions (social areas), as portrayed in Figure 7.28. In brief, these comprise the city core, the superior residential areas in the northern part of the native quarter and east of the Maidan, the mixed commercial residential areas of the southern half of the native quarter and the southeast of the city core, and the broad ring of peripheral wards. The legacy of the past remained strongly imprinted in the social space of the transitional city of 1961.

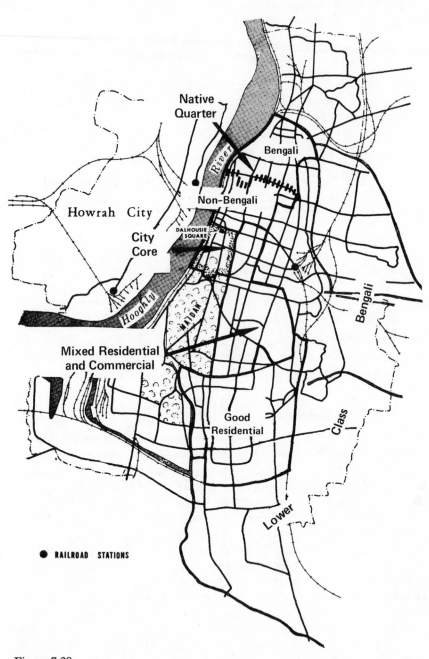

Figure 7.28
Calcutta's social areas.

Discussion

The factorial ecology, however, belies the assumption made in a social area analysis, *sensu stricto*, of Hyderabad (1966): that the social space of an Indian city can be studied simply using Indian equivalents of the two principal Shevky-Bell constructs of social rank and urbanization. According to the Hyderabad research workers, the two constructs are constituted in the Indian case as follows:

1. Social rank: (a) literacy among the general population; (b) literacy among females; (c) relative proportion of the population in scheduled castes, which varies inversely with (a) and (b).
2. Urbanization: (a) proportion of workers in manufacturing industries; (b) workers in commerical activities; (c) workers in other services.

Using these indicators, the Hyderabad analysts found that areas of lowest social rank formed a peripheral ring and also occupied the city's industrial zones, whereas the areas of highest social rank were found in the economic core of the city. But they also found that the residential quarters of workers in the three occupational groups used to define urbanization were highly segregated, almost mutually exclusive, so that the three constituent variables bore little relationship to each other. Nonetheless, they crossed the ratings of areas on the two constructs to develop four "types" of social areas.

In Calcutta, literacy and proportion of the population of scheduled castes are inversely related on Factor 4, which separates the good-quality high-status residential areas from the others. The structure of this factor is similar to the Hyderabad social rank construct. But Factor 1, the land use and familism gradient, is the direct equivalent of the Shevky-Bell urbanization (family status) dimension. To confirm that this is so, we can compare the distribution of social areas in the social status–family status space in Chicago (Figure 7.29) with a similar diagram for Calcutta, in which Factor 4 scores are used to locate wards on the ordinate and Factor 1 scores to provide positioning with respect to the abscissa (Figure 7.30). Similar segments of the social space are occupied by analogous social areas. In both instances, the high-status residential areas front superior amenities (the lake in Chicago and the Maidan in Calcutta). On the other hand, the geographic pattern of the other three social areas is inverted (Figure 7.31).

Calcutta's social ecology still thus contains many traditional elements. But if, as Bose argues, regional or ethnic groups in Calcutta are coming to bear the same relationships to each other as do the castes in India as a whole, the ethnic bases of the communities are being supplemented by strong functional correlates. There is increasing functional differentiation of land use, alongside regional occupational differentiation. And consistent with the notion of a city in transition, alongside these bases of differentiation are a strong land use and familism gradient, comparable in many ways to

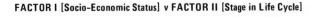

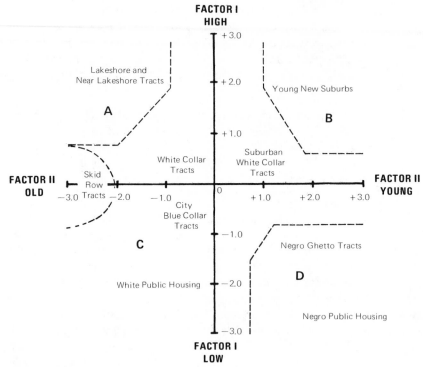

Figure 7.29
Chicago's social space.

"modern" American formulations of urbanization. On the other hand, several dimensions are required to index status, because many of the ethnic linkages between occupation, region of origin, and caste remain—Factor 6 differentiates the areas occupied by the Bengali "middle" commercial castes from those of the scheduled castes; Factor 5 is needed to define the middle-status non-Bengali commercial zones; Factor 4 separates the high-status residential areas but does not completely differentiate the commercial wards from the periphery; and Factor 3 reveals the familiar "Western" Hoyt-like axiality, but restricts it to literacy. Female employment is not part of the familism pattern, as in the Shevky-Bell model, but is related to the differences between Hindu and Muslim.

In terms of the factor models outlined in the fourth section of this chapter then, Calcutta conforms most closely to Combination 3, in which there is a separate family-status dimension, but socioeconomic status and minority group membership are linked. It is of some interest that the closest resemblances of Calcutta's ecology are to the ecology of

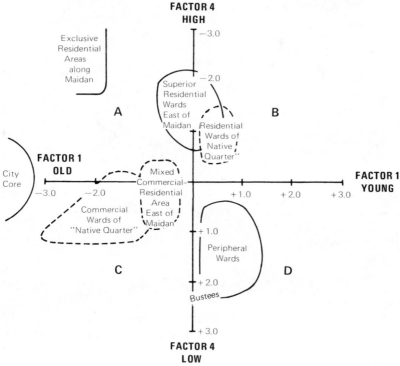

Figure 7.30
Calcutta's social space.

cities in the American South, where traditionally one found links between race and status in a system of caste. One description, written by Davis, Gardner, and Gardner in 1941, reads:

> Life in the communities in the Deep South follows an orderly pattern. The inhabitants live in a social world clearly divided into two ranks, the white caste and the Negro caste. The colored castes share disproportionately in the privileges and obligations of labor, school, and government.

This particular factor combination raises serious questions about relative emphasis placed on different dimensions in the choice model outlined at the beginning of this chapter. Bose noted the interrelationships between scarcity and ethnicity in Calcutta. The resulting ethnic base of much decision making can be embodied in the residential choice model with appropriate modifications, however. To the two choice dimensions of individual social space discussed in the American context can be added a third dimension of ethnicity (the regional relativity of the "American" model must by now also be evident). The dimensions can be reordered,

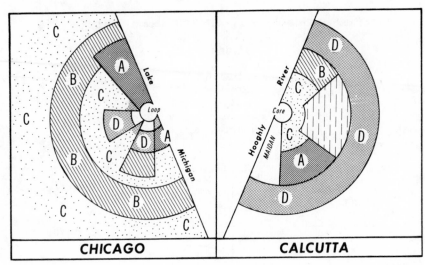

Figure 7.31
Generalized social areas.

with ethnicity placed at the head and altered in length; this makes ethnicity the most important element in the choice process.

Thus, in Calcutta, a new Rajasthani or Punjabi in-migrant usually seeks the safety of his ethnic enclave. Because most non-Bengali ethnic groups also occupy particular occupational niches, the variance in status within ethnic groups is substantially less than that between ethnic groups. In effect, ethnicity overrides choice based upon status, for the status and ethnic dimensions are collinear, with the latter more fundamental in defining the social dimensions within which choices are made. The exceptions are within the Bengali community, where occupationally derived differences corresponding to traditional caste differentiation still persist.

Similar less-than-economic decisions are involved in the creation of bustees, emphasizing the parallel relativity of location of many of the ethnic enclaves in physical space. Around the periphery of the city are found the poorest residents of and migrants to the city: scheduled castes, refugees from East Bengal, and lower-caste non-Bengalis. The process at work in the creation of the bustees they occupy was not one of orderly choice in an ethnically governed market framework but rather one of often illegal occupation of "empty" land at the urban periphery by poor families. There they erect a shack and attempt to survive in incredibly overcrowded and unhealthy living conditions. Abrams (1964) has vividly described the sequence of events:

As in a military campaign, some would bivouac during the night with their stock of materials behind a newly placed billboard. Next day, the horizon would be dotted with new rows of hovels, to which others would be added shack by shack, until the expansion was checked by a road, by a canal or by an owner prepared to spill blood.

With the passage of time, the bustee changes from a frontier community into a "regular," market-involved urban slum. Residences are bought and sold, and changing status is reflected in either physical improvement of the dwelling or a move to a new place of residence. Such changes are constrained, however, by the low status accorded the bustee dwellers by their occupations.

As the transitional process continues, however, one might reasonably expect the socioeconomic and ethnic bases of differentiation to separate. The expectation is, therefore, that differing urban ecologies related to differing factor combinations can be arranged along a scale of urban development from pre- to post-industrial forms. Testing of this notion now awaits systematic comparative ecological analysis.

PART IV
METROPOLITAN EXPANSION AND STRUCTURAL CHANGE

Chapter 8
The territorial sources of metropolitan growth

Now let us return to the American scene, and to emergent postindustrial metropolitan forms. Whether one wishes to call it a new urbanization or merely an extension of demographic trends, population distribution in the United States is changing. The roots of the changes lie in the emerging post-industrial economy; their expression is the postcity metropolitan region. How the new patterns evolved demographically through the dynamics of population growth and territorial expansion is the subject of this and the next chapter. Chapters 10 and 11 will go on to deal with the implications of the demographic shifts, Chapter 12 with the accompanying occupational selectivity, and Chapter 13 with the relationships between changes in in-dustrial location and the postindustrial transformation of the urban ecology in the 1960s and 1970s.

To be sure the concept of urbanization currently embraces more than "the multiplication of points of concentration and the increase in the size of individual concentrations" (Tisdale, 1942, p. 311). But it has been at least that during the twentieth century, and if one purports to grasp the structural bases of urban settlement, sufficient knowledge of its territorial evolution is essential. Marked by an enlarged scale of social and economic integration, metropolitan areas have grown to constitute today the primary ecological units subsuming population growth and concentration in the United States; but the units themselves are undergoing further dramatic transformation.

Two levels of ecological organization among metropolitan areas have been identified. At one level, the metropolis encompasses a circumscribed unit of territory discernible by commuting patterns, density levels, and socioeconomic interaction. The classic works of Gras (1922), McKenzie (1933b), and Hawley (1950) establish this aspect of metropolis as community. At another level, metropolitan areas fit into a national system of urban places (Berry, 1972; Lampard, 1968). This system assumes a multifaceted hierarchy of metropolitan centers, each operating within broad regional contexts closely tied to internal economic functions (Duncan et al., 1960; Duncan and Lieberson, 1970; Vance and Sutker, 1954). In this and the following five chapters, we shall focus on demographic and ecological relationships at the metropolitan community level. Analysis of the national system will be the concern of Part V.

Our specific task in the present chapter is the territorial decomposition of metropolitan growth in the United States from 1900 to 1970, by regional location and internal distribution patterns. The units of analysis are standard metropolitan statistical areas (SMSAs) as defined by the Office of Management and Budget. In short, an SMSA is an economically and socially integrated territorial unit with a recognized large population nucleus, that is, a central city (or multiple cities) of at least 50,000 inhabitants. Its boundaries are delimited by the county containing the central city and all contiguous (adjoining) counties that are found to be metropolitan in character and economically and socially integrated with the county of the central city. (See Appendix for complete definition).

To account for metropolitan population expansion beyond fixed county jurisdictions, the outer boundaries of SMSAs are semiflexible, permitting enlargement by adding outlying county units when contiguous county metropolitan criteria are met. Similarly, when separate urban places (cities) and their surrounding counties meet minimum size and compositional requirements, they also attain SMSA status. Total metropolitan population increase over any period can therefore be decomposed into population growth within previously defined metropolitan territory, the number of inhabitants in territory (contiguous counties) added to existing areas, and the population enumerated in newly designated SMSAs.

But a full explanation of the territorial evolution of metropolitan areas requires that additional analysis be undertaken. Population increases taking place within central cities and suburban rings also need to be decomposed by areal units. Total metropolitan central-city population increase over any period consists of growth within prior defined political boundaries of central cities, municipal annexation, inclusion of additional central cities to SMSA titles, and the enumeration of cities in new SMSAs. Similarly, aggregate suburban growth consists of population increases within existing areas, from redefined territory (that is, from adding outlying counties), population enumerated in rings of newly designated SMSAs, and population loss via annexation by the central cities.

Decomposition of the territorial sources of central-city and suburban-ring growth requires estimation of the aggregate amount of population annexed by central cities for decades before 1950. Previous historical studies of metropolitan expansion (Thompson, 1947; Bogue, 1953; Hawley, 1956; Schnore, 1957; Taeuber, 1972) were not able to make comprehensive adjustment for the aggregate amount of population annexation because complete data on number of inhabitants annexed are not available for most metropolitan areas prior to 1950. Both the 1960 and 1970 censuses of population, however, report population residing in territory annexed by all metropolitan central cities during the previous decade. We employed these data along with historical census information on municipal land area additions and enumerated suburban populations to estimate (by means of conventional regression techniques) population annexed by central cities for the decades between 1900 and 1950.

Our first step was to construct a linear multiple regression model for the 1950 and 1960 decades, treating the number of inhabitants annexed during each decade as a function of the amount of land area annexed, regional location (measured by four dummy variables), age of the central city (measured by the number of decades since the city attained metropolitan status), and the size of the population in the potentially annexable area (the suburban-ring population at the prior census date). In the regression equation neither central-city age nor region exhibited a significant association with number of inhabitants annexed, indicating that central-city age and region may prove to be more effective predictors of *whether* annexation occurs than of the *number* of inhabitants added via annexation. Also, as we shall show in the next chapter, it is invalid to use current age and region to predict past population annexation because, over time, the frequency and degree of annexation shift across age and regional categories.

Using data from the 1960 and 1970 censuses for all metropolitan central cities (as of 1973) that annexed at least one square mile of territory between 1950 and 1970, we regressed the mean ten-year population annexed, 1950 to 1970, on the mean ten-year land increments and mean suburban-ring population of 1950 and 1960. This yielded unstandardized b coefficients (partial slopes) indicating the unique contribution of each factor to annexed population.

The amount of land area added was most closely associated with the number of persons annexed ($r = 0.75$). The size of the population in the suburban ring is less strongly correlated, but statistically significant ($r = 0.20$). The regression equation predicting annexed population is $Y = 8826 + 577.57X_1 + 0.1329X_2$, where Y is the predicted population annexed over a ten-year period, X_1 is the decennial land-area change in square miles, and X_2 is the suburban-ring population. Two hundred and nine cities entered into the regression model.

Population annexation by central cities during the 1940s and each earlier decade was estimated by applying the derived regression parameters to known land-area changes and enumerated suburban populations for all central cities annexing one square mile or more of territory during the decade. Other cities are assumed to have annexed no population. The reported central-city and suburban-ring decennial population counts were then adjusted for altered boundaries by subtracting the estimated number of persons annexed from the city's enumerated population at the end of the decade and adding the same number to the suburban-ring population.

To assess the reliability of this method for calculating past annexation, we correlated the estimates derived from the regression equation for the decade 1940–1950 with actual figures reported by Bogue (1957) for 92 metropolitan central cities with population of more than one hundred thousand. Bogue assembled his data at the Bureau of the Census by carefully allocating 1950 enumeration district populations to detailed map overlays of areas annexed by the cities between 1940 and 1950 (personal communication). The resulting zero-order correlations between the actual population annexed between 1940 and 1950 and the predicted population annexed was .86, indicating 74 per cent of the variance explained.

Before proceeding with our results, a brief comment on the use of SMSAs as analytical units is helpful. While most agree that the appropriate statistical unit for measuring urbanization must be one that captures the complex and spatially broad character of large urban settlements, doubts have been expressed whether SMSAs validly delimit "metropolitan" areas (Berry et al., 1968; Feldt, 1965; Rosenwaike, 1970). Contrary to the stated purpose of the SMSA concept as operationally defining an integrated economic and social unit, Berry et al. (1968: 19–24) offer evidence that

1. The areas economically and socially integrated with the central cities are far more extensive than the formal boundaries of SMSAs.
2. In the least densely populated parts of the nation's settled area, regionally important commuting fields gravitate around urban centers of less than fifty thousand people.
3. At the other extreme, particularly in the manufacturing belt, labor markets overlap in elaborate ways. The urban regions of the "megalopolises" are highly complex, multicentered entities.

We call attention to these findings based on commuting patterns in order to recognize the limitations of the SMSA concept, and at the same time establish its utility for our purposes. The objective of this study is investigating the territorial sources of *aggregate* metropolitan growth rather than the otherwise important demographic and ecological characteristics of individual SMSAs. It is thus not critical to the present analysis that, adhering to set criteria, certain cities or counties should be reassigned to different SMSAs. We are also aware that central-city and SMSA boundaries cut across social and economic gradients rather than demarcating distinct entities, and that they often merge into diffuse megalopolises (Gottmann, 1961) and urban fields (Friedman and Miller, 1965). (The next section of this book will address itself to the regional system concept.) Nevertheless, official metropolitan areas approximately demarcate virtually all major urban regions in the United States and, particularly between 1945 and 1970, areas of growth and concentration versus areas of out-migration and population decline.

Following the Office of Management and Budget's guidelines (National Bureau of Standards, 1973), we dated the inception of SMSAs and the addition of outlying counties and cities to existing SMSA territory back to 1900. Cognizant of the charge that SMSAs often underbound actual metropolitan zones, we considered as much of the total territory and population "metropolitan" as validly allowable by formal criteria and at the earliest census date that all criteria were met. In 1970, for example, we treat as metropolitan all areas in the coterminous United States that had attained SMSA status prior to that date as well as all additional areas (cities or counties) designated "metropolitan" by the postcensus review (National Bureau of Standards, 1973). For analytical comparability New England SMSAs are reconstructed to their county-based equivalents, rather than traditional town units. Redrawing New England SMSA boundaries reduced slightly the number of metropolitan areas in that region.

Areal decomposition of metropolitan growth

Table 8.1 presents the territorial sources of aggregate metropolitan growth from 1900 to 1970. With the exception of the Depression decade, total metropolitan population increased monotonically during the first half of this century. Note, however, that until 1950 the bulk of these numerical increases occurred within prior defined SMSA boundaries. Metropolitan growth from adding outlying counties to existing SMSAs and from the inception of new SMSAs remained relatively stable.

Since 1950, aggregate metropolitan population increases have been substantially greater, due in large part to the unprecedented territorial expansion of existing SMSAs and the inception of numerous new SMSAs. Absorption of outlying counties by expanding SMSAs added nearly nine million persons to the metropolitan population between 1950 and 1970. During the same period, 81 new SMSAs have been designated, contributing more than 12,500,000 persons to the U.S. metropolitan population. In the 1960s alone, urban areas passing SMSA thresholds added over 7,200,000 people to those classified as metropolitan residents.

The bottom panel of Table 8.1 indicates the relative contributions of the three areal sources of aggregate metropolitan population increase. Observe that while population growth within existing SMSA boundaries has been larger in absolute numbers over the past three decades than in previous decades, its percentage of total SMSA increase has steadily declined. Conversely, the relative contributions of both SMSA territorial expansion and the inception of new SMSAs have consistently increased. Additions of new counties to 57 and 76 SMSAs, respectively, during the

Table 8.1

Sources of metropolitan population growth, 1900–1970

Decade	Total metropolitan population increase	Existing SMSAs	Added counties to SMSAs	Number of SMSAs enlarged	In new SMSAs	Number of SMSAs added
				Source of increase		
				Numerical increase (in thousands)		
1900–1910	10,385	7,331	145	(2)	2,909	(22)
1910–1920	11,227	8,436	264	(3)	2,526	(25)
1920–1930	16,014	12,245	836	(12)	2,931	(26)
1930–1940	6,010	5,075	178	(3)	755	(7)
1940–1950	18,098	14,354	902	(17)	2,841	(22)
1950–1960	30,887	21,330	4,260	(57)	5,296	(34)
1960–1970	30,699	18,976	4,443	(76)	7,279	(47)

Decade	Total metropolitan population increase	Existing SMSAs	Added counties to SMSAs	In new SMSAs
		Percentage distribution		
1900–1910	100.0	70.6	1.4	28.0
1910–1920	100.0	75.1	2.4	22.5
1920–1930	100.0	76.5	5.2	18.3
1930–1940	100.0	84.5	3.0	12.6
1940–1950	100.0	79.3	5.0	15.7
1950–1960	100.0	69.0	13.8	17.1
1960–1970	100.0	61.8	14.5	23.7

past two decades accounted for 13.8 and 14.5 per cent of total metropolitan population increase. Likewise, the respective designation of 34 and 47 new SMSAs in the two decades since 1950 have accounted for 17.1 and 23.7 per cent of aggregate U.S. metropolitan growth. Overall, nearly 40 per cent of the total metropolitan population increase in the 1960–1970 decade resulted from population enumerated within new or expanded territory attaining metropolitan status for the first time.

Regional variations of metropolitan growth since 1900 are presented in Table 8.2. Examining all SMSAs (Table 8.2a), it is evident that over time, the Northeast has contributed much smaller shares of total metropolitan growth. In contrast, the South and West have assumed rising proportions. North Central SMSAs have declined in terms of relative contributions to total metropolitan growth, but that area's long-term percentage contribution has not changed as dramatically as in other regions of the nation. This pattern also describes increases occurring within constant boundaries (Table 8.2b), but designation of new SMSAs (Table 8.2c) amplifies the observed trends. Further analysis shows that the South has gradually come to embrace a larger share of metropolitan growth in pre-existing areas, largely because of continued emergence of new SMSAs. In every decade since 1910, additions of new SMSAs in the South have accounted for at

Table 8.2a

Population growth in existing and new SMSAs by region, 1900–1970

	Decade						
	1900–10	1910–20	1920–30	1930–40	1940–50	1950–60	1960–70
All SMSAs	*Numerical increase (in thousands)*						
Northeast	4,685	3,835	4,682	1,358	3,129	4,941	4,950
North Central	3,033	4,066	5,234	1,043	4,721	8,019	6,727
South	1,452	2,152	3,711	2,172	5,781	9,809	11,727
West	1,215	1,174	2,387	1,437	4,466	8,118	7,265
Total	10,385	11,227	16,014	6,010	18,098	30,887	30,699
	Percentage distribution						
Northeast	45.1	34.2	29.2	22.6	17.3	16.0	16.1
North Central	29.2	36.2	32.7	17.4	26.1	26.0	21.9
South	14.0	19.2	23.2	36.1	31.9	31.8	38.2
West	11.7	10.4	14.9	23.9	24.7	26.3	23.7
Total	100.0	100.0	100.0	100.0	100.0	100.0	100.0

Table 8.2b

Population growth in existing SMSAs by region, 1900–1970

	Decade						
	1900–10	1910–20	1920–30	1930–40	1940–50	1950–60	1960–70
Existing SMSAs	*Numerical increase*						
Northeast	3,976	3,464	4,444	1,358	2,996	4,941	3,702
North Central	2,806	3,201	4,492	963	3,798	7,660	5,408
South	459	1,065	2,115	1,817	4,593	7,280	8,702
West	955	971	2,032	1,117	3,869	5,709	5,578
Total	7,476	8,701	13,082	5,254	15,257	25,590	23,420
	Percentage distribution						
Northeast	53.2	39.8	34.0	25.8	19.6	19.3	15.8
North Central	27.9	36.8	34.3	18.3	24.9	29.9	23.1
South	6.1	12.2	16.2	34.6	30.1	28.4	37.2
West	12.8	11.2	15.5	21.3	25.4	22.3	23.8
Total	100.0	100.0	100.0	100.0	100.0	100.0	100.0

Table 8.2c

Metropolitan population increase from designation of new SMSAs by region, 1900–1970

	Decade						
	1900–10	1910–20	1920–30	1930–40	1940–50	1950–60	1960–70
New SMSAs	*Numerical increase*						
Northeast	709	371	238	0	133	0	1248
North Central	947	865	742	80	923	359	1319
South	993	1087	1596	355	1188	2529	3025
West	260	203	355	320	597	2409	1687
Total	2909	2526	2932	756	2841	5297	7279
	Percentage distribution						
Northeast	24.4	14.7	8.1	0.0	4.7	0.0	17.1
North Central	32.6	34.2	25.3	10.6	32.5	6.8	18.1
South	34.1	43.0	54.4	47.0	41.8	47.7	41.6
West	8.9	8.0	12.1	42.3	21.0	45.5	23.2
Total	100.0	100.0	100.0	100.0	100.0	100.0	100.0

least 40 per cent of this facet of metropolitan population increase in the U.S.

If we combine geographic regions into the traditional heartland-hinterlands division (Northeast–North Central versus South–West) we obtain a marked indication of how the balance of metropolitan growth has shifted since 1900. In 1900–1910, 81 per cent of growth in constant boundaries and 57 per cent of new SMSA growth occurred in the heavily industrialized (North and North Central) regions of the nation. By the 1960–1970 decade, the dispersion of metropolitan population reached the level at which 60 per cent of constant SMSA area growth and 65 per cent of new SMSA growth took place *beyond* the traditional industrial belt. Changing economic structure, the rise of new types of industry, and in some cases, the importance of favorable climate have all been postulated to contribute to the transfer of the regional locus of growth (Karp and Kelley, 1971; Schwind, 1971; Greenwood and Sweetland, 1972; Perloff et al., 1960).

The regional dispersion of metropolitan population has had its parallel in the demographic decentralization process taking place within SMSAs. Table 8.3 decomposes total metropolitan growth into central-city and suburban-ring portions.

Several features of urban centrifugal drift become salient. First, aside from the depression decade, the total SMSA population increase occurring in central cities has remained relatively stable since 1900 (about 7 million per decade). Large absolute increases on the other hand have occurred in the suburban sector: the last decade saw a sixfold increase over 1900 to 1910. Second, considering percentage distributions of central-city growth, a steadily larger proportion of the total has been contributed by the addition of new central cities (40 per cent by 1960–1970), rather than, as some might expect, from within the sizeable number of existing cities. Further elaboration of this trend will be offered shortly. Third, different from the total central-city pattern but complementing it, suburbs within existing SMSAs at each date have taken increasingly larger shares of total suburban growth, the last decade being only a slight exception. Absolute increases from the suburban rings of new SMSAs have risen, but gains within established SMSA rings have far outpaced them. Finally, the relative proportions of growth in city and ring sectors have shifted in favor of the outer zone. Over four fifths of the total increase to existing SMSAs took place in the suburban periphery in 1960–1970 compared to 27 per cent during the 1900–1910 decade. In new SMSAs, roughly 60 per cent of the population resided outside principal cities, compared to less than 45 per cent in 1900–1910. The centrifugal drift in the locational base of population growth constitutes another reversal of patterns since the turn of the century.

As perhaps the most significant demographic aspect of metropolitanization, the process of decentralization requires further assessment. We have noted that roughly 40 per cent of total central-city population increase in the 1960–1970 decade resulted from the addition of cities in new SMSAs. A growing source of population, therefore, derives from the inclusion of new area. The figures in Table 8.4 reveal that enumeration of population from enlarged (annexed) territory has come to account for virtually all central-city increase in prior-defined central cities as well. In steadily increasing proportions, annexation of suburban territory and population provides the main source of growth. Not only has population growth within cities present at initial decades substantially declined, but, held to constant municipal boundaries, internal growth makes up barely 1 per cent of the 1960–1970 decade's total. Bear in mind also that these figures reflect national aggregations. We shall show in the following chapter that an examination of regional patterns of urban growth shows that in many instances cities have been declining for several decades. These places, which tend to be older, larger, and located in the North, are also those least likely to undertake annexation to offset internal losses. We are left then with a picture of aggregate central-city *growth* essentially contingent upon areal expansion.

Another salient feature documented in Table 8.4 is the increasing share of total central-city population growth derived from the addition of cities to existing SMSAs. The trend has been for metropolitan areas to become internally differentiated, to the point of having more than one large nucleus. The need for the Bureau of the Budget to add cities to official titles reflects this pattern. By our count, in 1900 less than 10 per cent of existing SMSAs were multicentered. In 1970 the comparable figure was

Table 8.3

Population growth in existing and new SMSAs by central-city–suburban-ring location, 1900–1970

Decade	Increases in all SMSAs			Increases within existing SMSAs			Increases from new SMSAs		
	SMSA total	Central cities	Suburban rings	SMSA total	Central cities	Suburban rings	SMSA total	Central cities	Suburban rings
				Numerical increase (in thousands)					
1900–1910	10,385	7,068	3,317	7,476	5,455	2,020	2,909	1,613	1,296
1910–1920	11,227	7,766	3,461	8,701	6,138	2,563	2,526	1,628	898
1920–1930	16,014	8,916	7,098	13,802	7,185	5,987	2,932	1,731	1,201
1930–1940	6,010	2,492	3,518	5,254	2,095	3,158	756	396	360
1940–1950	18,098	7,614	10,484	15,257	6,118	9,138	2,841	1,496	1,347
1950–1960	30,887	7,688	23,200	25,590	5,024	20,567	5,297	2,663	2,633
1960–1970	30,699	6,915	23,784	23,420	4,099	19,320	7,279	2,816	4,464

Percentage distributions

Period	Total			Total			Total			Total				
1900–1910	100.0	68.1	31.9	100.0	73.0	27.0	100.0	55.4	44.6	100.0	72.2	22.8	60.9	39.1
1910–1920	100.0	69.2	30.8	100.0	70.5	29.5	100.0	64.4	35.6	100.0	79.0	21.0	74.1	25.9
1920–1930	100.0	55.7	44.3	100.0	54.9	45.1	100.0	59.0	41.0	100.0	80.6	19.4	83.1	16.9
1930–1940	100.0	41.5	58.5	100.0	39.9	60.1	100.0	52.4	47.6	100.0	84.1	15.9	89.8	10.2
1940–1950	100.0	42.1	57.9	100.0	40.1	59.9	100.0	52.6	47.4	100.0	80.4	19.6	87.2	12.8
1950–1960	100.0	24.9	75.1	100.0	19.6	80.4	100.0	50.3	49.7	100.0	65.3	34.6	88.7	11.3
1960–1970	100.0	22.5	77.5	100.0	17.5	82.5	100.0	38.7	61.3	100.0	59.3	40.7	81.2	18.8

Table 8.4

Sources of metropolitan central-city growth, 1900–1970

Decade	Total increase	Increase within existing cities			Added cities to SMSA titles	Cities in new SMSAs
		Subtotal	Constant boundaries	Annexation		
		Numerical increase (in thousands)				
1900–1910	7,068	5,429	4,862	567	26	1,613
1910–1920	7,766	6,057	5,272	785	81	1,628
1920–1930	8,916	7,125	6,166	959	60	1,731
1930–1940	2,492	2,045	1,709	336	50	396
1940–1950	7,614	5,644	4,256	1,388	474	1,496
1950–1960	7,698	4,551	468	4,083	473	2,663
1960–1970	6,915	3,593	66	3,527	506	2,816
		Percentage distribution				
1900–1910	100.0	76.8	68.8	8.0	0.4	22.8
1910–1920	100.0	78.0	67.9	10.1	1.0	21.0
1920–1930	100.0	79.9	69.2	10.7	0.7	19.4
1930–1940	100.0	82.1	68.6	13.5	2.0	15.9
1940–1950	100.0	74.1	55.9	18.2	6.2	19.6
1950–1960	100.0	59.2	6.1	53.1	6.1	34.6
1960–1970	100.0	52.0	1.0	51.0	7.3	40.7

roughly 25 per cent, and of those created at that date, approximately 40 per cent contained more than one central city. No doubt the extent of multinodality would be greater still if arbitrary SMSA boundaries were discarded in numerous cases of contiguous urbanization.

Turning to the areal components of suburban growth presented in Table 8.5, we observe that, with the exception of the 1930s, every decade in this century has witnessed a steady climb in total suburban population growth. A rise in the proportion of growth occurring within the defined area at the beginning of each decade continued until 1930–1940, whereupon a gradual decline commenced. This latter trend is due to the increasing number of new SMSAs and, particularly important, county additions.

Nonetheless, we find territorial redefinition to have artificially concealed total suburban population increase as much as it has enhanced it. That is, where central cities gained population by municipal annexation, the suburban rings lost. The total enumerated suburban population increased by 23.2 million during the 1950s and 23.8 million during the 1960s. Had annexed suburban population not been transferred to central cities, the absolute suburban population increase in each decade would have been 27.3 million. In previous decades, total annexation was smaller, but no less significant in distorting the real distribution of central-city and suburban-ring population growth.

Table 8.5

Sources of metropolitan suburban-ring growth, 1900–1970

Decade	Total increase	Increase within existing suburbs			Added counties	Suburban ring of new SMSAs
		Subtotal	Constant boundaries	Annexation		
		Numerical increase (in thousands)				
1900–1910	3,317	1,901	2,468	−567	119	1,296
1910–1920	3,461	2,379	3,164	−785	184	898
1920–1930	7,098	5,120	6,079	−959	777	1,201
1930–1940	3,518	3,030	3,367	−336	128	360
1940–1950	10,484	8,710	10,098	−1,388	428	1,347
1950–1960	23,200	16,779	20,862	−4,083	3,788	2,633
1960–1970	23,784	15,643	19,170	−3,527	3,677	4,464
		Percentage distribution				
1900–1910	100.0	53.7	74.4	−17.1	3.6	39.1
1910–1920	100.0	68.7	91.4	−22.7	5.3	26.0
1920–1930	100.0	72.1	85.6	−13.5	10.9	16.9
1930–1940	100.0	86.1	95.7	−9.6	3.6	10.2
1940–1950	100.0	83.1	96.3	−13.2	4.1	12.8
1950–1960	100.0	72.3	89.9	−17.6	16.3	11.3
1960–1970	100.0	65.8	80.6	−14.8	15.5	18.8

Overview

Before we move on to examining the role of annexation in differential patterns of city and suburban growth, let us briefly summarize the territorial components of U.S. metropolitan growth since the turn of the century. With only brief pauses, aggregate metropolitan population growth has continued to climb steadily since 1900. In the past two decades, growth has geographically circumscribed substantially wider territory. Regionally, the locus of metropolitan growth has shifted to the more recently urbanized West and South. Internally, metropolitan growth exhibits another form of dispersion. Aggregate central-city growth is now almost wholly dependent upon territorial annexation and the designation of new central cities. By far the largest amounts of population increases continue to occur within the suburban zones of metropolitan areas. And here, too, rapid absorption of outlying territory and the formation of new SMSAs have been an essential source of growth in recent decades.

Put in historical perspective, population distribution figures for the first and latest decades are informative. In 1900, less than one third of the U.S. population resided in metropolitan areas. Within these metropolitan areas, 70 per cent of the residents lived in the central cities, the remainder in nearby suburbs. The nation grew by 16 million during the first ten years of the century; of this growth, 63 per cent was within metropolitan areas, the greater proportion (again 70 per cent) in central cities. By 1970 the pattern had clearly reversed. Over two thirds of all persons in the coterminous United States resided in SMSAs, with the bulk (55 per cent) residing *outside* the central cities.

We know that suburbanization is not a new phenomenon. In relative terms, the edge of urban settlement has traditionally grown more rapidly than the center, and, as we shall observe shortly, large-scale suburban dispersion is more than fifty years old. But until recent decades, expanding metropolitan populations have not noticeably exceeded central county boundaries. In fact, before 1950, even many single county SMSAs were overbounded, frequently containing large segments of rural territory.

Since the 1950s we have seen not only the filling up of area that was open at the start of each decade, but a sizeable augmentation of new suburban territory. The growth experience of a few rapidly expanding SMSAs is instructive. In 1970, by including new counties, the Atlanta, Georgia, SMSA added 2,590 square miles of territory and 207,652 inhabitants; the Houston, Texas, SMSA increased land area by 5,071 square miles and population by 257,358; and the Minneapolis–St. Paul, Minnesota–Wisconsin, SMSA added 2,539 square miles and 151,512 new inhabitants. Additions between 1960 and 1975 of more than two hundred outlying counties to previously designated SMSA territory give testimony to the pervasiveness of this component of metropolitan growth.

In concluding this chapter, we should point out that many of the newly designated outlying metropolitan counties are at least two county tiers removed from the central county and contain vast segments of thinly populated territory. The evolving character of metropolitan structure is thus one of increasing expansiveness. Indeed, indications are that we are on the threshold of a radically changed, unbounded metropolitanization that makes urban-nonurban distinctions meaningless. At present, it may be premature to discard the concept of metropolitan community; but, as we shall soon observe, the emergence of a new urban form is unmistakably upon us.

APPENDIX
Criteria for delineating standard metropolitan statistical area

1. Each standard metropolitan statistical area must include at least:
   (a) one city with 50,000 or more inhabitants, or
   (b) a city having a population of at least 25,000 which, with the addition of the population of contiguous places, incorporated or unincorporated, have a population density of at least 1,000

persons per square mile, and which together constitute for general economic and social purposes a single community with a combined population of at least 50,000, provided that the county or counties in which the city and contiguous places are located has a total population of at least 75,000.[1]

2. A contiguous county will be included in a standard metropolitan statistical area if
   (a) at least 75 per cent of the resident labor force in the county is in the nonagricultural labor force,[2] and
   (b) at least 30 per cent of the employed workers living in the county work in the central county or counties of the area.[3]

3. A contiguous county that does not meet the requirements of Criterion 2 will be included in a standard metropolitan statistical area if at least 75 per cent of the resident labor force is in the nonagricultural labor force and it meets two of the following additional criteria of metropolitan character and one of the following criteria of integration.
   (a) Criteria of metropolitan character:
      (1) At least 25 per cent of the population is urban.
      (2) The county had an increase of at least 15 per cent in total population during the period covered by the two most recent Censuses of Population.
      (3) The county has a population density of at least fifty persons per square mile.
   (b) Criteria of integration:
      (1) At least 15 per cent of the employed workers living in the county work in the central county or counties of the area, or

---

[1] In New England, the cities and towns qualifying for inclusion in a standard metropolitan statistical area must have a total population of at least 75,000.

[2] *Nonagricultural labor force* is defined as those employed in nonagricultural occupations, those experienced unemployed whose last occupation was a nonagricultural occupation, members of the Armed Forces, and new workers.

[3] In applying Criteria 2 and 3, the central county or counties of the area are defined as:
   (a) The county or counties which contain the urban part of a central city of a standard metropolitan statistical area.
   (b) For areas having central cities qualifying on the basis of Criterion 1(b), the county or counties containing the urban part of the city of 25,000, and the contiguous places which together with the city constitute the community of 50,000.
   (c) In a standard metropolitan statistical area of 250,000 or more, a county containing
      (1) A city of at least 50,000 which has a contiguous boundary with a central city (cities) of the standard metropolitan statistical area, and
      (2) A population which is at least 80% urban and has nonagricultural employment of at least 50,000.

A city whose name is included in the title of a standard metropolitan statistical area is a central city.

The *urban part* of a central city means that part of a central city identified as "urban" by the Bureau of the Census.

(2) the number of people working in the county who live in the central county or counties of the area is at least 15 per cent of the employed workers living in the county, or

(3) the sum of workers commuting to and from the central county or counties is equal to 20 per cent of the employed workers living in the county.

Area titles

4. The following guidelines are used for determining titles for standard metropolitan statistical areas:

(a) The title of the standard metropolitan statistical area always includes the name of the largest city.

(b) There shall be no more than three city names in the title of any standard metropolitan statistical area. The addition of up to two city names may be made in the area title on the basis of the following:

(1) For those areas where the largest city has a population of 50,000 or more inhabitants (Criterion 1a), the additional city or cities must have a population equal to one third or more of that of the largest city and a minimum population of 25,000, provided that the name of each additional city having a population of at least 250,000 will be included in the title.

(2) For those areas where the largest city has a population of at least 25,000 but less than 50,000 (Criterion 1b), the additional city or cities must have a population equal to one third or more of that of the largest city and a minimum population of 15,000.

(c) Area titles which include the names of more than one city shall start with the largest city and list other cities in order of their size according to the most recent census, except that the names of cities qualifying an area as an SMSA under Criterion 1b shall precede those of any other qualifying city.

(d) In addition to city names, the area titles will contain the name of the state or states included in the area.

Loss of designation

5. A standard metropolitan statistical area shall lose such designation if it fails to meet the criteria for defining an SMSA as measured by information reported in two successive censuses. A contiguous county (city or town in New England) included in a standard metropolitan statistical area shall be excluded from that standard metropolitan statistical area if it

fails to meet the criteria for inclusion in an SMSA as measured by information reported in two successive censuses.[4]

The names of cities which are included in the title of a standard metropolitan statistical area by the application of Criterion 4b shall be dropped from the title if they fail to meet Criterion 4b as measured by information reported in two successive censuses.

The 1970 Census shall be the first census used for the application of this criterion.

Special provisions: New England

In New England, the city and town are administratively more important than the county, and data are compiled locally for such minor civil divisions; towns and cities are the units used in defining Standard Metropolitan Statistical Areas. In New England, because smaller units are used, and more restricted areas result, a population density criterion of at least one hundred persons per square mile is used as the measure of metropolitan character. The concept of a central core is used to describe the central areas in a standard New England metropolitan statistical area. The central core of a standard metropolitan statistical area in New England shall consist of the central city (cities) of the area and those cities and towns immediately adjacent to it that have a population of at least one hundred persons per square mile and that qualify for inclusion in the standard metropolitan statistical area by virtue of their integration with the central city.[5]

6. A city or town adjacent to a central city (cities) of a standard metropolitan area in New England will be designated a part of the central core of the standard metropolitan statistical area if it has a population density of at least one hundred persons per square mile, and if

   (a) at least 15 per cent of the employed workers living in the city or town work in the central city (cities) of the area, or

   (b) the number of people working in the city or town who reside in the central city (cities) is equal to at least 15 per cent of the employed workers living in the city or town, or

---

[4] The provisions of Criterion 5 shall not apply in cases in which
   (1) the application of Criterion 1 or Criterion 8 would lead to the designation of a new standard metropolitan statistical area, or
   (2) the application of Criterion 9 would lead to the inclusion of a contiguous county in a different standard metropolitan statistical area.

[5] It is recognized that in addition to areas defined along city and town boundaries, there is a need for a county version of New England areas for purposes of metropolitan data comparisons with areas in other regions. In addition, because some statistics are not readily available except at the county level, users must at times generalize the New England areas to county lines in order to use statistics for them. Therefore, a standard set of metropolitan counties will be designated in New England, on a basis comparable with definitions used in the rest of the country, as a supplement to the primary version of the areas defined along town and city boundaries.

    (c) the sum of the number of workers commuting to and from the central city or cities is equal to 20 per cent of the employed workers living in the city or town.

7. A contiguous city or town in New England will be included in a standard metropolitan statistical area if it has a population density of at least one hundred persons per square mile, and if

    (a) at least 15 per cent of the employed workers living in the city or town work in the central core of the area, or

    (b) the number of people working in the city or town who reside in the central core is equal to at least 15 per cent of the employed workers living in the city or town, or

    (c) the sum of the number of workers commuting to and from the central core is equal to 20 per cent of the employed workers living in the city or town.

Qualifying cities in adjacent counties

8. (a) If two or more adjacent counties each have a city qualifying under Criterion 1a or 1b and the cities are within twenty miles of each other (city limits to city limits), they will be included in the same area unless there is definite evidence that the two cities are not economically and socially integrated.

    (b) If two or more adjacent counties each have a city qualifying under Criterion 1a or 1b and the cities are more than twenty miles from each other (city limits to city limits), they will not be included in the same area unless there is definite evidence that the two cities are economically and socially integrated.

Contiguous county eligible for inclusion in more than one SMSA

9. In the application of Criteria 2 and 3, a contiguous county (city or town in New England) eligible for inclusion in more than one standard metropolitan statistical area will be included in that standard metropolitan statistical area with which it is most closely integrated as measured by the application of Criterion 3b.

Chapter 9
The pattern and timing of suburbanization

We observed in Chapter 8 that from the start of this century metropolitan growth has increasingly taken place in the suburban rings. So pervasive has been the centrifugal drift of urban population during the past two decades that it has become the dominant feature of America's changing population distribution. In this chapter we shall continue our examination of urban centrifugal drift, with special emphasis on the role of annexation in city and suburban growth patterns.

Several studies noted earlier offer detailed historical descriptions of metropolitan city and suburban growth (Thompson, 1947; Bogue, 1953; Hawley, 1956; Taeuber, 1972). Although these efforts provide useful demographic analyses, they repeatedly exhibit one major limitation in their respective treatments of central city and suburban growth: no accounting is made of the distortion that municipal annexation leaves on reported rates of population increase. This constitutes a serious shortcoming because municipal acquisitions of suburban territory and population introduce a systematic bias that artificially inflates central-city growth and understates suburban growth.

One study that acknowledges this problem redistributes annexed population from city to suburban-ring categories for the 1950–1960 decade (Schnore, 1962). The resulting adjustments accurately reverse apparent centralizing trends. Unfortunately, due to the paucity of annexation data before 1950, comprehensive extension of similar corrections to historical growth rates of all central cities and suburban rings has not been done.

Our purpose here is to reassess patterns of central-city and suburban growth from 1900 to 1970, controlling for all metropolitan boundary changes, including suburban population annexed by central cities. As described in Chapter 8, conventional multiple regression techniques applied to historical data on municipal land area additions and enumerated suburban populations permit us to calculate reasonable approximations of population annexation before 1950. Reported central-city and suburban-ring decennial population counts are adjusted for altered political boundaries by subtracting from each central-city population the estimated number of persons annexed during the decade, and adding the same figure to the suburban-ring population. (During the 1950–1960 and 1960–1970 decades, actual annexation figures are used in adjusting central-city and suburban-ring populations.) Growth in the two metropolitan sectors is then examined by ten-year intervals, making certain at each decade that: (1) only metropolitan areas are compared that meet SMSA definitional criteria *at the prior census year*, and (2) county and city components of each SMSA are held constant *as of the prior census year*. These two precautions ensure that metropolitan growth due to the creation of new SMSAs or the addition of cities or counties to existing areas does not enter into the results.

The first part of our analysis is devoted to elaborating the differential occurrence of annexation by regional location, historical period, and, most importantly, by city growth stage. Annexation and other boundary adjustments are then incorporated to arrive at corrected rates of population increase and changing density for all metropolitan central cities and suburban rings. The resulting adjustments provide long-needed refinements upon our information on patterns of urban expansion as well as the timing of suburbanization.

Preliminary considerations

Before we proceed, clarification of terms is in order. The process of urban expansion has been described in numerous ways ranging from the organic analogies of early human ecology to recent prophecies about eventual abandonment of the city. In this chapter we make reference to *expansion, suburbanization,* and *centrifugal drift* to describe the overall tendency for large urban areas to grow outwardly. *Decentralization* is reserved for specific usage to denote the process whereby a suburban ring grows faster, either in relative percentage rates or in absolute numbers, than the central city. *Deconcentration,* on the other hand, refers to a decline in settlement density of the population residing in the urban center. *Suburbanization* in its totality is the enlargement and spread of a functionally integrated population over an increasingly wider expanse of territory. Hence consideration in turn of both total population growth and patterns of dispersion offers the best demographic measurement of suburbanization—more general than past treatments of suburbanization that have relied on single measures. Thompson (1947), Bogue (1953), and Hawley (1956), for example, examined only total changes in city and suburban population, or what we call decentralization. Winsborough (1963b), on the other hand, presented an "ecological theory of suburbanization" that considers only density patterns—what we call deconcentration.

In our analysis the *city* is operationally defined as the formally recognized central-city population of Standard Metropolitan Statistical Areas. The *suburban-ring population* refers to that population residing outside the central city but within the SMSA. This is in keeping with common research approaches, but it is by no means a universal practice. Past literature has differed greatly on how geographical differentiation of metropolitan populations may be accomplished. Previous research strategies and their findings, upon which this chapter builds, can briefly be considered.

Bogue (1953) decomposed intercensal population changes of metropolitan areas from 1900 to 1950 into central-city, urban-ring, and rural-ring components. His study was among the first to show that decentralization, or a faster rate of population increase in the suburban ring than in the city, began (for metropolitan areas as a whole) during the 1920s. The analysis is unable to control for shifting central-city boundaries, although the author readily acknowledges that this shift influences the size of population change in city and ring categories.

Hawley (1956) conducted a more thorough investigation of decentralization aimed at measuring growth rates in varying concentric zones around the metropolitan center. His data also establish that population centralized up to the benchmark year of 1920 and decentralized thereafter. Again, since political boundaries serve to delineate central areas from satellite areas, the vagaries of annexation bias the results. An indication of the effect this might have had on reported rates of centralization is given by zonal variations in growth. Hawley observed that the most rapid rates of population increase occur within central cities from 1900 to 1910, in the zero-to-five-mile zone adjacent to cities in 1910–1920, and in the five-to-ten-mile zone after 1920. It seems more than likely that as growth spread in this outward manner the observed buildup at the center before 1920 was enhanced by cities' capturing peripheral population through annexation.

McKenzie (1933b) offered further evidence that annexation accounted for a considerable amount of city growth before 1920. From close inspection of three cities, the tracing of the number of inhabitants by areal boundaries reveals that decennial increases from 1830 to 1930 were largely dependent upon annexation. For example, the city of Cleveland as delimited by its 1830 area began losing population in 1870. Yet with continued extension of official boundaries, it maintained increases at every census up to 1930. The same inflation of city growth rates characterize St. Louis and Boston as well. Summary data gathered by McKenzie of land area annexations for 35 large cities over this period establishes the pervasiveness of the phenomenon. Recognizing this, Schnore (1959) attempts to take large known annexations into consideration in marking the onset of decentralization, but his corrections are purely judgmental. The importance of 1920 as a turning point in population distribution remains substantiated for most metropolitan areas, although older places are shown to have decentralized earlier. Still, as Schnore admits, failure accurately to measure annexed population causes the timing to occur later than if it were known.

Several studies of suburbanization employ methods that avoid the political boundary dilemma. Because these approaches deal with a number of smaller geographic units, they render more precise examination of historical patterns of centrifugal drift. The gain in analytical power, however, must be weighed against inability to obtain a complete cross-sectional picture of metropolitan growth. One technique, cohort analysis of census tracts, follows changes in density levels for small areas aggregated by the date in which they attain a moderate development level (Duncan et al., 1962; Guest, 1973). This type of analysis finds that older areas of an urban center, particularly ones fully developed before 1920, experience a general decline in density with each decade. Population concentration in more recently developed areas gradually climbs over time, but never approaches former levels. Since age is highly associated with centrality, the expansion process can readily be described as one in which the urban core deconcentrates while the periphery grows.

Comparative studies of density functions complement these findings. Guest (1973) reports from a sample of 37 metropolitan areas that high population growth before 1880 is far more important than later growth in explaining either current levels of city density or rates of decline away from the center. He attributes this to the introduction of modern transportation, which enabled emerging population settlements to expand over wider areas. Winsborough (1963b) finds that less concentrated populations are characteristic of cities that had transit systems in 1890. These same older cities are relatively more concentrated by 1950, indicating the dispersed character of cities which developed after the advent of short-distance transport improvements. Finally, one longitudinal analysis (Mills, 1970) finds gradual ten-year declines in density gradients as far back as 1910. Although the observed trend sets an early timing for deconcentration, the limited sample of six older cities prohibits establishment of a general pattern.

This brief overview indicates the variety of methods used to investigate suburbanization. This chapter continues the mode of analysis employed by Bogue, Hawley, and Schnore; namely, longitudinal and cross-sectional comparison of central-city and suburban growth rates. A criticism has at times been made that the central-city–suburban dichotomy hinders accuracy because of wide differences in cities' respective proportions of the metropolitan areas. Mills (1970:5), for example, states that

> the central-city–suburb dichotomy does not provide a fixed measure of suburbanization, since the part of the metropolitan area that is included in the central city differs greatly from one metropolitan area to another. A five-point change

in the central city has a different meaning in a metropolitan area in which the central city contains one-third of the residents than in one in which it contains three-quarters.

We hasten to point out, however, that the distinction is not only methodologically acceptable, but valuable, for several reasons.

First, the objection of different metropolitan proportions assumed by central cities may pose an obstacle to comparing individual SMSAs, but presents less difficulty for analyzing groups of metropolitan areas. On the whole, cities take up similar proportions of the encompassing urban region, as regularities in their diminishing share of population attest. As always, caution must be exercised in comparing percentage change in cases in which population bases differ. Despite this caveat, percentages still provide the simplest and most readily interpretable measure of population change.

Second, the city-suburban boundary roughly corresponds to an essential division of the metropolitan community. As an ecological unit, the metropolis undergoes expansion by peripheral growth, which is typically accompanied by a development of centralized coordinative activities to maintain integration. The complete process of expansion, therefore, requires both centrifugal and centripetal movement—the former occurring on the edge of development, and the latter at the core. The central-city boundary, albeit not the precise delimiter of the organizational nucleus of the community, offers the best approximation across a large number of metropolitan areas. Thus, one consequence of delineating growth by central cities and suburban rings is the shedding of additional light on the changing structure of metropolitan communities.

Third, one should not overlook the fact that political jurisdictions, however fortuitous or erratic their origins, constitute real factors compounding many urban problems. The "myth of suburbia" as a single entity may have been dispelled, but the fiscal, social, and economic disparities between most large cities and their surrounding municipalities are well documented. The suburban "wall" is neither illusory nor, it appears, a temporary unintended by-product of urban growth.

The role of annexation in city and suburban growth

To provide an overview of annexation activity in SMSAs at different time periods and in different geographic regions, Table 9.1 details percentage increases in central-city population caused by municipal boundary extensions since 1900. Historically, annexation was a frequent event. Before 1900 it provided an essential source of city growth by acting as an automatic response to, and co-optation of, peripheral development. In the heyday of civic boosterism and intercity rivalries, virtually all metropolitan cities added large populated areas (McKenzie, 1933b; Jackson, 1972). By the turn of the century, however, incorporation of outlying areas, the development of autonomous suburban public services, and the scattering of population beyond contiguous municipalities led to widespread defeat of annexation proposals. Popular resistance to extending political boundaries continued until the late 1940s, when annexation activity regained increased usage,

Table 9.1

Percentage increase of metropolitan central-city population resulting from annexation

	Decade						
	1960–70	1950–60	1940–50	1930–40	1920–30	1910–20	1900–10
Date of metropolitan area inception [a]							
1860 and before (N = 15)	0.9	0.4	0.5	0.5	0.6	1.2	1.3
1870 (N = 7)	1.1	2.5	2.0	0.3	4.2	4.5	6.8
1880 (N = 7)	19.2	8.1	2.0	0.6	4.5	4.2	6.0
1890 (N = 18)	9.1	7.2	2.0	1.0	5.3	7.2	7.8
1900 (N = 16)	2.1	9.1	6.6	1.4	2.9	6.0	8.3
1910 (N = 22)	10.7	20.6	11.9	2.4	9.6	9.5	
1920 (N = 25)	10.6	20.8	4.3	0.5	9.5		
1930 (N = 26)	12.4	24.8	10.1	2.2			
1940 (N = 7)	12.8	85.7 [c]	37.7				
1950 (N = 22)	11.4	15.1					
1960 (N = 34)	14.6						
All cities [b]	6.1	8.1	3.3	0.8	3.1	3.3	3.4
Geographic region [b]							
Northeast	0.0	0.0	0.8	0.4	1.1	1.7	2.0
South	12.8	20.7	11.2	1.4	6.8	4.6	3.3
North Central	5.8	5.8	1.5	0.8	3.4	3.4	3.6
West	6.7	13.9	3.3	1.4	5.4	10.6	16.9
Percentage of total city growth resulting from annexation	98.4	90.0	25.0	15.7	13.6	12.8	10.3
Number of areas at initial date	199	165	143	136	110	85	63

[a] Date of inception is census year in which central city first reached 50,000 or, for multi-city SMSAs, when they first met current definitional criteria.

[b] The sample size for each intercensus period is the number of urban areas qualifying as SMSAs (current definition) at the initial date or earlier.

[c] The amount of central city growth due to annexation seems abnormally high for the 1940 cohort during the 1950–1960 decade (85.7%). Coincidentally, large annexations were undertaken in 6 of the 7 metropolitan areas during this period. The mean population annexed by these 7 cities is 74,008 compared to 24,746 for all metropolitan central cities, 1950 to 1960. The individual annexations are Phoenix, Arizona, 332,398; Amarillo, Texas, 45,383; Corpus Christi, Texas, 34,638; Columbus, Georgia, 43,495; Jackson, Mississippi, 33,364; Stockton, California, 18,810; Waterloo-Cedar Falls, Iowa, 1,004.

though not on its former scale. From 1900 to 1940, total city growth by annexation declined steadily from 3.4 per cent to 0.8 per cent. In the 1950s, the extent of central-city growth resulting from adding new areas rose to 8.2 per cent, and in the 1960s registered 6.2 per cent.

Data in Table 9.1 indicate some important specifications of the overall trend. A distinct pattern prevails for each intercensus period: the younger a metropolitan area, the more its central city grows by enumerating new population from surrounding areas. Annexation, then, is clearly a more important factor in earlier stages of city development, and successively decreases in the course of urban maturation. Regional differences support this observation. Older central cities in the Northeast register small increments from annexation in the earliest decade shown, and these amounts decline to virtual insignificance by 1970. Both North Central and Western cities show a gradual tapering up to 1950, as expected. But it should be noted that at the turn of the century cities in the West, which were then in their initial periods of growth, increased inhabitants substantially by widening their corporate area. In the South, population additions enhanced city growth relatively little in the 1900–1910 decade (about the national metropolitan average). Rather, for this newest regional grouping, annexation increased steadily for every ten-year period up to 1960, pausing only in the sluggish Depression years.

In short, annexation has been a vital component of central-city growth in the initial decades of metropolitan development. As urbanization has spread from the Northeast to the West and South, the process has been repeated for newly emergent metropolitan centers. It might further be noted that the recent upsurge in annexation has reached the point where population added in this manner offsets overall central-city decline. The next to the bottom row of Table 9.1 shows that by the 1950s annexation accounted for 90 per cent of all city growth. In the last decade, the contribution was virtually total. These figures evidence the importance that municipal boundary extensions have had for artifically prolonging aggregate central-city growth.

Correcting central-city and suburban population for annexation refines the timing of metropolitan decentralization. Previous analyses that focus on differential rates of population increase concur in setting the initial shift to greater suburban growth at the period of the 1920s. In an urban text, Hawley (1971) again examines contrasting city and suburban growth rates, and reiterates this conclusion. Similar figures of percentage increase are presently reconsidered in Table 9.2. Giving no heed to changing city boundaries, the nation's suburban rings, as anticipated, first show higher rates of population growth during the 1920–1930 decade—36.7 per cent compared to 22.8 per cent for central cities. In the Northeast and West, decentralization began ten years earlier, with the latter region showing much more rapid overall growth. The altered balance of growth appeared "on time" in North

Central cities, and a decade later in the South. When the distribution of population increase is adjusted for annexation, however, *faster rates of growth in the suburban ring are shown to have occurred in every decade since 1900 for every regional grouping*. It is evident that central cities maintained their

Table 9.2

Mean increase and percentage increase of SMSA, central-city, and suburban-ring population, unadjusted and adjusted for annexation, by region

	Decade						
Regional Component	1960–70	1950–60	1940–50	1930–40	1920–30	1910–20	1900–10
All Areas							
SMSAs							
Mean increase	95,358	129,273	100,380	37,319	111,322	99,249	116,366
Per cent	16.3	25.0	21.3	8.3	27.1	24.8	31.1
Central cities							
Unadjusted increase							
Mean	18,137	27,583	39,471	15,038	64,775	71,258	86,185
Per cent	6.2	9.0	13.2	5.1	22.8	25.7	33.0
Adjusted increase							
Mean	3	2,683	29,761	12,564	56,059	62,026	77,182
Per cent	0.0	0.8	10.0	4.3	19.7	22.4	29.6
Suburban rings							
Unadjusted increase							
Mean	77,221	101,690	60,909	22,281	46,547	27,991	30,181
Per cent	26.4	47.9	35.4	14.4	36.7	22.7	26.6
Adjusted increase							
Mean	95,355	126,590	70,619	24,755	55,263	37,223	39,184
Per cent	32.5	59.6	41.1	16.0	43.7	30.2	34.5
Geographic region							
Northeast							
SMSAs							
Mean increase	98,650	126,165	87,270	39,947	134,183	119,459	159,020
Per cent	9.5	14.1	10.5	5.0	19.3	18.8	29.0
Central cities							
Unadjusted increase							
Mean	−10,806	−16,489	22,717	14,171	62,487	70,880	104,184
Per cent	−2.3	−3.4	4.8	3.1	14.7	18.1	30.8
Adjusted increase							
Mean	−10,897	−16,620	19,158	12,107	57,949	64,174	97,427
Per cent	−2.3	−3.4	4.0	2.6	13.6	16.4	28.8
Suburban rings							
Unadjusted increase							
Mean	109,456	142,654	64,553	25,776	71,696	48,571	54,836
Per cent	19.7	35.0	18.2	7.8	26.6	19.9	26.1
Adjusted increase							
Mean	109,547	142,785	68,112	27,840	76,234	55,285	61,593
Per cent	19.7	35.1	19.3	8.5	28.3	22.7	29.3

Table 9.2—*continued*

Regional Component	Decade						
	1960–70	*1950–60*	*1940–50*	*1930–40*	*1920–30*	*1910–20*	*1900–10*
South							
SMSAs							
Mean increase	77,123	105,142	88,802	40,493	58,186	56,040	38,306
Per cent	21.3	34.0	35.2	18.0	29.2	27.8	19.3
Central cities							
Unadjusted increase							
Mean	31,001	47,788	45,975	21,296	46,503	48,957	33,910
Per cent	15.5	24.7	26.6	13.1	30.7	31.7	21.7
Adjusted increase							
Mean	5,461	7,706	26,567	18,866	36,274	41,822	28,745
Per cent	2.7	4.0	15.4	11.6	23.9	27.1	18.4
Suburban rings							
Unadjusted increase							
Mean	46,122	57,354	42,827	19,197	11,683	7,091	4,396
Per cent	28.6	49.5	53.8	30.0	24.4	15.0	10.5
Adjusted increase							
Mean	71,662	97,436	62,235	21,627	21,912	14,226	9,561
Per cent	44.5	84.1	78.1	34.3	45.6	30.0	22.8
North Central							
SMSAs							
Mean increase	69,960	100,840	75,434	20,948	110,945	104,376	97,055
Per cent	12.7	23.2	17.9	5.1	31.2	32.0	30.2
Central cities							
Unadjusted increase							
Mean	1,787	11,236	29,047	4,674	68,536	84,460	82,452
Per cent	0.6	4.0	9.8	1.6	24.7	33.0	31.4
Adjusted increase							
Mean	−14,584	−5,124	24,632	2,346	58,861	75,490	73,014
Per cent	−5.1	−1.8	8.3	0.8	21.2	29.5	27.8
Suburban rings							
Unadjusted increase							
Mean	68,173	89,604	46,387	16,274	42,409	19,916	14,603
Per cent	25.4	57.6	36.9	14.6	54.2	28.5	25.0
Adjusted increase							
Mean	84,544	105,964	50,802	18,602	52,804	28,931	24,041
Per cent	31.5	68.1	40.4	16.7	67.5	41.4	41.4

relative growth edge over suburban areas until the 1920s by annexing adjacent populations.

Although the actual *rate* of population increase has been higher on the urban periphery for a longer time than suspected (most likely, since before 1900), decentralization in an absolute sense did not occur generally until the 1930s. This is true of all metropolitan areas taken together, but regional

Table 9.2—*continued*

	Decade						
Regional Component	*1960–70*	*1950–60*	*1940–50*	*1930–40*	*1920–30*	*1910–20*	*1900–10*
West							
SMSAs							
Mean increase	190,848	310,265	244,540	77,870	199,011	110,007	159,133
Per cent	27.8	44.8	50.2	17.3	57.2	38.2	87.8
Central cities							
Unadjusted increase							
Mean	52,621	104,424	89,727	33,281	112,617	77,708	128,190
Per cent	16.6	28.1	30.4	11.2	43.5	34.0	88.3
Adjusted increase							
Mean	31,332	52,799	79,877	29,067	98,711	53,419	103,603
Per cent	9.9	14.2	27.0	9.7	38.1	23.4	71.4
Suburban rings							
Unadjusted increase							
Mean	138,227	205,841	154,813	44,589	86,394	32,299	30,943
Per cent	37.5	64.2	80.9	29.1	97.4	54.5	85.7
Adjusted increase							
Mean	159,516	257,466	164,663	48,803	100,300	56,588	55,530
Per cent	43.3	80.2	86.0	31.9	113.2	95.4	153.8

variations should be noted. Mean increases in population, which can be alternatively interpreted as differential proportions of SMSA growth in cities and rings, were greater in the suburban sector of Western metropolitan areas as early as 1910. Annexation, however, conceals this decentralization until 1930. Both Southern and North Central cities were outpaced in total numbers by their outlying areas in the 1930s. The former region, which was pointed out above to have increasingly relied on annexation to bolster city growth, would not seem to exhibit this pattern of decentralization until 1950 if annexation were overlooked. Suburbs in the Northeast began taking a larger share of metropolitan growth in the 1920–1930 decade, regardless of boundary changes. The extent of distortion that arises from ignoring enlarged city limits depends, of course, on the size of the population added.

Because annexation declines in importance with advanced phases of a city's growth history, population counts in newer metropolitan cities are more sensitive to its effects. Likewise, regional aggregations, which also reflect age groupings, tend to compound inaccuracies. The largest disparity between adjusted and unadjusted population changes appears at the turn of the century in the West, and although this continues somewhat unabated to the present, Southern SMSAs currently exhibit the most "error."

Analysis of changing density patterns completes our chronicle of urban centrifugal drift. Tables 9.3 and 9.4 present mean levels of population concentration in central cities and suburbs at each census year both before and after adjustments for annexation during the previous decade. In Table 9.3, the area included by the suburban ring has been held constant by retaining all counties that are part of the most recent SMSA boundaries (National Bureau of Standards, 1973). But the actual area for each suburban ring shrinks with decennial annexation losses, which must be controlled for. The effect of this adjustment on suburban density is to raise slightly, for each age cohort and census year, the level of population concentration. Main trends are left undisturbed.

Table 9.3 also demonstrates that population per square mile in the

Table 9.3

Mean population per square mile in suburban rings of metropolitan areas (unadjusted and adjusted for annexation)

Date of metropolitan area inception	Year							
	1970	1960	1950	1940	1930	1920	1910	1900[a]
A. *Unadjusted*								
1860 and before	835	669	471	370	339	255	211	167
1870	575	465	311	245	217	160	125	101
1880	325	259	172	139	128	99	96	84
1890	328	260	180	137	122	94	82	71
1900	164	123	89	69	60	52	48	43
1910	170	144	112	87	81	72	63	
1920	143	115	84	65	58	50		
1930	136	106	75	56	48			
1940	63	58	51	37				
1950	97	81	61					
1960	96	61						
1970	106							
B. *Adjusted*								
1860 and before	835	673	474	373	341	259	217	167
1870	579	476	318	246	227	169	136	101
1880	345	268	174	140	133	112	99	84
1890	340	269	182	139	125	97	86	71
1900	166	130	94	71	62	56	50	43
1910	179	160	116	89	87	77	68	
1920	152	129	87	66	62	55		
1930	145	124	80	57	55			
1940	73	81	64	37				
1950	104	90	68					
1960	106	81						
1970	115							

Note: The adjusted figures record the population at each year within the boundaries of the prior census date divided by the land area at the prior census date.

[a] Population per square mile in 1900 not adjusted for prior annexation.

suburban ring is higher the earlier an SMSA came into being, and climbs regularly for all areas from 1900 to the present. These relationships are straightforward, but some caution must be exercised in interpreting absolute suburban density levels. Because of the use of county units, suburban-ring density is depressed in some cases by inclusion of large land areas beyond the actual edge of urban development.

More interesting are the data presented in Table 9.4, which describe

Table 9.4

Mean population per square mile in central cities of metropolitan areas (unadjusted and adjusted for annexation)

Date of metropolitan inception[a]	Year							
	1970	1960	1950	1940	1930	1920	1910	1900[b]
A. *Unadjusted*								
1860 and before	11,290	11,987	12,825	12,058	11,845	10,925	10,360	9,526
1870	7,759	8,857	10,098	10,455	10,287	11,552	11,523	11,476
1880	5,306	6,934	8,144	7,665	7,755	8,008	7,968	6,646
1890	5,207	6,806	8,145	8,250	8,280	8,233	7,658	7,167
1900	3,875	4,417	5,270	5,181	5,090	4,731	4,430	3,526
1910	3,751	4,799	5,775	5,668	5,745	6,226	5,511	
1920	3,774	4,737	5,830	5,864	5,955	5,298		
1930	3,244	3,980	5,473	5,452	5,022			
1940	2,389	6,281	4,716	5,423				
1950	3,284	3,851	4,203					
1960	2,877	3,419						
1970	3,147							
B. *Adjusted*								
1860 and before	11,315	12,107	12,908	12,090	12,345	12,009	11,559	9,526
1870	8,015	9,493	10,905	10,348	13,641	15,435	14,992	11,476
1880	6,746	8,166	8,319	7,764	9,021	9,902	8,775	6,646
1890	6,206	7,727	8,689	8,537	9,036	9,073	8,549	7,167
1900	4,187	5,291	5,758	5,221	5,285	5,471	4,946	3,526
1910	4,593	5,773	6,353	5,910	7,401	7,491	7,270	
1920	4,521	5,970	6,464	6,149	6,773	7,366		
1930	4,094	5,462	6,235	5,483	6,179			
1940	6,968	4,942	6,182	5,822				
1950	3,885	4,635	6,170					
1960	3,818	4,866						
1970	4,040							

Note: The adjusted figures record the population at each year within the boundaries of the prior census date divided by the land area at the prior census date.

[a] Date of inception is the census year in which central city first reached 50,000 inhabitants, or, for multiple-city SMSAs, when they first met current multiple-city definitional criteria.

[b] Population per square mile in 1900 not adjusted for prior annexation.

density patterns in metropolitan central cities from 1900 to 1970. The top panel of Table 9.4 presents central-city densities in terms of the actual population and land area enumerated at each census date. The bottom panel adjusts the density figures for population and land area annexed by the central cities between decennial censuses. Adjustment is made by subtracting the land area and population annexed by cities between census dates from the enumerated land area and population at the latter census date. This procedure provides the population at the latter census date residing in central city boundaries (area) of the former date; that is, it excludes annexed area and population during the decade. Because most annexation occurs at the less densely settled urban periphery, adjustment for annexation systematically raises central-city density. Interestingly, it also systematically raises suburban density (see Table 9.3), because suburban areas annexed by cities tend to have higher population density than the remainder of the suburban ring.

Table 9.4 highlights how important the period in which a city evolved is to the intensity of its settlement density. At each decade, new cohorts of metropolitan centers tend to possess lower densities than older centers. Closer inspection uncovers three general age groups which exhibit remarkable intracohort consistency in their historical density levels. Cities which attained metropolitan size before 1880 have regularly possessed density levels approximately twice that of cities which did not attain metropolitan status until 1900 or later. Cities that attained metropolitan status in 1880 or 1890 fall consistently in between.

It seems no mere coincidence that these dates reflect frequently cited and important transportation eras in our urban history. Places that grew before 1880 evolved in the period of premodern transportation, and necessarily were dense pedestrian cities. The latter half of the nineteenth century witnessed the introduction of rudimentary public transportation, which greatly facilitated cross-city movement. By 1900, extensive electric-powered mass transit systems appeared, allowing new housing to be constructed over a greater area at lower density and the increased separation of homes from workplaces (Warner, 1962; Tarr, 1972; Jackson, 1973, 1975). Consequently, central cities evolving in 1900 or later never developed the intense land use stages of the oldest metropolitan centers.

What about the timing of deconcentration (that is, actual density declines) in metropolitan central cities? Hawley (1971) constructs a table similar to the top panel of Table 9.4, which shows absolute declines in city density occurring intermittently before 1950. After that date, deconcentration characterizes all metropolitan centers, regardless of age. Hawley draws attention to the fact that density reductions were taking place in newer as well as older cities. His conclusion sums up the emergent pattern:

> Perhaps never again will urban centers have reason to attain great sizes and densities before deconcentration begins. Nor is there any reason to believe that

the growth experience of the older metropolitan areas will be repeated by the newer ones. (Hawley, 1971:161)

While this conclusion is intuitively appealing, more detailed examination of our data suggests that it may be premature. Recall that earlier we showed that annexation tends to be a major component of the growth of a central city during its initial decades of metropolitan development. Can it be, then, that reported declines in density of newer central cities are simply an artifact of widespread annexation? Apparently so, and because the corporate boundaries of cities often differ between census dates, even the bottom panel of Table 9.4, when used alone for longitudinal comparisons, can be misleading. Accurate detection of the timing of deconcentration requires simultaneous use of both panels of Table 9.4.

To obtain *changes* in central-city densities between decennial censuses within constant city boundaries (excluding annexation), one subtracts the density figures for age cohorts at each census year, presented in the top panel, from the corresponding adjusted density figures ten years later presented in the bottom panel. For example, to compute changes of population density in central cities between 1960 and 1970 *within 1960 central-city boundaries*, the unadjusted densities in 1960 for each age cohort are subtracted from their corresponding adjusted densities in 1970.

Table 9.5 presents *decennial changes* in population density between 1900 and 1970 for all age cohorts of central cities within central-city boundaries

Table 9.5

Mean decennial changes in population per square mile in central cities of metropolitan areas occurring within central-city boundaries at prior census date

Date of metropolitan area inception[a]	Decade						
	1960–70	1950–60	1940–50	1930–40	1920–30	1910–20	1900–10
1860 and before (N = 15)	−672	−718	850	245	1420	1649	2033
1870 (N = 7)	−842	−605	450	61	2089	3912	3516
1880 (N = 7)	−188	22	654	9	1013	1934	2129
1890 (N = 18)	−600	−418	439	257	803	1415	1382
1900 (N = 16)	−230	21	577	131	554	1041	1420
1910 (N = 22)	−206	−2	685	165	1175	1980	
1920 (N = 25)	−216	140	600	194	1475		
1930 (N = 26)	114	−11	783	461			
1940 (N = 7)	687	226	759				
1950 (N = 22)	34	432					
1960 (N = 34)	399						

[a] Date of inception is the census year in which central city first reached 50,000 inhabitants, or, for multiple-city SMSAs, when they first met current multiple-city definitional criteria.

at the prior census date. When we measure deconcentration in terms of density changes within central-city boundaries existing at the beginning of each decade, we observe that not until the 1950 decade do density reductions appear for any of the age cohorts of central cities. Furthermore, these density reductions are restricted primarily to central cities that attained metropolitan status in 1920 or earlier. With only one exception, every central-city cohort attaining metropolitan status after 1920 has experienced *increased* population density to 1970.

Table 9.5 further indicates that even during decades of rapidly advancing short-distance transportation technology (1900 to 1930), older as well as newer cohorts of metropolitan central cities exhibited substantial increases in population density within existing corporate limits. Moreover, with the advent of mass automobile usage (1920 to 1950), central-city densities continued to increase across all age cohorts. Thus, while the streetcar, electric trains, and automobile may be credited with lower initial density levels in cities that developed during those vehicles' respective technological eras, they were not responsible for "emptying out" or reversing concentration within existing central-city boundaries.

Summary

This chapter can be summed up by first remarking that suburbanization is a multifaceted phenomenon which has not been fully and appropriately measured in previous empirical studies. Three indices of expansive city growth were examined herein. The first, *relative decentralization*, or higher rates of percentage population increase in the suburban ring as compared to the city, quantifies the rapidity of growth at the city's rim. It is the least reliable indicator of suburbanization. Because new growth almost always takes place on the urban periphery, whether or not it falls within city boundaries is somewhat arbitrary. In this sense, cities have been "decentralizing" since at least 1900; that is, once-annexed city population is redistributed to the suburban ring.

A second index, *absolute decentralization*, is free of percentage base differences and therefore provides clear interpretation of raw population increases and the respective proportion in city and suburb. Correcting for annexation, we find that almost all suburban rings have assumed a larger share of total metropolitan growth since 1930, and many a decade or so earlier. It has been the accumulation of disproportionate total growth that is primarily responsible for the larger balance of population now living in the suburban ring.

Our third measure of suburbanization assesses *the degree to which metropolitan centers have deconcentrated as their peripheries expanded*. The history of urban density patterns is one of successively lower levels for cities that grew during periods of faster, more efficient local transportation. Nevertheless, with the exception of some of our older central cities, deconcentration, as measured by decennial density declines within central-city boundaries at the beginning of each decade, is a post-1950 phenomenon. In decades prior to 1950, city annexation of suburban territory concealed much of the real increase of population density within earlier central-city boundaries. Even after 1950, central-city density declines were selective, occurring primarily in older cities. Population density continued to increase in most cities that attained metropolitan status in 1940 or later, once the artificial effects of annexation were removed.

Consistent with historical patterns of urban expansion, annexation has played a major role in the growth of newer central cities, while it has also concealed increases in their population density. Thus, without longitudinal controls for annexation, the data convey the impression that newer cohorts of central cities are deconcentrating when, in fact, population density within previous city boundaries is actually increasing.

It should be reiterated, however, that overall increases in population size and density that did take place in many central cities during the most recent decade were offset by absolute declines in others. The impact that depletion of future energy sources, economic conditions, or technological innovations may have on urban settlement patterns has yet to be assessed. But the suburbanization of the U.S. population that began in significant numbers over fifty years ago is now firmly established, and not likely to be reversed by developments currently anticipated.

Chapter 10
Metropolitan expansion and central-city organization: a test of the theory of ecological expansion

The previous chapter's analysis documented the dynamic yet disproportionate growth that has occurred in metropolitan areas since the turn of the century. We observed that during the most recent decades almost all metropolitan growth has been in the suburban rings, whereas most central cities have grown very little and many have experienced an absolute population loss. As a result of urban centrifugal drift, more people are now residing in the suburban rings than in the central cities.

Concurrent with this population redistribution, there has been a significant alteration in the ecological structure of our urban areas. The compact city of nineteenth-century America has been replaced by the diffuse metropolitan area, as entire communities have become territorially specialized and dependent on one another.

The general purpose of this chapter is to examine the relationship between population redistribution and organizational structure at the level of the metropolitan community. Its specific purpose is to provide an empirical test to the theory of ecological expansion. In essence, this theory stipulates that *population growth in peripheral areas of a system will be matched by an increase in organizational functions in its nucleus to ensure integration and coordination of activities and relationships throughout the expanded system.*

Roderick D. McKenzie introduced the theory to the field of human ecology in 1933, explaining that expansion "connotes movement outward from a spatially determined center of settlement without loss of contact with that center. It presupposes a sufficient development of the center or core of settlement to insure reciprocal relations within an ever-widening range of territory" (1933a: 20).

Amos Hawley, a student of McKenzie's, further developed and elaborated upon McKenzie's original formulation of expansion in his 1950 treatise *Human Ecology: A Theory of Community Structure.* In this work Hawley views expansion as a progressive absorption of more or less unrelated populations into a single organization. According to Hawley, expansion

> begins at an intersection of heavily traveled routes, e.g., a city, and proceeds with the extension and improvement of transportation and communication facilities. The phenomenon involves centrifugal and centripetal movements. Centrifugal movements are the process by which new lands and populations are incorporated into a single organization. The centripetal movements make possible a sufficient development of the center to maintain integration and coordination over the expanding complex of relationships. (1950:369)

The centrifugal and centripetal movements result in an expanded community having two loci. The first is an inner locus, or city, which is conceived as the organizational nucleus, and the second is an outer locus, or suburban area, whose population and economic institutions are integrated and coordinated through the auspices of the central city.

Ecological expansion at the level of the local community is predicated on the reduction of time and cost involved in short-distance movements. Where no efficient means of short-distance transportation exist, the friction of space is high, restricting both the people of the community and their employment and service institutions to an area fixed by pedestrian movement. This produces a segmented pattern of relatively autonomous, territorially heterogeneous settlement units throughout a region. Growth in these units occurs primarily through increases in population density or through cellular-like fission (characteristic of the preindustrial city), and not through centrifugal and centripetal processes of ecological expansion.

Growth without expansion is well illustrated in the American case: cities grew to be quite large before they began to interact routinely with their outlying settlement units to form a single interdependent system. In the early nineteenth century, the introduction of the railroads and barge canal systems, both adapted for long-haul movement, released cities from a previous dependency on their immediate hinterland and the vagaries of natural waterways for the provision of agricultural commodities. With the importation of food surplus and other raw materials from distant inland territories, cities grew rapidly in size. Their growth, however, occurred to a large extent through increases in population density; the scope of community interdependence remained essentially constant. Outlying towns and villages continued to function as semi-independent entities, each providing its own specialized services and community facilities.

It was not until the second half of the nineteenth century that the centrifugal and centripetal processes of ecological expansion began to operate. The major innovations that instigated the expansion processes were the horse-drawn streetcar and the intraurban electric trains. The horse-car extended the distance that could be traversed in an hour's time from approximately three miles to nearly five miles, and electric trolleys further enlarged the sixty-minute radius of travel to seven or eight miles. Rapid-transit electric trains added from ten to twenty miles to the radius of convenient daily movement to and from the central city (Hawley, 1950: 406–407). These electric trains were responsible for bringing many of the outlying towns and villages into daily contact with the central city and for creating radial strings of new suburban settlements.

The adaptation of rail transportation to intraurban movement signaled the beginnings of widespread population decentralization and functional specialization of local land areas. No longer was it necessary to reside within the three-mile distance to place of work that was imposed by pedestrian movement. Places of employment and places of residence became increasingly separated. Similarly, the differentiation of local areas was made possible, so that professional, administrative, clerical, financial, and other specialized activities could each be efficiently concentrated in a median location rather than distributed throughout the community. The central business district developed as an organizational hub, and as population dispersed outward along the rail lines the entire community began to acquire a new structure.

The electric rail carriers were inflexible, however. They operated on fixed time schedules and restricted the expansion of the community to those areas located near the rail arteries. It was the development of the motor vehicle in the late nineteenth century and its widespread use by 1930 that overcame these limitations and substantially accelerated the expansion process. Motor vehicle registration increased from 8,000 in 1900 to 26,351,999 in 1930, or from one in every ten thousand persons to one in every five persons (U.S. Bureau of the Census, 1960). During the same period, highway patterns evolved that wove a tight web throughout the interstitial suburban areas that had been isolated from the main rail arteries. The speed and flexibility of the automobile, together with the hard-surfaced road, fostered an encompassing zone of daily interaction that reached distances of up to fifty miles from the urban center (Hawley, 1950:410).

The early twentieth century also witnessed dramatic improvements in short-distance communication. Household and business telephones became commonplace, radio broadcasts began originating from the urban centers, and the central-city–based newspapers were made available daily to the outlying settlements. All contributed significantly to the social and economic integration of the expanding community.

As expansion progressed, the formerly independent towns and villages scattered around the central city lost their specialized services and institutions to the center. Their stores could not compete with the diversified units located in the central city, and their professional workers and other specialists migrated to the inner locus to take advantage of the external economies afforded by a central location. Many of their local institutions were displaced by chain outlets and branch offices, managed and controlled by administrative headquarters concentrated in the central city (Hawley, 1971:148).

A large number of these outlying towns and villages became residential suburbs whose new function was to house commuting workers. A smaller number became commercial satellites of the central city, providing standardized goods and services to the nearby population. Still others were transformed into predominantly industrial suburbs, as manufacturing industries discovered that they could remove their production facilities from the congested center but continue to maintain administrative offices there. Expansion thus eliminated both the autonomy and the heterogeneity of the surrounding towns and villages, absorbing them into a single diffuse community composed of territorially specialized units whose functions were integrated and coordinated through the auspices of the central city. This larger and altogether new type of community became known as the metropolitan community.

In this chapter we use standard metropolitan statistical areas (SMSAs) as the units of analysis, as we examine the effects of metropolitan expansion on the organizational structure of central cities. If the centrifugal and centripetal forces of metropolitan expansion have operated in the manner postulated above, we should expect to find consistent positive relationships between the size of metropolitan areas and the development of organizational functions in the central cities. More important, very strong relationships should exist between the size of the outer locus and the development of coordinative and integrative functions in the central city, even when controls are introduced for other pertinent variables including the size of the central city, SMSA age, income and racial composition of central-city residents, and distance between metropolitan centers.

Data and method

The data employed in this chapter pertain to 157 monocentered SMSAs with populations of 100,000 or more in 1960. To obtain measures of organizational structure, data were gathered from the journey-to-work section of the 1960 Census of Population (U.S. Bureau of the Census, 1963) on the number of SMSA residents working in the central cities and in the suburban rings for each of three occupational and four industrial categories that perform integrative or coordinative functions in the metropolitan area. These categories are:

1. Professional, technical, and kindred.
2. Managers, officials, and proprietors.
3. Clerical and kindred.
4. Transportation, communication, and public utilities.
5. Finance, insurance, and real estate.

6. Business and repair services.
7. Public administration.

When the number of central-city employees in each of the seven categories is divided by the central-city population size, we obtain seven detailed components of central-city organization. This method of obtaining components of central-city organization is very similar to the procedure carried out by sociologists examining the relative size of the administrative and supportive components in formal organizations (Pondy, 1969; Rushing, 1967). To determine the extent of administrative intensity within each organization, the number of employees performing administrative or supportive functions is divided by the total number of personnel. Using these same data, we shall also construct indices of organizational development in the central cities and suburban rings and derive two summary measures of the total organizational component of central cities.

Population data were obtained from the 1960 Census of Population. The demographic variables include population in the central city, population in the urbanized area, population in the suburban ring, population in the SMSA, and proportion of the SMSA population in the suburban ring. Although organization and population are generally regarded as functionally interdependent factors in ecological analysis (that is, the organization requires a population of sufficient size to staff its functions, but there cannot be a larger population than the organization can sustain), the organizational components will be treated as dependent variables and the demographic factors as independent variables.

Correlation and standardized partial regression coefficients (Beta weights) will be used to measure the relationships between the organizational and the population variables and their levels of significance. Additional variables, including SMSA age, per capita income of central-city residents, nonwhite percentage in the central-city population, and distance between metropolitan centers will be introduced as statistical controls.

Results

As a preliminary step to analyzing the relationship between metropolitan expansion and central-city organization, it is useful to examine briefly the means and standard deviations of the organizational components of central cities for the total sample and for different size categories of central cities and suburban rings. Table 10.1 presents these data.

Looking first at the organizational components on the basis of central-city size, we observe that only three of the seven organizational components show consistent increases with central-city size. Such is not the situation when we group the organizational components of central cities on the basis of the size of the suburban rings. As the theory of ecological expansion would

Table 10.1
Means and standard deviations[a] of organizational components of central cities (central-city employees per 1000 central-city residents), by size of central city and suburban rings, 1960

Organizational components of central city	Total sample (N = 157)	Size of central city[b]			Size of suburban ring[c]		
		Small (N = 37)	Medium (N = 83)	Large (N = 37)	Small (N = 29)	Medium (N = 96)	Large (N = 32)
Professional, technical, kindred	52 (17)	51 (21)	51 (15)	55 (16)	41 (10)	53 (17)	59 (16)
Managers, officials, and proprietors	40 (10)	43 (13)	39 (10)	41 (7)	35 (7)	41 (11)	43 (10)
Clerical and kindred	79 (25)	71 (17)	77 (26)	92 (25)	60 (14)	78 (21)	99 (28)
Transportation, communication, and public utilities	35 (13)	36 (15)	32 (13)	40 (11)	28 (11)	34 (13)	43 (12)
Financial, insurance, and real estate	24 (12)	21 (8)	24 (14)	29 (9)	17 (6)	24 (8)	34 (9)
Business and repair service	12 (5)	12 (5)	12 (6)	14 (3)	9 (2)	13 (5)	15 (4)
Public administration	25 (22)	20 (15)	24 (18)	30 (34)	19 (9)	23 (17)	34 (38)

[a] Standard deviations are in parentheses.
[b] Central-city size: *Small*: population less than 75,000; *Medium*: population between 75,000 and 300,000; *Large*: population over 300,000.
[c] Suburban-ring size: *Small*: population less than 50,000; *Medium*: population between 50,000 and 300,000; *Large*: population over 300,000.

suggest, there is a monotonic increase in the mean value of each of the components of central-city organization as the size of the suburban population becomes larger. It is also noteworthy that the gradient of each central-city component is consistently steeper when the mean values are computed on the basis of the size of the outer locus.

Table 10.2 presents the means and standard deviations of the organizational components of central cities on the basis of the size of the SMSA and the relative size of the suburban rings. We see that every organizational component except one exhibits a monotonically increasing gradient as the SMSA size categories change from small to medium to large.

When the organizational components of central cities are grouped in terms of the relative size of the suburban rings, the mean values of all organizational components again exhibit the expected positive gradients. These data, together with the data presented in Table 10.1, provide tentative support to the thesis that expansion of the outer locus of metropolitan areas has been matched by a development of organizational functions in the central

Table 10.2

Means and standard deviations of organizational components of central cities by size of SMSA and by proportion of SMSA population in suburban ring, 1960

Organizational components of central city	Size of SMSA[a]			Relative size of suburban ring[b]		
	Small (N=17)	Medium (N=104)	Large (N=36)	Small (N=28)	Medium (N=116)	Large (N=13)
Professional, technical, and kindred	42 (7)	53 (18)	55 (16)	41 (9)	53 (17)	63 (19)
Managers, officials, and proprietors	35 (6)	41 (11)	41 (11)	36 (6)	40 (9)	51 (17)
Clerical and kindred	60 (14)	78 (24)	91 (26)	61 (16)	81 (23)	102 (33)
Transportation, communication, and public utilities	28 (12)	34 (13)	41 (12)	29 (9)	36 (14)	43 (12)
Financial, insurance, and real estate	16 (6)	24 (13)	29 (10)	20 (8)	25 (12)	33 (11)
Business and repair service	8 (2)	12 (5)	14 (4)	11 (5)	12 (5)	16 (5)
Public administration	15 (7)	24 (18)	30 (34)	20 (8)	24 (23)	36 (32)

[a] Size of SMSA: *Small*: population less than 125,000; *Medium*: population between 125,000 and 600,000; *Large*: population greater than 600,000.

[b] Relative size of suburban ring: *Small*: less than 30 per cent of SMSA population in suburban ring; *Medium*: between 30 and 70 per cent of SMSA population in suburban ring; *Large*: over 70 per cent of SMSA population in suburban ring.

cities. Much more extensive analysis is necessary, however, before any conclusions can be drawn.

The next step in testing the theory of expansion is to determine the degree of organizational development that exists within the inner and outer loci of metropolitan areas. Organizational development may be measured in central cities and suburban rings by the degree to which organizational functions performed within each central city and suburban ring exceed the expected amount, that is, the average for all metropolitan areas (Bogue, 1949:33; Carroll, 1963; Feldt, 1965). Indices for organizational development in central cities can be derived using the formula: $I = (o_c/p_c)/(O_m/P_m)$ where o_c is the number of SMSA residents performing a given organizational function within the central city, p_c is the population size of the central city, O_m is the total number of SMSA residents performing the same organizational function within all SMSAs, and P_m is the total population of all SMSAs. It should be recognized that the numerator in this formula is the organizational component of the central city, and that the denominator (which is a constant) is the average organizational component for all metropolitan areas.

Indices of organizational development in the suburban rings may also be constructed by placing the organizational components of the suburban rings (the ratio of SMSA residents performing organizational functions within the suburban rings to the population size of the suburban rings) in the numerator of the above formula. By computing separate indices of organizational development for central cities and suburban rings, we will be able to determine the impact of population decentralization on the development of organizational functions within each locus, as well as arriving at comparative figures on the differential in organizational development between the central cities and their suburban rings.

Table 10.3 shows the mean indices of organizational development for the entire sample and on the basis of the size of the SMSA. It is immediately apparent that for the total sample and for each SMSA size class, the indices of organizational development are substantially greater in the central cities than in the suburban rings. In fact, for every organizational category except those of professional, technical, and kindred, and of public administration, the mean index of development in the central cities is more than twice that of the suburban rings. This large differential holds not only for the total sample but within each SMSA size class.

The impact of the size of the metropolitan area on organizational development in central cities is likewise apparent in Table 10.3. Although all organizational categories show greater degrees of development in central cities of large SMSAs than in central cities of small SMSAs, the most pronounced increments are in the industrial categories. Central-city finance, insurance, and real estate functions increase from 12 per cent *below* the expected amount in small SMSAs to over 57 per cent *above* the expected

Table 10.3

Mean indices of organizational development in central cities and suburban rings, by size of SMSA, 1960

Organizational category and locus of employment	Total sample		Size of SMSA[a]		
	Mean	Std. dev.	Small	Medium	Large
Professional, technical, and kindred					
Central cities	1.215	0.395	0.975	1.231	1.281
Suburban rings	0.631	0.272	0.586	0.595	0.758
Managers, officials, and proprietors					
Central cities	1.323	0.388	1.146	1.347	1.339
Suburban rings	0.617	0.440	0.517	0.619	0.659
Clerical and kindred					
Central cities	1.285	0.406	0.973	1.268	1.481
Suburban rings	0.428	0.182	0.360	0.399	0.544
Transportation, communication, and public utilities					
Central cities	1.325	0.509	1.069	1.285	1.563
Suburban rings	0.484	0.272	0.476	0.476	0.512
Financial, insurance, and real estate					
Central cities	1.130	0.413	0.879	1.293	1.574
Suburban rings	0.281	0.160	0.157	0.251	0.426
Business and repair services					
Central cities	1.242	0.485	0.857	1.244	1.416
Suburban rings	0.467	0.431	0.342	0.463	0.537
Public administration					
Central cities	1.226	1.118	0.768	1.207	1.498
Suburban rings	0.700	0.905	0.503	0.687	0.833

[a] See Table 10.2 for size categories.

amount in large SMSAs. Business and repair services in the central cities increase from approximately 14 per cent *below* the expected amount in small SMSAs to over 41 per cent *above* the expected amount in large SMSAs, whereas central-city public administrative functions increase from 23 per cent *below* the expected amount to nearly 50 per cent *above* the expected amount when the SMSA size categories move from small to large.

Having observed that the central cities are far more developed in organizational functions than their suburban rings and that the indices of organizational development in the central cities generally rise as the size of the SMSA increases, the important question becomes: What is the relationship between expansion of the outer locus of metropolitan areas and organizational development in central cities? This question is answered in Table 10.4, which presents the mean indices of organizational development in central cities on the basis of both the absolute size and the relative size of the outer locus.

Table 10.4

Mean indices of organizational development in central cities and suburban rings, by absolute size and by relative size of suburban rings, 1960

Organizational category and locus of employment	Size of suburban ring[a]			Relative size of suburban ring[b]		
	Small	Medium	Large	Small	Medium	Large
Professional, technical, and kindred						
Central cities	0.965	1.234	1.381	0.965	1.245	1.485
Suburban rings	0.568	0.601	0.779	0.694	0.615	0.640
Managers, officials, and proprietors						
Central cities	1.147	1.346	1.413	1.192	1.313	1.691
Suburban rings	0.573	0.619	0.652	0.676	0.600	0.646
Clerical and kindred						
Central cities	0.970	1.268	1.619	0.988	1.313	1.669
Suburban rings	0.357	0.407	0.555	0.477	0.419	0.468
Transportation, communication, and public utilities						
Central cities	1.055	1.305	1.629	1.100	1.347	1.691
Suburban rings	0.529	0.465	0.504	0.620	0.453	0.469
Financial, insurance, and real estate						
Central cities	0.903	1.272	1.806	1.080	1.320	1.750
Suburban rings	0.174	0.262	0.435	0.246	0.280	0.366
Business and repair services						
Central cities	0.895	1.264	1.490	1.134	1.221	1.653
Suburban rings	0.357	0.473	0.549	0.587	0.418	0.647
Public administration						
Central cities	0.930	1.165	1.680	0.999	1.217	1.799
Suburban rings	0.767	0.699	0.644	1.153	0.593	0.687

[a] See Table 10.1 for size categories.
[b] See Table 10.2 for size categories.

Focusing first on the indices of organizational development in central cities on the basis of the absolute size of the suburban rings, we observe that the mean values of all seven indices monotonically increase as the outer locus expands. Organizational development also generally occurs in the suburban rings as they expand, but at a much slower rate. For example, finance, insurance, and real estate functions in the suburban rings exhibit the largest increments as the suburban rings expand, but these increments are quite small when compared with the development of the same functions in the central cities as the outer rings expand. Whereas the mean index of development of financial, insurance, and real estate functions in the suburban rings increases from nearly 83 per cent below the expected amount in small suburban rings to approximately 66 per cent below the expected amount in large suburban rings, the mean index of development of these same functions in central cities increases from approximately 10 per cent *below* the expected amount in central cities with small suburban rings to over 80 per cent *above* the expected amount in central cities with large suburban rings. It may be inferred from the differential slopes in the gradients of organizational development exhibited by central cities and suburban rings that increases in the size of the outer locus have had a much larger impact on organizational development in the center than in the outer locus.

Table 10.4 also presents the mean indices of organizational development in the central cities and suburban rings on the basis of the relative size of the suburban rings. Again we observe that all seven indices of organizational development in central cities monotonically increase as the relative size of the suburban rings become larger. Thus, whether the outer locus is measured in terms of its absolute size or its relative size, organizational functions in the central city become more developed as the outer locus expands.

The data presented in Tables 10.1 through 10.4 lend credence to the theory of ecological expansion; yet by presenting the organizational variables in terms of broad categories and without statistical controls, we have not exploited the data to their fullest. In order to take advantage of the interval nature of the data to examine more exactly the effects of peripheral expansion on organizational development in the central cities, we shall utilize correlation and regression techniques; first to obtain zero-order correlations between the demographic and the organizational variables, and second to obtain standardized partial regression coefficients which will give us the direct effects of population increments in the outer locus on the organizational components of the central cities.

If expansion of metropolitan areas has been matched by a development of organizational functions in the central cities, we should find not only positive relationships between SMSA size and each of the organizational components of central cities, but also increases in the magnitude of the correlation coefficients as successively larger portions of the outlying population are correlated with the organizational components of the central cities. The zero-order correlation coefficients presented in Table 10.5 reveal that this is generally the case. All the correlations between SMSA size and the components of central-city organization are in the hypothesized direction, and all but two of the coefficients are significant at the .001 level. Also as expected, a positive gradient of correlation coefficients exists for each of the organizational components as the independent variable successively changes from exclusively central-city size to exclusively suburban-ring size. Each component of central-city organization is positively related beyond the .001 level of significance to the size of the suburban rings.

The last column in Table 10.5 presents the zero-order correlations between the relative size of the suburban rings and the organizational components of central cities. Of the seven organizational components, six

Table 10.5

Zero-order correlations between organizational components of central cities and the population size (log) of portions of SMSAs, 1960

Organizational components of central city	Central city	Urbanized area	Suburban rings	SMSA	Percentage of SMSA population in suburban rings
Professional, technical, and kindred	.07	.24**	.36**	.24**	.43**
Managers, officials, and professionals	−.03	.13	.28**	.15	.45**
Clerical and kindred	.27**	.45**	.57**	.46**	.49**
Transportation, communication, and public utilities	.19*	.31**	.39**	.32**	.33**
Financial, insurance, and real estate	.25**	.35**	.46**	.40**	.36**
Business and repair services	.21*	.21*	.42**	.35**	.36**
Public administration	.14	.25**	.26**	.23*	.21*

* Significant at .01.
** Significant at .001.

are positively related to the relative size of the suburban rings at the .001 level of significance and one at the .01 level. Manager, official, and proprietor functions and clerical functions performed in the central cities exhibit particularly high relationships with the relative size of the suburban rings.

To examine the overall effect of increases in metropolitan size on the organizational structure of central cities, it is useful to develop a summary measure of the organizational component in central cities similar to that of the administrative/personnel (A/P) ratio in formal organization theory (Pondy, 1969; Rushing, 1967). By adding the number of central-city employees in the three mutually exclusive occupational categories and dividing the sum by the population size of the central city, we obtain such a measure based on organizational employment by occupation. Likewise, by adding together the number of central-city employees in the four mutually exclusive industrial categories and dividing this by the population of the central city, we obtain a total organizational component of central cities based on its industrial employment. It is enlightening to examine the relationships between the two composite measures of central-city organization and the population size of the different territorial units of the metropolitan areas.

Table 10.6 shows the zero-order correlations between the total organizational component of central cities, measured separately by occupational and industrial data, and the population size of portions of the metropolitan areas. The positive gradient that exists as successively larger portions of the

Table 10.6

Zero-order correlations between total occupational and industrial components
of central-city organization and the population size (log) of territorial
portions of SMSA, 1960

Total components of central city organization	*Central city*	*Urbanized area*	*Suburban rings*	*SMSA*	*Percentage of SMSA population in suburban rings*
Organizational component by occupation	.17	.36**	.50**	.37**	.52**
Organizational component by industry	.26**	.40**	.48**	.41**	.39**

** Significant at .001.

suburban population are correlated with the two summary central-city
measures demonstrates the close relationship between the size of the outer
locus and organizational development in the metropolitan center. Noteworthy is the fact that both measures of the total organizational component
of central cities exhibit higher correlations with SMSA size and with the
absolute and relative size of the suburban rings than with central-city size.
Clearly, than, the central city is not the appropriate ecological unit for which
the central-city organizational functions serve.

Summing up the results thus far, we have shown that central cities are
much more developed in organizational functions than are the suburban rings,
and that positive and highly significant relationships exist between increments in both the absolute and the relative size of the suburban rings and the
development of organizational functions in the central cities. We have yet to
determine, however, if the size of the outer locus *directly* influences organizational development within the metropolitan center. The final step in the
present analysis will be to measure the direct effects of the suburban rings
on organizational development within the metropolitan center.

In determining the direct influence of the suburban rings, it is important
to control for a number of variables that may also be influencing the organizational structure of central cities. For example, the literature on the systems
of cities (Berry and Pred, 1961; Berry and Horton, 1970; Berry, 1972;
Duncan, et al., 1960; Lampard, 1968) suggests that higher-order cities
perform more supralocal administrative and service functions than do
lower-order cities. Since, as we shall show in Part V of this volume, higher-
order cities are typically larger cities, it is crucial to control for the size of the
central city when measuring the direct influence of the suburban ring on
organizational development within the central city. Other ecological studies
(Bogue, 1949; Hawley, 1941; Meyer, 1972; Schnore, 1957; Schnore, 1959)
suggest that age of the SMSA, per capita income of the central-city residents,

nonwhite percentage in the central city, and distance to the nearest other metropolitan center likewise influence the organizational structure of central cities. These variables, along with central-city size, were therefore selected as independent (control) variables in the regression analysis used to determine the direct effects of size of the outer locus.[1]

Table 10.7 presents the standardized partial regression coefficients between the organizational components of the central cities and the six independent variables. The partial regression coefficients indicate that the size of the outer locus has a strong direct influence on the development of organizational functions in the central cities. Manager, official, and proprietor functions and business and repair services in the central city are very strongly influenced by the size of the suburban ring. Also noteworthy is the finding that when other relevant variables are held constant, an inverse relationship emerges between central-city size and each organizational

Table 10.7

Standardized partial regression coefficients (Beta weights) between organizational components of central cities and selected metropolitan variables, 1960

Organizational components of central city	Size of central city (log)	Size of suburban rings (log)	Age of SMSA	Per capita income of central-city residents	Percentage of nonwhite in central city	Distance to nearest other SMSA
Professional, technical, and kindred	−.34*	.72**	−.16	.14	−.04	.07
Managers, officials, and proprietors	−.44**	.81**	−.28*	.05	.16*	.34**
Clerical and kindred	−.43**	.74**	.21*	.14*	.02	.06
Transportation, communication, and public utilities	−.35*	.55**	.19	−.09	.05	.17*
Finance, insurance and real estate	−.24	.59**	.06	.03	.09	.11
Business and repair service	−.16	.81**	−.31**	.06	.01	.37**
Public administration	−.10	.27*	−.01	.09	.22*	.05
Total central-city component by occupation	−.45**	.83**	−.01	.14*	.03	.14
Total central-city component by industry	−.28*	.64**	.04	.03	.17*	.17*

* Coefficient is greater than twice its standard error.

** Coefficient is greater than three times its standard error.

[1] The age of the SMSA is operationalized as the number of decades that have passed since the census first reported its central city as having at least 50,000 residents. Distance to nearest other metropolitan center is measured in terms of highway mileage and includes distance to any other metropolitan center, even if that center is not part of our sample. The source of the mileage data is the Household Goods Carriers Bureau (1967).

component. This suggests that there may be economies of scale in administering and servicing city populations.

Regression analysis was also conducted by substituting the relative size of the outer locus (the per cent of the SMSA population residing in the suburban ring) for the absolute size of the outer locus in the regression model. The results were just as conclusive. With central-city size as well as the other four metropolitan variables held constant, the standardized partial regression coefficients between the relative size of the outer locus and the organizational components of central cities ranged from +.57 for managers, officials, and proprietors to +.21 for public administration. Standardized partial regression coefficients between the relative size of the outer locus and the total occupational and industrial components of central-city organization were +.59 and +.46, respectively. These strong positive relationships between the organizational components of central cities and both the absolute and the relative size of the suburban rings clearly reflect the balance that exists between the size of peripheral areas of ecological units and the extent of organizational development in their nuclei.

Conclusions

The results presented thus provide empirical support for the contention that increases in the peripheral areas of metropolitan communities have been matched with a development of organizational functions in their centers. Categoric and multivariate analyses indicate that central cities are far more developed in coordinative and integrative functions than their suburban rings and that the degree of organizational development in central cities increases as the outer locus expands. In fact, the size of the outer locus has substantially greater direct effects on organizational development within the central city than does either size of the central city, age of the SMSA, per capita income of central city residents, nonwhite percentage in the central city, or distance between metropolitan centers. If the present centrifugal drift of population, manufacturing activity, and establishments providing standardized goods and services continues, we may expect to find the central cities becoming even more territorially specialized in future years.

Chapter 11
The impact of suburban population growth on central-city service functions: a question of externalities and spillovers

Our discussion of metropolitan expansion described the process by which central cities and their outlying territory become inseparable parts of a single ecological system known as the metropolitan community. So interdependent have central cities and their suburbs become that only analytically can we separate their demographic and organizational structures. Without the commercial units and service industries located in the central cities, the large suburban population could not be supported. Conversely, the commercial units and service industries located in the central cities are dependent on the suburban population to staff many of their functions, particularly middle and upper management, and draw on their goods and services.

With the continuing expansion of metropolitan areas, more and more suburbanites have been making daily use of central-city facilities. As a commercial hub and service center, the city's facilities are utilized intensely by a large part of the suburban population for their employment, shopping, recreation, professional services, and other needs.

This chapter examines the central city as a commercial hub and service center for the suburban population. More specifically, it addresses two fundamental issues: (1) What is the effect of the large and growing suburban population on central-city commercial activity? and (2) What impact, if any, have recent increases in the suburban population had on public services provided by central-city governments? Primary attention is given to the latter issue, and its implications for central city planning and metropolitan financing are discussed.

The metropolitan communities to be examined are all SMSAs that had attained metropolitan status by 1950 ($N = 168$). Data representing four broad, but distinct, categories of service functions performed in central cities were obtained from the 1948, 1958, and 1967 Censuses of Business and from the 1950, 1960, and 1968–1969 Compendia of City Government Finances (U.S. Bureau of the Census). The categories are: (1) retail trade, (2) wholesale trade, (3) business and repair services, and (4) public services provided by central-city governments. Sales and receipts for central-city retail trade, wholesale trade, and business and repair services are used as indicators of the magnitude of these three categories of service functions, whereas annual operating expenditures for noneducational services provided by central-city governments are used as indicators of the magnitude of the public-service functions. For cross-sectional analysis, the sales, receipts, and operating expenditures are expressed in terms of "amount per central-city resident"; for longitudinal analysis, these data are converted to constant dollars and expressed in terms of first differences.

Demographic data were obtained from the 1950, 1960, and 1970 Censuses of Population. The suburban population is defined as all population residing outside the central city but within the metropolitan area for those SMSAs that contain a single central city, and as all population residing outside the largest central city but within the metropolitan area of SMSAs that contain more than one central city.

The primary method used is path analysis, which provides an algorithm for decomposing the total correlation between the dependent and independent variables so that the direct effects of each independent variable on the dependent variables may be ascertained. For example, if central-city service functions vary with both the size of the central-city population and the size of the suburban population then a path model mapping the relationship appears, as shown in Figure 11.1.

In a multivariate path model such as this, the zero-order correlation between an independent variable and the dependent variable is the sum

Figure 11.1
X_1, central-city service variable; X_2, central-city population; X_3, suburban population size; p_{12} and p_{13}, path coefficients; r_{23}, zero-order correlation coefficient; e_1, e_2, and e_3, error terms.

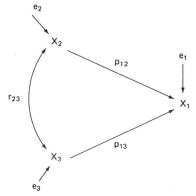

of its direct effect via the path from that independent variable to the dependent variable and its indirect association through its correlation with the other independent variable (Land, 1969, pp. 12–15). In terms of the coefficients,

$$r_{12} = p_{12} + r_{23} \, p_{13},$$
$$r_{13} = p_{13} + r_{23} \, p_{12}.$$

The path coefficients (p_{12} and p_{13}) measure the variance in X_1 for which each independent variable (X_2 or X_3) is directly responsible, when the other independent variable is held constant. (In simple recursive models such as ours, the path coefficients are equivalent to standardized partial regression coefficients as presented in Chapter 10.) Therefore, by computing path coefficients, we will be able to determine the direct effect of variation (or change) in both the central city and the suburban population on each central-city service function, as well as the direct effects of other independent variables.

Suburban population and central-city services

Does the size of the suburban population have any bearing on service functions performed in the central city? Table 11.1 presents path co-

Table 11.1
Path coefficients between per capita measures of central-city service functions and population size (log) of central cities and suburban areas, 1960 and 1970

	Central-city service functions			
Population size	*Retail trade (1958)*	*Wholesale trade (1958)*	*Business and repair services (1958)*	*Public services (1960)*
1960:				
Central city	−.76*	.33*	.11	.01
Suburban area	.83*	.27*	.54*	.39*
	Retail trade (1967)	*Wholesale trade (1967)*	*Business and repair services (1967)*	*Public services (1969)*
1970:				
Central city	−.83*	.16	.04	.06
Suburban area	.61*	.36*	.53*	.41*

* Significant at .001.

efficients between each of the four general categories of central-city service functions and the population sizes of the central city and suburban areas. (For cross-sectional analysis, the logs of the size of the central-city and suburban populations were regressed on the per capita service measures to normalize the skewed distribution caused by the very largest cities and suburbs.)

Looking first at the direct effects of central-city size on the central-city service functions, we observe that only central-city wholesale trade in 1958 exhibits a significant positive relationship with central-city size. Business and repair services, along with public services, show little relationship with central-city size. Also interesting is that, when suburban population size is held constant, a strong inverse relationship emerges between retail trade per capita in the central city and the population size of the central city. This, of course, does not mean that central-city retail trade declines with increases in central-city size but rather that, controlling for the effects of the suburban population, increments in central-city population are associated with less than proportionate increases in central-city retail trade.

On the other hand, all four categories of central-city service functions have highly significant positive relationships with the size of the suburban population, in both 1960 and 1970. Moreover, with the exception of wholesale trade at the former point in time, suburban population has a larger direct (positive) effect on every category of central-city service functions than does central-city population.

Having noticed the close cross-sectional association between central-city service functions and population size of the suburban areas, the question now becomes: To what extent are changes over time in these service functions influenced by, or related to, changes in the suburban population size? This is determined in Table 11.2, which shows the path coefficients between changes in the four categories of central-city service functions and changes in the population of central cities and suburban areas during the two most recent decades.

The longitudinal results indicate that changes in the size of the suburban population have had highly significant direct effects on all four categories of central-city service functions. The proportion of variance in the four service categories for which changes in the suburban population were directly responsible ranged from 22 per cent to 51 per cent during the 1950–1960 decade, and from 10 per cent to 38 per cent during the 1960–1970 decade.[1]

It is again important to observe that changes in the suburban population exerted much larger direct effects on every central-city service category,

[1] The squared path coefficients measure the proportion of variance (or change) in the dependent variable for which the determining variable is directly responsible (Land, 1969, p. 10). The zero-order correlations between changes in central-city and suburban populations were .11 during the 1950–1960 time period and .27 during the 1960–1970 time period.

Table 11.2

Path coefficients between changes in central-city service functions and changes in the population size of central cities and suburban areas, 1950–1960 and 1960–1970

	Central-city service functions			
Population change	Retail trade (1948–58)	Wholesale trade (1948–58)	Business and repair services (1948–58)	Public services (1950–60)
1950–1960:				
Central city	.54**	.41**	−.15	−.18
Suburban area	.47**	.71**	.69**	.65**
	Retail trade (1958–67)	Wholesale trade (1958–67)	Business and repair services (1958–67)	Public services (1960–69)
1960–1970:				
Central city	.74**	.27**	.22*	.11
Suburban area	.31**	.62**	.51**	.45**

* Significant at .01.
** Significant at .001.

with the exception of retail trade, than did changes in the central-city population.

The path coefficients presented in Tables 11.1 and 11.2 clearly indicate that the suburban population has a strong influence on central-city service functions. Perhaps the most notable finding is that virtually no relationship exists, either cross-sectionally or longitudinally, between noneducational public services provided by the central-city government and its population, while positive and strong relationships exist both cross-sectionally and longitudinally between increases in the suburban population and central-city public services. The obvious and important inference suggested by this finding is that the suburban population has at least as great an impact on public service provided by central-city governments as does the central-city population itself. The remainder of the chapter will examine this issue in more detail.

The public-service issue

One of the most serious problems facing our central cities in recent years has been the growing service-resource gap that has developed from a disproportional increase in public services provided by central-city governments, exceeding that of their resources. Either explicit or implicit in most discussions addressing this problem has been the suggestion that the major force behind the increasing demand for public services is the changing composition of the central-city population. Much less emphasis has been given to the increased demand for public services in the central cities created by the rapidly expanding suburban populations.

It would be difficult to deny that the changing composition of the central-city population has increased the need for certain municipal services, such as public welfare and housing. However, we should not overlook the fact that increases in suburban populations have created a large demand for many other central-city services. For example, the suburban population makes regular use of central-city streets, parks, zoos, museums, and other public facilities; its routine presence in the central city increases sanitation department problems and contributes to the costs of fire protection; the daily in-and-out movement of the large commuting population requires services that constitute a large portion of the operating budget of both the police and highway departments (Hawley, 1957, p. 773). These are only some of the costs experienced by central-city governments as a result of services they provide to their suburban neighbors. Just what has been the relationship between growth of suburban populations and services commonly provided by the central city?

In Table 11.3 are presented the path coefficients between the sizes of the central-city and suburban populations and the annual per capita expenditures for six common central-city service functions.[2]

We see that the suburban population exhibits a higher positive relationship with every central-city service function than does the central-city population itself. Moreover, although the direct effects of the suburban population are significant at the .001 level on every central-city service function, except highway services in 1960, not a single significant positive relationship exists between the central-city population and the central-city service functions when suburban population size is held constant. These

[2] *Police service* includes police patrols and communications, crime-prevention activities, detention and custody of persons awaiting trial, traffic safety, vehicular inspection, and the like. *Fire-prevention services* include inspection for fire hazards, maintaining fire-fighting facilities such as fire hydrants, and other fire-prevention activities. *Highway services* include maintenance of streets, highways, and structures necessary for their use, snow and ice removal, toll highways and bridge facilities, and ferries. *Sanitation services* include street cleaning, collection and disposal of garbage and other waste, sanitary engineering, smoke regulation, and other health activities. *Recreation facilities and services* include museums and art galleries, playgrounds, play fields, stadiums, swimming pools and bathing beaches, municipal parks, auditoriums, auto camps, recreation piers, and boat harbors. *General control* includes central staff services and agencies concerned with personnel administration, law, recording, planning and zoning, municipal officials, agencies concerned with tax assessment and collection, accounting, auditing, budgeting, purchasing, and other central finance activities (U.S. Bureau of the Census, 1970, pp. 64–67).

Table 11.3

Path coefficients between per capita expenditures for common central-city service functions and population size (log) of central cities and suburbs, 1960 and 1970

Central-city service function	Population, 1960		Population, 1970[a]	
	Central city	Suburb	Central city	Suburb
Police	.15	.56**	.09	.61**
Fire	−.22*	.43**	−.35**	.57**
Highway	−.19*	.11	−.28**	.25**
Sanitation	−.01	.31**	−.15	.56**
Recreation	−.12	.24**	−.03	.37**
General control	.06	.33**	−.02	.41**

[a] Central-city service data for 1969.
* Significant at .01.
** Significant at .001.

data provide support, in the cross-sectional case, to the contention that increases in the size of the suburban population are a major contributor to increased expenditures for common service functions performed by central-city governments.

The next step is to examine the direct effects of changes (over time) in the size of the central-city and suburban populations on changes in expenditures (in constant dollars) for the six central-city services. Table 11.4

Table 11.4

Path coefficients between change in expenditures (in constant dollars) for common central-city service functions and change in the central-city and suburban populations, 1950–1960 and 1960–1970

Central-city service function	Population Change, 1950–1960		Population Change, 1960–1970[a]	
	Central city	Suburb	Central city	Suburb
Police	−.12	.74**	.13	.56**
Fire	.01	.79**	.17	.54**
Highway	−.13	.61**	.17	.51**
Sanitation	−.03	.72**	.18	.62**
Recreation	−.10	.80**	.23*	.54**
General control	−.10	.51**	.20*	.58**

[a] Central-city service function change 1960–1969.
* Significant at .01.
** Significant at .001.

lists the path coefficients between changes in each service function and changes in the central-city and suburban populations between 1950 and 1960, and between 1960 and 1970.

We observe that the impact of the suburban population on the six central-city service functions is even larger longitudinally than it was cross-sectionally. The amount of change in these service functions for which changes in the suburban population are directly responsible range from 26 per cent to 64 per cent during the 1950–1960 interval, and from 26 per cent to 38 per cent during the 1960–1970 interval. Only during the 1960–1970 interval do we find changes in central-city population size exhibiting consistent positive relationships with changes in the city services. Even during this time period, however, the direct effects of changes in the suburban population were much larger than the effects of changes in the central-city population.

It was shown in Chapters 8 and 9 that there was a good deal of suburban annexation by central cities between 1950 and 1960. When this occurs, city services are generally provided to the annexed population. There may be a lag, however, between the time at which the suburban area is annexed and the period at which many city services begin to be provided to those residing in the annexed area. Therefore, as a check against the possibility that city expenditure data are not being related to the "appropriate" city population, controls for annexation were introduced. The path coefficients from the cross-sectional and longitudinal analyses of the relationship between service and expenditures and the populations of the central cities and the suburbs, controlling for annexation, are presented in Table 11.5.

Table 11.5

Path coefficients between population size (log) of central cities and suburbs and per capita expenditures for selected city services, 1960, and changes in city and suburban sizes and city service expenditures, 1950–1960, controlling for annexation

Central-city expenditure category	Static (1960) (N = 198)		Dynamic (1950–1960) (N = 157)	
	Central city	Suburb	Central city	Suburb
All expenditures	−.27**	.52**	−.31**	.69**
Police	−.35**	.68**	−.25**	.78**
Fire	−.50**	.53**	−.11	.83**
Highway	−.36**	.25**	−.25**	.65**
Sanitation	−.46**	.55**	−.18	.76**
Recreation	−.38**	.44**	−.22*	.84**
General control	−.33**	.48**	−.21*	.55**

* Significant at .01.
** Significant at .001.

The controls for annexation cause the path coefficients between central-city population and all service factors to become negative, in both the cross-sectional and the longitudinal analyses. The direct effects of the suburban population, controlling for annexation, remain positive and strong cross-sectionally, and actually increase in the longitudinal analysis. Evidently, by controlling for annexed population, we have taken away from the central city residents who are actually being provided with city services, and have given their impact to the suburban sector. The result is that in the cross-sectional case, the *per capita* expenditures increase (statistically) when annexed population is subtracted from the central city population, even though the central-city population decreases. In the longitudinal case, central-city expenditures increase partially because public services are provided to the annexed population, yet that additional population is deleted from the central-city population when annexation controls are introduced. Thus, by controlling for annexation, we actually increase the impact of the suburban population.

The next step in examining the effects of centrifugal population drift on central-city services is to compare the impact of suburban population growth on central-city services with its impact on its own (suburban) services. Because a large portion of the suburban population makes regular use of central-city facilities and services and often, for one reason or another, their governments do not provide full public services to their own area, we would anticipate that the impact of suburban population growth on central-city services would be as large as its impact on its own services.

To compare the impact of suburban population growth on services provided by the central city with its impact on services provided by suburban jurisdictions, data for both metropolitan and central-city operating expenditures for government-supplied services were gathered from the 1957 and 1967 Census of Governments. With these data we derived the operating expenditures for services provided in the suburban area by subtracting the operating expenditures for those services provided in the central city from the operating expenditures for the corresponding services for the entire SMSA. With data on central-city and suburban service expenditures at two points in time, we can compare the impact of the suburban population on central-city and suburban area services cross-sectionally (at 1957 and at 1967) and longitudinally (from 1957 to 1967).

Table 11.6 describes the direct effects of size of the suburban population on expenditures per capita for central-city services and on expenditures per capita for services provided in suburban rings during the years 1957 and 1967. In 1957, size of the suburban population is positively and significantly related to per capita expenditures for each central-city service. Regarding its own per capita expenditures for services, however, the size of the suburban population is negatively related both to total per capita operating expenditures and to four of the six detailed service categories. That is, as the size of the

Table 11.6

Path coefficients between population size of suburbs (log) and per capita expenditures for selected central-city and suburban services, 1957 and 1967

Public service function	1957 (N = 162) Suburban population[a] with:		1967 (N = 195) Suburban population[b] with:	
	central-city expenditures	suburban expenditures	central-city expenditures	suburban expenditures
All services	.34**	−.64**	.40**	−.34**
Police	.60**	−.27**	.48**	.11
Fire	.38**	.28**	.41**	.45**
Highway	.23*	−.73**	.12	−.53**
Sanitation	.28**	.25**	.34**	.44**
Recreation	.24**	−.09	.32**	−.06
General control	.31**	−.82**	.40**	−.62**

[a] 1960 population.
[b] 1970 population.
* Significant at .01.
** Significant at .001.

suburban population increased, the amount expended per capita for these services in the suburban rings declined. This is particularly the case for operating expenditures on highways and for general administrative control. Even for fire and sanitation services in suburban areas, to which increases in the suburban population have positive and significant direct effects, the direct effects of those increases on the same services in the central city are larger.

In 1967, the relationships are similar, if not quite as emphatic. Again, there is an inverse relationship between total per capita expenditures for services in the suburban ring and population size of the suburban ring, whereas there is a positive relationship between size of the suburban population and total per capita expenditures for central-city services. The only deviations from the 1957 relationships are that central-city highway expenditures per capita were not significantly related to size of the suburban population, and that for both sanitation and recreation services the suburban population exerted a larger direct effect on suburban-area services than on central-city services.

A reasonable explanation for this increased positive relationship between suburban operating expenditures for fire and sanitation services and the size of the suburban population is that a lag almost always occurs from the time of population growth or incorporation of suburban areas to the time at which adequate service begins to be supplied. The smaller negative relationship between suburban populations and suburban services in 1967 than in 1957 may be explained in a like manner. Perhaps if, or when, public services in suburban areas catch up to population growth in those areas, all negative relationships will disappear. On the other hand, there may be significant economies of scale in the financing of these suburban services.

Examining the longitudinal relationships between 1957 and 1967, we observe in Table 11.7 that changes in the suburban populations had a larger direct effect on changes in expenditure (in constant dollars) for every central-city service, with the exception of sanitation service, than did changes in the central-city population. Conversely, there is not one service factor in suburban areas for which changes in the central-city population exerted a direct effect that is even one half the magnitude of the direct effects of changes in the suburban population on that same service factor. It may be inferred from this finding that the impact of the suburban population on central-city services is not reciprocated by a proportional impact of the central-city population on suburban services.

It is also important to note that changes in the suburban population exhibit larger direct effects on changes in police, fire, highway, and administration expenditures in the central city than on changes in expenditures for the same services in the suburban areas. Two very large differentials in the impact of change in the suburban population on change in central-city and suburban-area service expenditures are police expenditures (.58 as opposed to .34) and fire expenditures (.57 as opposed to .06).

The smaller impact of growth in the suburban population on its own public services than on those in the central city provides aggregate level support to the findings of Hawley and Zimmer (1956a, 1956b) that many

Table 11.7

Path coefficients between change in central-city population and change in service expenditures in the central city and suburbs and path coefficients between change in suburban population and change in service expenditures in the central city and suburbs, 1957–1967   (N = 157)

Public service function	Δ Central-city population[a] with:		Δ Suburban population[b] with:	
	Δ central-city expenditures	Δ suburban expenditures	Δ central-city expenditures	Δ suburban expenditures
All services	.13	.24**	.49**	.53**
Police	.16	.12	.58**	.34**
Fire	.24**	−.01	.57**	.06
Highway	.27**	.06	.65**	.45**
Sanitation	.43**	−.26**	.41**	.53**
Recreation	.41**	.09	.61**	.72**
General control	.21*	.25**	.62**	.57**

[a] 1960 to 1970.
[b] 1960 to 1970.
* Significant at .01.
** Significant at .001.

suburbanites desire a minimum of local public services because they fear that additional public services will lead to additional taxes and force them out of the homeowning category. Many suburbanities are able to afford their homes only by keeping taxes low; hence they will oppose public expenditures for additional or higher-quality services. It therefore follows that while population growth may occur in those suburban areas, concurrent growth in public services is by no means assured.

Other causal factors

Having demonstrated through both static and dynamic analyses that the growth of common central-city service functions is strongly related to population increases in suburban areas, we now test for the possibility that some third variable or combination of variables is responsible for the strong relationships between suburban populations and central-city public services. First, we must account for the effects of the age of the city. We know that suburban populations in the older SMSAs are generally larger than those in younger SMSAs. It has also been found that "older cities bearing the stamps of obsolescence, high density, high industrialization, and aging inhabitants, generate higher expenses than their size alone might have led one to suspect" (Vernon, 1960, p. 172).

Another important variable is personal income of central-city residents. Cities whose residents have higher personal incomes are usually able to provide more and better-quality services than those whose residents have lower personal incomes. It has also been long known that city-to-suburb migration is closely related to personal income of central-city residents. Residents with higher incomes have a much larger choice of residential locations than do low income residents, who are often economically and socially confined to the inner city.

Finally, we must control for the racial composition of the central city. One may suspect that certain municipal expenditures are either directly or indirectly related to the racial composition of the central-city population. At the same time, the "flight to the suburbs" has been greatest in those cities that have experienced the largest influx of nonwhite migrants in the past 25 years.

To discover if each of the above three variables were positively related to both per capita operating expenditures for services provided by central-city governments and population size of suburban areas, zero-order correlations were computed.[3]

[3] The correlation and path coefficients presented in Tables 11.8, 11.9, and 11.10 apply to 157 SMSAs in 1960, for which complete data were available on central-city age (operational as the number of decades since the central city first attained a population size of 50,000 or more), per capita income of central-city residents, nonwhite percentage in the central city, and the number of suburban residents who commute to work in the central city.

Table 11.8

Zero-order correlation coefficients between selected characteristics of central cities and per capita operating expenditures of central-city governments and size of suburban population, 1960

	Central-city characteristic		
	Age	Per capita income	Nonwhite (%)
Per capita operating expenditures in central city (all services)	.43**	.14*	.17*
Size of suburban population (log)	.65**	.16*	.28**

* Significant at .05.
** Significant at .001.

As may be observed in Table 11.8, all relationships are positive. Per capita operating expenditures for central-city services were found to be positively related at the .001 level of significance to the age of the central city, but positively related at only the .05 level of significance to per capita income of central-city residents and the percentage of nonwhites in the central-city population. Similarly, the size of the suburban population was found to be positively related at the .001 level of significance to both the age of the central city and the percentage of nonwhites in the central-city population, but less ($p < .05$) to per capita income of central-city residents.

These results raise an important question: What is the relationship between suburban population size and per capita operating expenditures for central-city public services when we control, not only for central-city size, but also for age of the central city, per capita income of the central-city residents, and the nonwhite percentage in the central-city population? In addition, we might ask: What is the relationship between per capita expenditures for central-city services and each of the three central-city variables (age, income, nonwhite percentage), when the remaining two central-city variables as well as central-city size and suburban size are held constant? Table 11.9 answers these questions.

The crucial finding in Table 11.9 is that the impact of the suburban population remains strong and in the hypothesized direction when controls are introduced for central-city size, age, per capita income, and nonwhite percentage. With these variables held constant, the direct effects of the size of the suburban population on the central-city service functions are all positive, with significance at the .001 level for every service function except highway services.

Central-city size exhibits a negative relationship to all service functions except sanitation services when the other variables are held constant. The significant negative relationships found between central-city size and a

Table 11.9

Path coefficients between per capita expenditures for central-city services and selected variables, 1960

Central-city service function	Central-city population (log)	Suburban population (log)	Age	Per capita income	Nonwhite (%)
All services	−.24**	.38**	.34**	.10	.09
Police	−.04	.60**	.02	.15	.16
Fire	−.51**	.45**	.30**	.20*	−.03
Highway	−.29**	.08	.19*	.19*	−.04
Sanitation	.05	.28**	−.19*	.14	.27**
Recreation	−.19*	.27**	.04	.10	.12
General control	−.22*	.33**	.23*	.18	.22*

* Significant at .01.
** Significant at .001.

number of the services indicate that economies of scale may operate in the provision of these services.

Examining the direct effects of central-city age, per capita income, and percentage nonwhite on the service functions, we observe that age has a substantial direct effect on the total operating expeditures and on expenditures for fire and highway services, as well as on general control. Also as expected, personal income of the central-city residents exerts a positive effect on all services, with significant direct effects on fire and highway services. The percentage of nonwhites exerts significant direct effects on sanitation services and general control but, in contrast to the effects of age and personal income, exhibits essentially no relationship with fire and highway services.

In sum, Table 11.9 indicates that size of the suburban population, rather than size or composition of the central-city population, is the most important determinant of central-city expenditures per capita for public services. It may also be inferred from Table 11.9 that, *ceteris paribus*, the overall per capita operating expenditures for central-city services increase with the age of the central city and decline with increases in its size.

The commuting population

In an effort to refine the above analysis and determine the impact on central-city public services of suburban residents who utilize central-city services daily, data were obtained from the journey-to-work reports of the 1960 Census of Population on the total number of people in each SMSA who reside in the suburbs and commute to work in the central city. Path

Table 11.10

Path coefficients between per capita expenditures for central-city services and selected variables, 1960

Central-city service function	Central-city population (log)	Suburban commuters (log)	Age	Per capita income	Nonwhite (%)
All services	−.23*	.35**	.35**	.11	.08
Police	−.02	.52**	.06	.17	.16
Fire	−.52**	.44**	.32**	.21*	−.04
Highway	−.40**	.28**	.14	.18	−.08
Sanitation	.02	.31**	−.18	.14	.26**
Recreation	−.18	.24**	.03	.11	.12
General control	−.21*	.28**	.25**	.20*	.22*

* Significant at .01.
** Significant at .001.

coefficients were again computed, substituting the number of commuters for suburban population size in the multiple regression model. The results are presented in Table 11.10.

We observe that the number of suburbanites who commute to work in the central city has a direct impact at the .001 level of significance on the total per capita operating expenditures for central-city services, as well as on per capita expenditures for each individual central-city service. Recalling that highway services were the only central-city function to which size of the suburban population was not significantly related (Table 11.9), it is noteworthy that, when number of commuters is used as the independent variable, a highly significant positive relationship emerges.

The fact that the overall results in table 11.10 are so similar to those in Table 11.9 indicates that the number of suburban residents who commute to work in the central city corresponds closely with the size of the suburban population. When the zero-order correlation between size of the suburban population and the number of suburban residents who commute to the central city was computed, it was found to be .95. Regression analysis showed the unstandardized slope between suburban size and number of commuters to be .105. In other words, an almost perfect linear relationship exists between suburban population size and the number of suburban residents who commute daily to the central city, with each increase of 1,000 suburban residents leading to an additional 105 commuters. Furthermore, the ratio of suburban residents who work in the central city to the central-city resident population increases with the size of the suburban population. A correlation coefficient of .46 exists between suburban population size (log) and the ratio of commuters to central-city residents. Thus, as the size of the suburban population increases, not only do larger numbers of suburban residents daily utilize central-city public services, but, more important, the proportion of suburban residents relative to central-city residents who utilize city services also increases. On the average, there are 132 commuters using central-city services per thousand central-city residents. These findings, along with the results provided in Tables 11.3 through 11.10, offer empirical support to the argument that the rapid growth of suburban populations has contributed greatly to the increased demand, and hence increased expenditures, for common central-city public services.

Summary and implications

Although most social scientists acknowledge that the suburban population influences the service structure of central cities, the degree of that influence appears to have been underestimated. Both cross-sectional and longitudinal analysis indicate that the suburban population has a large impact on central-city retail trade, wholesale trade, business and repair services, and public services provided by central city governments. More detailed examination of the public sector shows that the suburban population in general, and the commuting population in particular, exert strong effects on police, fire, highway, sanitation, recreation, and general administrative functions performed in the central cities. The impact of the suburban population remains strong when controls are introduced for central-city size and age, annexation, per capita income of central-city residents, and the percentage of the central-city population that is nonwhite.

The empirical evidence that has been presented in the last two chapters substantiates not only our basic reseach hypotheses on urban organization but also the contention that metropolitan growth is a special case of ecological expansion. During the past twenty years, large numbers of people have moved outward from the central cities into the suburban areas while maintaining daily contact with the central city. At the same time, organizational functions have developed in the central city to ensure integration and coordination of the expanding complex of relationships. The high correlations shown between the population in the suburban areas and the organizational components and service functions in the central cities reflect their close interdependence.

As metropolitan areas continue to expand, an increasing number of suburban residents make routine use of central-city services and facilities. The additional activity in the central city created by suburban residents has also been reflected in the operating expenditures of central-city governments. Hence, per capita expenditures for central-city services have been found to be at least as sensitive to the size of the suburban population as the size of the central-city population itself.

Expansion, then, may be considered to be at the root of the service-resource problem facing our central cities. The forces of expansion have increased the demand for public services in the central cities, but at the same time have redistributed beyond central-city boundaries those best able to pay for the additional services. Left behind in the centrifugal drift have been poorly educated minorities, hard-core unemployed, the aged, social misfits, and others who are more likely to be a liability than an asset to the central-city fiscal base.

Exacerbating the service-resource problem facing the central cities has been the fact that suburban population growth has not been matched by a proportional growth of suburban public services. The much larger impact of suburban population growth on central-city public services compared to its impact on suburban public services suggests that suburban residents may be unwilling to bear the costs of "urban" services in their own area, and that they will continue to utilize central-city public facilities as long as doing so is less expensive than providing their own. Evidently, it is not economically rational for suburban-area residents to build and operate their own libraries, large parks, zoos, museums, or other public facilities if they have ready and free access to those in the central city.

Suburban populations may thus be exploiting central cities to the extent that, by regularly using central-city services and public facilities, they are imposing marginal costs on the central cities without providing full repayment for those costs. Although partial payments are made by suburbanities to central cities in the form of income taxes, user changes, and sales taxes levied by some central cities, limited research has shown that these sources do not generate the necessary revenue to cover the additional costs. Neenan (1970), for example, in his study of benefit and revenue flows between Detroit and six of its suburban municipalities, shows that the suburban communities enjoy a considerable net gain from the public sector of Detroit. His analysis indicates that Detroit's net subsidy to its suburbs ranges from $1.73 per capita for Highland Park, a low-income industrial suburb, to $2.58 per capita for Birmingham, a higher-income residential and commercial suburb.

A more subtle, yet just as important, means by which suburban populations exploit central cities is by not bearing their fair share of the welfare costs in the metropolitan areas. Through zoning restrictions and discriminatory practices, the suburban populations have been able to ensure that most of the low-income and poorly educated people in the metropolitan area are confined to the central cities. Suburban areas are therefore able to avoid the costs of public housing, public health, and other welfare expenses that impose a heavy burden on the operating budgets of many central cities.

What implications does this study have for future planning and policy in metropolitan areas ? First, our findings suggest that central-city officials and planners should be particularly attentive to trends in the population growth of their suburban areas when projecting future demand for central-city services. As long as areal specialization continues to increase within the metropolitan community, we can expect the impact of the suburban population on the central-city facilities and services to grow.

A second implication is that the suburban population, through daily use of central-city facilities, substantially raises the costs of municipal services. This general finding for metropolitan areas, together with Neenan's detailed cost-benefit analysis of Detroit, provides a rationale for consolidating the politically autonomous suburban units with the central city in the form of a metropolitanwide government. With a single jurisdiction controlling the services and resources, not only would the tax load for the provision of services be spread in a more equitable fashion throughout the metropolitan community, but economies of scale might also be realized. Heavy resistance on the part of suburban populations to political reorganization, however, makes the outlook for consolidation in the near future quite pessimistic. For the time being, then, the only recourse open to the central city is increased federal assistance for the provision of municipal services. Perhaps suburban resistance to consolidation will recede only when the circuitious flow of taxes from suburb to Washington to central city increases to an extent that the service-resource gap begins to favor the central city.

Chapter 12
The changing occupational structure of the metropolis: selective redistribution

The growth and decentralization of metropolitan population documented in the previous chapters has been accompanied by significant changes in the ecological structure of not only the central cities but also of the suburban rings. The dispersion of retail establishments and standard consumer services closely followed the centrifugal drift of metropolitan population. According to the International Council of Shopping Centers, some 14,000 shopping centers were constructed in this country between 1954 and 1974, most of them to serve expanding suburban populations. During the same period there has been a substantial drift of manufacturing activity to the suburban rings. Between 1947 and 1967, the central cities of our 23 largest and oldest SMSAs lost an average of 17,370 manufacturing positions, whereas their suburban rings gained an average of nearly 85,000 positions.

Along with the centrifugal drift of manufacturing activity and establishments offering standard consumer goods and services, there has been a centripetal buildup in the cities of professional, financial, communications, and specialized business services. The specialized nature of many of these functions require them to have a median location (Chinitz, 1964; Thompson, 1965; Vernon, 1960). These specialized facilities have been accumulating in the central business districts and have replaced standard goods and service establishments that were unable to withstand the increasing land value of a central location.

Another prominent centripetal movement has been in the administrative sector (Armstrong, 1972; Cowan et al., 1969). Administrative headquarters rely on large pools of clerical workers and a complement of financial, legal, professional, and technical services that are often available only in the central business districts. The growing demand for centralized office space has stimulated the construction of an increasing number of downtown office complexes. In some of our larger metropolitan areas, the central business district office building binge has reached staggering proportions (Carruth, 1969; Manners, 1974; McQuade, 1973). As we shall observe later, many of our central cities are becoming occupationally specialized in professional, technical, and clerical office functions.

No doubt the growth of administrative, professional, and specialized business services in the central cities, together with the dispersion of population, manufacturing activity, and establishments providing standard (consumer) goods and services, has transformed the structure of the metropolitan community. Through these centripetal and centrifugal movements an extensive network of territorial interdependences has been fostered within each SMSA. But the expansion of the metropolitan community has also created serious problems for the viability of our central cities. These problems, which include high unemployment, rising welfare rolls, shrinking tax bases, and growing public-service demands, are proposed to be direct consequents of movements inherent in the expansion process. Let us elaborate.

The centrifugal and centripetal forces of expansion operate on two fundamental sectors of the metropolitan community—the residential sector and the employment sector. In the residential sector, it is well documented that persons with better educational backgrounds, who generally hold white-collar jobs, have been dispersing to the suburban rings (Guest, 1971; Haggerty, 1971; Pinkerton, 1969; Schnore, 1972; Smith, 1970). Conversely, persons with lower educational backgrounds, who generally hold blue-collar positions, continue to drift to the urban centers (Hawley and Zimmer, 1970). The implication of these movements, of course, is that our central cities are becoming the domicile of disproportionately large numbers of less educated, blue-collar workers while the suburban rings are developing as the residential locus of the better educated, white-collar worker.

Unfortunately, conflicting movements have been occurring in the employment sector. We noted that jobs having lower education requirements, such as those in manufacturing and standard consumer trade, are dispersing to the suburban rings; meanwhile, jobs requiring higher education, such as those in administration, the professions, and specialized business services, are accumulating in the central cities.

The changing distribution of employment opportunities in the metropolis has been especially detrimental to large numbers of lower-class residents of the central cities, particularly poor minority groups. The educational backgrounds of these people are not appropriate for the new (white-collar) functions accumulating in the central cities, and many cannot afford the growing cost of commuting by automobile to blue-collar jobs dispersed throughout the suburban rings. Moreover, suburban zoning restrictions on low-cost housing and discriminatory practices prevent the vast majority of the urban poor from obtaining inexpensive residential sites near expanding suburban industries (see Downs, 1973; Kain, 1968; National Committee Against Discrimination in Housing, 1970). The outcome is that central-city unemployment is now more than twice the national average, and even higher among city residents who have traditionally found employment in blue-collar industries, which are now caught in the suburban drift (Friedlander, 1972; Harrison, 1972; Hoskin, 1973).

High unemployment rates (and concomitant rising welfare rolls) among central-city residents are only two of the deleterious implications of the changing structure of metropolitan areas. The centrifugal and centripetal movements of expansion also have been instrumental in eroding the tax base of central cities, while creating an even greater need for public services in the city. Most of our central cities are plagued with rising debt service, with no apparent reversal of the trend in sight.

It is suggested that four components of the expansion process are at the root of the widening service-resource gap afflicting the central cities. These are: (1) the exodus to the suburban rings of middle- and upper-income families; (2) the large influx into the central cities of poor minority groups, mainly black and brown; (3) the centrifugal drift of commerce and industry beyond the taxing jurisdiction of the central cities and beyond the reach of the urban poor; (4) the daily flow into the central cities of large numbers of suburbanites who make routine use of central-city public services as part of their journey-to-work, recreation, and shopping activities. As a consequence of these four interrelated movements, central-city governments are finding themselves in the difficult position of having to meet increased public service demands with shrinking resource bases. The fundamental problem is not insufficient metropolitan resources to support municipal services, but the redistribution of most of these resources, in the process of urban expansion, to the politically autonomous suburban rings.

In this chapter we shall examine the nature and extent of this redistribution of economic activity in metropolitan America since World War II and assess its implications for the separation of home from workplace of different occupational and racial groups. Our guiding hypothesis is that the centrifugal and centripetal movements of metropolitan expansion have exacerbated current urban problems by eroding central-city tax bases and by spatially removing employment opportunities from those who need them most.

Data and methods

All data that we shall analyze come from publications of the U.S. Bureau of the Census. To examine the centrifugal drift of manufacturing employment, data for 245 SMSAs were gathered from the 1947 and 1967 Censuses of Manufacturers. For longitudinal comparability, SMSAs were adjusted to constant areas according to their 1973 SMSA boundaries (National Bureau of Standards, 1973). Since SMSAs are composed of counties (except in New England, where towns and cities are used), adjustments for peripheral boundary changes were made by adding or deleting appropriate county data at a particular date. In New England, SMSAs were reconstructed in terms of county units to make them definitionally comparable to metropolitan areas in other regions of the nation.

It was also necessary to adjust these data to correct for intrametropolitan boundary changes. Annexation and consolidation systematically understate suburban-ring manufacturing growth and overstate central-city manufacturing growth. Adjustments for central-city annexation of manufacturing employment were made by employing a correction factor based on the proportion of suburban population annexed by central cities between 1950 and 1960 and between 1960 and 1970. Using this correction factor and conventional techniques of extrapolation and interpolation, it is possible to reconstruct all metropolitan manufacturing data in terms of 1950 central-city boundaries.[1]

Along with employment data from the Censuses of Manufacturers, place-of-work data by occupation for the labor force residing in 101 longitudinally comparable SMSA central cities and suburban rings were obtained from the Detailed Characteristic state reports of the 1960 and 1970 Censuses of Population. Our sample includes all of the SMSAs of one hundred thousand or more population in 1960 that neither added nor deleted central cities or outlying counties to their definition between 1960 and 1970.[2]

---

[1] The 1967 data were adjusted to 1950 boundaries by a two-step procedure. First the 1967 data were adjusted to 1960 boundaries using the formula

CCEMP6760 = CCEMP6767
$$- [(0.7 \times ANNEX6070/CCPOP6770) \times CCEMP6767]$$

where CCEMP6760 is 1967 central-city manufacturing employment in 1960 boundaries, CCEMP6767 is the reported 1967 manufacturing employment in 1967 central city boundaries, ANNEX6070 is the population annexed by the central city between 1960 and 1970, and CCPOP6770 is the 1967 central-city population in 1970 boundaries (obtained through linear interpolation). The 1967 central-city employment in 1960 boundaries is then adjusted to 1950 boundaries by using the formula

CCEMP6750 = CCEMP6760
$$- [(ANNEX5060/CCPOP6070) \times CCEMP6760],$$

where CCEMP6750 is the 1967 central-city employment in 1950 boundaries, ANNEX5060 is the population annexed by the central city between 1950 and 1960, and CCPOP6070 is the 1960 central-city population in 1970 boundaries. Since SMSA boundaries remain constant, 1967 suburban-ring manufacturing employment in 1950 suburban boundaries is obtained by subtracting 1967 central-city manufacturing employment in 1950 boundaries from reported 1967 SMSA employment. We further assumed that central-city annexation of territory containing manufacturing establishments in 1948 and 1949 was negligible.

[2] As a further control, SMSAs whose central cities annexed population between 1960 and 1970 that exceeded 20 per cent of their 1960 population were excluded. Statistical adjustments for occupational annexation by those cities in the sample which did annex population are described in the appendix.

The place-of-work data enabled us to construct and compare the occupational distributions within central cities and suburban rings in 1960 and 1970 and examine their changing composition. If our working assumptions thus far are correct, we should find central cities becoming increasingly specialized in white-collar functions as blue-collar functions shift to the suburban rings.[3]

Since metropolitan place of work data are cross-classified by residence (central city or suburban ring), we shall also examine changing commuting patterns of white-collar and blue-collar workers. Furthermore, where longitudinally comparable commuting data are available for both blacks and whites, their journey-to-work flows will be examined and compared to assess differential changes in separation of residence from workplace. In all cases in which longitudinal comparisons of journey-to-work flows are made, appropriate adjustments will be made for possible workplace, as well as for residential annexation. The techniques for adjusting place-of-work data for annexation by occupation, residence, and race are described in the appendix of this chapter.

Results

Let us begin with an examination of the growth and redistribution of manufacturing employment in SMSAs since World War II. Table 12.1 provides an overview.

It is apparent that, adjusting for annexation, recent growth in SMSA manufacturing employment has been due primarily to increases in manufacturing activity in the suburban rings. Between 1947 and 1967, central cities registered a net loss of 293,307 manufacturing jobs, or a 4 per cent overall decline. On the other hand, manufacturing employment in the

---

[3] Two caveats should be noted here. First, the place-of-work data reported in the Detailed Characteristic Tables refer to the resident labor force within each SMSA. Excluded are workers who live entirely outside the SMSA but commute to work in the central cities and suburban rings. Although the vast majority of metropolitan workers do reside within their SMSA of employment, two assumptions must hold in order that our analysis of these data be precise. First, the aggregate occupational distribution of workers in the central cities and suburban rings who commute from outside the SMSA must be the same as the aggregate distribution of metropolitan residents employed in the central cities and suburban rings. Second, the aggregate numbers of central-city and suburban-ring workers who live outside the SMSAs must not have substantially changed between 1960 and 1970. Since there are obviously central-city and suburban workers who do commute from outside the SMSA, the absolute figures we report cross-sectionally will be biased downward for both cities and rings. However, the percentage distributions, relative differences, and employment changes over time should be representative.

A second caveat is necessary because place-of-work data in the 1960 census were reported by occupation for employed metropolitan residents of age 14 and older, whereas, the 1970 place-of-work data were reported by occupation for employed metropolitan residents of age 16 and older. We do not believe, though, that the exclusion of 14- and 15-year-old metropolitan employees in 1970 will substantially influence our results.

Table 12.1

Manufacturing employment in 245 SMSAs, their central
cities, and their suburban rings (constant SMSA boundaries
with adjustments for annexation), 1947–1967

	1947	1967	Change 1947–1967
Central cities	7,356,733	7,063,426	−293,307
Suburban rings	4,141,704	8,044,030	3,902,326
SMSAs	11,498,437	15,107,458	3,609,021

suburban rings increased by 3,902,326, representing a 94 per cent increase in
suburban manufacturing employment over a two-decade period. The dramatic growth of manufacturing employment in the suburban rings together
with the absolute declines in central cities increased the suburban share of
total SMSA manufacturing employment from 36 per cent in 1947 to 53 per
cent in 1967.

What are the reasons for this remarkable shift of manufacturing
activity to the suburban rings ? One important factor is post–World War II
changes in production technology, especially capital intensification and new
assembly-line techniques that have large single-story space requirements.
Many existing central city factories are of the older multistory design, and
are thus inappropriate for today's mass production technology. Numerous
manufacturers have found it more efficient and less expensive to build
entirely new facilities on relatively cheap suburban land than to redesign and
convert their obsolete central-city structures. Likewise, entering industries
with extensive areal requirements found it exceedingly difficult to obtain
large tracts of central-city land at practical costs, and they also have turned
predominantly to open space in the suburban rings for their plant sites.

Another factor stimulating the suburban manufacturing drift has been
changing modes and improvements in short-distance transportation. The
widespread development of suburban highway systems since World War II,
pervasive automobile ownership, and increased reliance on trucking for
freight shipments have operated concurrently to attract manufacturers to the
suburban rings. Many manufacturers recognized that by locating outside the
congested centers yet near a suburban expressway, their transport costs
could be substantially reduced, an adequate metropolitan labor supply
could be tapped, and problems of limited employee parking space and
freight transfer areas could be solved. Traffic congestion and lack of sufficient
employee parking space have been particularly troublesome for large manufacturers located in older, more densely settled central cities, where street
designs were established well before the age of the automobile and truck.

A third yet no less important reason for the decentralization of industry
has been the spread of public services throughout the suburban rings and

the expansion of other external economies. Large industrial facilities require communication and power lines, water supplies, highway services, and police and fire protection. Before World War II, such public services were confined largely to the central cities and their adjacent built-up areas. With the extension of these services throughout the suburban rings, manufacturers have had much greater freedom of location within the metropolitan periphery. At the same time, urban expansion brought numerous subcontractors, local suppliers, and business and repair services to the suburban rings—external economies that, in the past, were available to manufacturers only within the confines of the central cities.

There is no reason to suspect, however, that manufacturing decentralization has exhibited a uniform pattern in all SMSAs. Plant obsolescence, lack of open space, and traffic congestion have been particularly prevalent in larger, older central cities located in the Northeastern and North Central states. When we examine average changes of manufacturing employment in central cities and suburban rings by SMSA size, age, and region we find such characteristics to have substantial effects (see Table 12.2).

By 1950 the large, old central cities in the Northeast and Midwest had filled up most remaining open space, leaving only the suburban ring

Table 12.2

Mean change in manufacturing employment within 245 SMSAs, their central cities and their suburban rings (constant SMSA boundaries with adjustments for annexation)

| Metropolitan characteristics | N | Mean change in manufacturing employment, 1947–1967 | | |
		SMSAs	Central cities	Suburban rings
Size				
Under 250,000	113	3,457	307	3,150
250,000–500,000	63	7,800	−619	8,419
500,000–1,000,000	36	12,741	−1,993	14,734
Over 1,000,000	33	68,739	−6,584	75,323
Inception date (age)				
After 1950	80	6,321	659	5,662
1930–1950	53	9,846	1,857	7,989
1900–1920	63	13,142	1,974	11,168
Before 1900	49	35,785	−11,609	47,394
Region				
Northeast	42	11,040	−8,216	19,256
North Central	68	11,350	−6,827	18,177
South	98	11,935	2,646	9,289
West	37	32,513	6,939	25,574
Total	245	14,731	−1,197	15,928

free for additional industrial development. Younger and smaller SMSA central cities and those located in the South and West typically had more open areas for industrial development. In addition, because most of the latter SMSAs were patterned during the automobile age, they are less congested and have better access routes for freight transfer. The large differences in changes between central-city and suburban manufacturing employment within the various types of SMSAs clearly reflect the influence of these factors.

Bearing in mind that, since World War II, the greatest absolute growth of lower-income population has been predominantly in large Northeastern and Midwestern central cities, we further cross-classified manufacturing employment changes, by region, for SMSAs of one million or more population. The results are striking. Adjusting for annexation, central cities of the 16 largest Northeastern and North Central SMSAs lost an average of 34,571 manufacturing positions between 1947 and 1967, whereas their suburban rings gained an average of 86,358 positions. Conversely, central cities of the 17 largest Southern and Western SMSAs gained an average of 19,756 manufacturing positions, and their suburban rings gained an average of 64,936 positions. The important figures, however, pertain to the 16 largest Northeastern and North Central SMSAs, where central-city manufacturing employment fell a total of approximately 550,000 positions, or an average *annual* decline of over 1,700 manufacturing jobs per city between 1947 and 1967.

Having focused thus far on the centrifugal drift of manufacturing employment, let us now broaden our analysis to cover recent shifts in the overall occupational structure of central cities and suburban rings. Table 12.3 presents the mean number of employed persons by occupation for 101 longitudinally comparable central cities and suburban rings in 1960 and 1970.

Central-city employment in all occupational categories, with the exception of professional, clerical, and service functions, declined between 1960 and 1970. The largest declines were in the blue-collar categories—craftsmen and operatives. On the other hand, suburban-ring employment expanded in all occupational categories, with the anticipated exception of farm workers. Like central cities, the largest suburban increases were in professional, clerical, and service functions. Unlike central cities, however, suburban rings exhibited substantial gains in craftsmen and operatives. Adjusting for occupational annexation, suburban rings of the 101 SMSAs gained an average of 30,264 employees, whereas the central cities registered an average net loss of 3,427 employees between 1960 and 1970.

It was noted earlier that recent decentralization patterns have been highly selective—removing large numbers of blue-collar functions from the central cities and replacing them with white-collar functions. Table 12.4 substantiates this contention.

Aggregating central-city and suburban employment into white-collar

Table 12.3

Mean number of employees by occupation in 101 longitudinally comparable SMSA central cities and suburban rings, 1960–1970

Occupational category	Mean number of central-city employees		
	1960	1970*	Change 1960–1970
Professional and technical	16,015	20,138	4,123
Managers and proprietors	12,458	11,354	−1,104
Clerical workers	26,915	30,002	3,087
Sales workers	11,113	9,813	−1,300
Craftsmen	18,041	16,043	−1,998
Operatives	25,078	19,838	−5,240
Laborers	5,676	4,668	−1,010
Service workers	15,669	15,713	44
Farm workers	232	203	−29

Occupational category	Mean number of suburban-ring employees		
	1960	1970*	Change 1960–1970
Professional and technical	9,392	16,203	6,811
Managers and proprietors	5,787	8,288	2,501
Clerical workers	10,066	18,411	8,345
Sales workers	5,209	7,921	2,712
Craftsmen	11,684	14,575	2,891
Operatives	14,263	17,501	3,238
Laborers	3,649	4,363	714
Service workers	8,393	12,630	4,237
Farm workers	2,691	1,506	−1,185

* Central-city and suburban employment figures adjusted for annexation between 1960 and 1970.

Table 12.4

Total number of white-collar and blue-collar employees (excluding farmers) working in 101 longitudinally comparable SMSA central cities and suburban rings, 1960–1970

Central cities	1960	1970*	Change 1960–1970
White-collar	6,716,618	7,202,065	485,447
Blue-collar	6,510,906	5,682,649	−828,257
Suburban rings			
White-collar	3,075,938	5,133,061	2,059,123
Blue-collar	3,836,919	4,956,053	1,119,134

* Central-city and suburban employment adjusted for annexation.

and blue-collar categories (excluding farm workers), we see that between 1960 and 1970 blue-collar employment in the 101 central cities declined by over 825,000 positions, yet white-collar employment increased by nearly 500,000 positions. This represents a 12.7 per cent reduction in all central-city blue-collar jobs during the decade, with a concurrent 7.2 per cent increase in central-city white-collar positions.

The suburban rings show huge increases in both white-collar and blue-collar functions. Interestingly, however, white-collar employment in the suburbs grew at more than twice the rate of blue-collar employment. Between 1960 and 1970 suburban white-collar jobs increased by 67 per cent, whereas blue-collar jobs increased by 29 per cent. As a result, in 1970, suburban rings were the locus of more white-collar employment than blue-collar employment.

It is also noteworthy that the overall occupational composition of the suburban rings in 1970 (51 per cent white-collar and 49 per cent blue-collar) was exactly the same as the central-city composition in 1960. Moreover, from 1960 to 1970, differences between central cities and suburban rings declined in terms of total employment, white-collar employment, blue-collar employment, and employment in each of the nine occupational categories listed in Table 12.3. These differences in occupational structure diminished however they were measured—in terms of absolute number of jobs, percentage of metropolitan jobs, or relative distributions of occupations in central cities and suburban rings. This structural convergence clearly manifests the urbanization of America's suburbs.

Examining selective changes in central-city white-collar and blue-collar employment by SMSA size, age, and region, we observe in Table 12.5 that the largest and oldest SMSA central cities had the greatest redistribution. Central cities of SMSAs larger than one million lost an average of 49,144 blue-collar jobs between 1960 and 1970, but gained an average of nearly 9,000 white-collar jobs. Likewise, central cities that reached metropolitan status before 1900 lost an average of 38,505 blue-collar positions and gained an average of 5,808 white-collar positions. Northeastern and North Central SMSA central cities lost averages of 20,325 and 10,158 blue-collar positions, respectively, compared with average net gains of 171 blue-collar jobs in Southern central cities and 1,649 blue-collar jobs in Western central cities.

The loss of substantial numbers of blue-collar job opportunities in large, old central cities, and in cities in the Northeastern and North Central states, signifies the roots of the employment problem facing inner city residents with limited educational backgrounds. In terms of total numbers of available jobs (and here we believe absolute amounts are the pertinent figures) in just one decade (1960 to 1970), the 16 largest central cities in our sample lost a total of 786,304 blue-collar jobs; the 21 oldest central cities lost a total of 808,605 blue-collar jobs, and the 59 central cities located in Northeastern and North Central states (the old industrial belt) lost a total of 1,798,497 blue-collar jobs. As noted earlier, it is precisely the same sets of

Table 12.5

Mean number of white-collar and blue-collar employees in 101 SMSA central cities, 1960 and 1970

Metropolitan characteristics	N	Central-city white-collar			Central-city blue-collar		
		1960	1970*	Change 1960–70	1960	1970*	Change 1960–70
Size							
Under 250,000	(44)	14,647	16,648	2,001	16,982	16,209	−773
250,000–500,000	(30)	25,649	30,520	4,871	26,603	26,436	−167
500,000–1,000,000	(11)	58,192	68,378	10,196	64,138	63,873	−265
Over 1,000,000	(16)	291,409	300,113	8,704	266,254	217,110	−49,144
Inception date (age)							
After 1950	(13)	15,035	19,760	4,725	14,640	15,468	828
1930–1950	(31)	23,407	28,067	4,660	23,719	23,705	−14
1900–1920	(36)	31,879	36,257	4,378	34,749	33,915	−834
Before 1900	(21)	221,327	227,135	5,808	206,397	167,892	−38,505
Region							
Northeast	(27)	119,118	124,202	5,084	112,399	92,074	−20,325
North Central	(32)	62,850	62,298	−552	65,676	55,518	−10,158
South	(30)	32,598	40,341	7,742	32,240	32,411	171
West	(12)	47,650	57,397	9,747	39,515	41,164	1,649
Total	(101)	66,501	71,308	4,807	64,464	56,264	−8,200

* Employment figures adjusted for annexation.

central cities that have experienced the largest influx of minority and other low-income groups since World War II.

Not unexpected is the finding in Table 12.6 that the largest growth in suburban blue-collar employment has also been in the outer rings of the largest and oldest SMSAs. Note, however, that these suburban rings have shown even greater increases in white-collar employment. Mean growth of white-collar employment in the suburban rings of the largest and oldest SMSAs is more than twice as great as blue-collar employment growth.

Comparison of the differential growth of white-collar employment in the central cities and suburban rings indicates that suburban areas in the largest and oldest SMSAs have been most effective in competing with their central cities for white-collar functions. The reasons for the disproportionate growth of white-collar employment in these suburban rings are many, including huge increases in their middle- and upper-income population bases attracting numerous retail establishments along with physicians, dentists, lawyers, and other professionals to serve the massive resident population; the desire of management personnel to work closer to their suburban homes; availability of less expensive land for smaller customized office complexes; accessibility to commercial airports; and the push factors of congestion, pollution, and high crime in large, old central cities.

Table 12.6

Mean number of white-collar and blue-collar employees in 101 SMSA suburban rings, 1960 and 1970

Metropolitan characteristics	N	Suburban-ring white-collar			Suburban-ring blue-collar		
		1960	1970*	Change 1960–70	1960	1970*	Change 1960–70
Size							
Under 250,000	(44)	4,393	6,978	2,585	7,168	8,479	1,311
250,000–500,000	(30)	11,099	17,933	6,834	16,141	21,160	5,019
500,000–1,000,000	(11)	26,414	43,958	17,544	39,199	51,743	12,544
Over 1,000,000	(16)	141,193	237,780	96,587	162,880	211,188	48,308
Inception date (age)							
After 1950	(13)	6,095	11,160	5,065	9,311	11,971	2,660
1930–1950	(31)	7,846	15,401	7,555	10,158	15,617	5,459
1900–1920	(36)	12,336	20,628	8,289	18,098	24,682	6,584
Before 1900	(21)	109,971	179,426	168,455	130,927	163,229	32,302
Region							
Northeast	(27)	63,789	91,257	27,468	79,350	86,657	7,307
North Central	(32)	25,572	49,336	23,746	31,925	48,809	16,884
South	(30)	11,693	23,194	11,501	15,494	23,982	8,488
West	(12)	17,689	37,237	19,548	20,007	32,053	11,976
Total	(101)	30,455	50,822	20,367	37,989	49,070	11,081

* Employment figures adjusted for annexation.

Another possible push factor is executive decisions predicated on the changing racial and ethnic composition of resident labor-force pools in our largest and oldest central cities. Apparently there is a tendency among some executives to move their companies to outlying suburbs in order to circumvent hiring workers from central-city resident labor pools that are becoming numerically dominated by persons of minority descent. To illustrate, a high-ranking Detroit insurance executive interviewed by *The New York Times* noted several major corporate moves from the central city because of institutional racism. The executive is quoted as remarking: "A vice-president of [he named a prominent organization] told me that they wanted to move for one reason—to get rid of lower echelon workers, like file clerks and typists. These days in Detroit those workers have to be black."

Ken Patton, New York City's economic development administrator, reports that studies conducted by his agency indicated that a major force behind corporate moves to the suburbs is what he termed "social-distance." To quote Mr. Patton: "The executive decision-maker lives in a homogenized ethnic and class community. Increasingly his employees in the city are from communities quite different [from his] in class and ethnicity.... The decision-maker can't relate to the city kid, that kid [that] doesn't look the same as him."

According to *The New York Times*, Mr. Patton's analysis was similar to the private views expressed by some city officials, including then Mayor John Lindsay. In discussions with reporters and editors, these officials complained that company presidents move their complexes to the suburbs when they discover that a large percentage of their office help is black or Puerto Rican.

The growing separation of home from workplace

What are the implications of the selective movements of population and economic activity for separation of residence from workplace of different occupation and racial groups ? As anticipated, Table 12.7 shows substantial declines between 1960 and 1970 of employees who live and work in the central cities, particularly among blue-collar workers. Concurrently, with the suburbanization of economic activity, commuting streams from city residence to suburban jobs increased across white-collar and blue-collar occupations.

The rapid growth of employment and housing in the suburban rings during the 1960s also contributed to large increments in the number of metropolitan residents living and working in the rings. Still, an increasing number of persons moved to the suburbs while retaining jobs in the central cities, as is evidenced by the expansion of blue-collar and white-collar commuting from suburban residence to central city jobs. Note also that

Table 12.7

Mean number and percent distribution of blue-collar and white-collar employees by place of residence and place of work for 101 longitudinally comparable SMSAs, adjusted for annexation, 1960 and 1970

		Blue-collar		White-collar	
Live	*Work*	*1960*	*1970*	*1960*	*1970*
Central city	Central city	50,973 (47.6)	40,096 (35.6)	49,115 (48.1)	47,126 (36.1)
Central city	Suburban ring	5,690 (5.3)	8,847 (7.9)	3,668 (3.6)	7,080 (5.4)
Suburban ring	Central city	13,492 (12.6)	15,856 (14.1)	17,387 (17.0)	23,739 (18.2)
Suburban ring	Suburban ring	32,300 (30.2)	40,563 (36.1)	26,787 (26.2)	44,085 (33.7)
Suburban ring	Outside SMSA	3,151 (2.9)	4,954 (4.4)	3,540 (3.5)	6,198 (4.7)
Central city	Outside SMSA	1,525 (1.4)	2,167 (1.9)	1,573 (1.5)	2,485 (1.9)

although there was an average increase of approximately 5,000 blue-collar workers in the 101 SMSAs between 1960 and 1970, there was an average net decline of approximately 2,600 blue-collar workers who both lived and worked in the same metropolitan zone.

Examination of the relative distribution of commuting streams provides further documentation of increased cross-cutting in journey-to-work flows. Observe that the percentage of metropolitan workers in all four nonlateral commuting streams (that is, C. C.–Ring, Ring–C. C., Ring–Outside SMSA, C.C.–Outside SMSA) has increased among white-collar as well as blue-collar workers. Overall, the percent of white-collar workers commuting non-laterally increased from 25.6 per cent in 1960 to 30.2 per cent in 1970. During the same period blue-collar employees living in different zones from those in which they worked increased from 22.2 per cent to 28.3 per cent of the blue-collar labor force. Table 12.7 thus indicates not only that growing numbers of metropolitan workers are commuting across city and suburban boundaries, but also that larger proportions of the metropolitan labor force are facing increased separation of home from workplace.

We contended in the initial section of this chapter that the suburban-ization of job opportunities has been particularly disadvantageous to minority groups who are residentially concentrated in the central cities. To examine this issue, place-of-work data for blacks and whites in 34 SMSAs were compared in 1960 and 1970.

The sample includes all SMSAs that are longitudinally comparable by criteria previously described, and that had comparable data on residences and workplaces of whites and blacks in 1960 and 1970. Because the 1960 census provided place-of-work data for whites and nonwhites, and the 1970 census for all workers and blacks, we examined only SMSAs where blacks constituted more than 90 per cent of the nonwhite population.

Table 12.8 provides comparative journey-to-work flows in the 34 SMSAs for blacks and whites in 1960 and 1970. Not surprisingly, the data reveal that blacks predominantly reside and work in the central cities. In 1960, 72.2 per cent of the black labor force both lived and worked in the central cities, compared with 46 per cent of the white metropolitan labor force. In 1970, 64.6 per cent of the black metropolitan labor force lived and worked in the central cities, compared to less than one third of the white metropolitan labor force.

The substantial drop during the 1960s of whites living and working in the central cities gives testimony to the massive centrifugal drift of residences and jobs occupied by whites. Even with expanding numbers and percentages of whites living in the suburbs and commuting to jobs in the central cities, total white employment in the 34 SMSA central cities dropped an average of 27,000 positions between 1960 and 1970.

Table 12.8 further reveals that the number and per cent of blacks involved in all four types of nonlateral commuting streams increased

Table 12.8

Mean number and percentage distribution of black and white employees by place of residence and place of work, adjusted for annexation, 34 selected SMSAs, 1960 and 1970

Live	Work	Blacks		Whites	
		1960	1970	1960	1970
Central city	Central city	37,451 (72.2)	39,715 (64.6)	182,988 (46.0)	143,715 (32.7)
Central city	Suburban ring	3,316 (6.4)	7,342 (12.0)	15,219 (3.8)	24,413 (5.6)
Suburban ring	Central city	1,854 (3.6)	2,792 (4.5)	58,735 (14.8)	70,899 (16.1)
Suburban ring	Suburban ring	7,957 (15.3)	9,011 (14.7)	122,809 (30.9)	172,924 (39.3)
Suburban ring	Outside SMSA	453 (0.9)	877 (1.5)	12,787 (3.2)	20,652 (4.7)
Central city	Outside SMSA	830 (1.6)	1,701 (2.8)	5,072 (1.3)	7,132 (1.6)

between 1960 and 1970. Of particular note is the fact that whereas the percentage of blacks reverse commuting from central-city residence to suburban jobs increased substantially during the 1960s, the percentage of metropolitan blacks living and working in the suburban rings actually declined from 1960 to 1970. It is apparent that housing opportunities for blacks in the suburbs are falling far short of job opportunities in those areas.

Although some might argue that the lack of appropriate housing for blacks near suburban employment is largely a class rather than a racial issue, the data presented in Table 12.9 belie this position. Regardless of occupational status, a disproportionate number of blacks employed in the suburban rings commute from the central cities. Overall, 49 per cent of blacks employed in the suburbs commute from the central city, compared to only 13.7 per cent of white suburban workers. Without question, these figures reflect the intense segregation of blacks in the central cities and the corresponding predominantly white racial composition of the suburban rings. Nevertheless, the fact that such high proportions of black suburban workers in all occupational categories commute from the city attests to the growing logistical problems they are facing with the suburbanization of economic opportunity in metropolitan America.

Conclusions

Although any predictions about future structural changes in the metropolis are necessarily speculative, the trend appears well established. That is,

Table 12.9

Relative amount of reverse commuting from central cities to suburban rings of black and white workers, by occupation for 84 selected SMSAs, 1970

| | Percentage of suburban workers commuting from central cities | | |
Suburban-ring occupation	Blacks	Whites	Total
Professional	45.4	13.5	14.7
Managerial	41.6	12.3	12.9
Clerical	51.0	13.3	15.2
Sales	45.7	12.2	12.9
Crafts	54.4	15.5	17.7
Operative	54.7	16.6	21.7
Laboring	47.7	14.9	21.2
Service	43.4	11.3	17.8
Farming	8.8	2.2	3.1
All occupations	49.0	13.7	16.7

we may expect increased suburbanization of most types of economic activity along with middle- and upper-income population. This will be matched by continued declines in central-city blue-collar jobs, and their partial replacement by more highly skilled white-collar functions. At the same time, the resident population of many central cities will become increasingly dominated by minority groups, the aged, and other poor.

If the above trends continue, as is anticipated, unemployment and poverty in our central cities will obviously worsen. Furthermore, since most central cities are currently stretched to their fiscal limits in providing public-service jobs, little promise exists for additional expansion of this sector. In fact, many central cities are already reducing the number of employees on municipal payrolls; the recent experience of New York is not an isolated circumstance, but the tip of an iceberg. The only feasible solution at this time appears to be large and continuous infusions of state and federal funds for inner-city jobs. These funds, no doubt, will increasingly come (indirectly at least) from the wallets and portfolios of the new suburban masses.

APPENDIX
Methods of adjusting SMSA place-of-work data by residence, occupation, and race for annexation, 1960–1970

Annexation tends to artificially inflate central-city employment growth and understate suburban-ring employment growth when longitudinal analyses are based on metropolitan place-of-work data. Employment totals for central cities and suburban rings in 1970 must therefore be adjusted for annexation occurring during the 1960–1970 decade. The method of adjustment can be described in several steps:

1. An estimate of total "annexed employment" is made from published census figures of total annexed resident population. The assumption is made that suburban jobs within the SMSA are transferred to the central city by annexation in the same proportion that residents are transferred. The formula for estimating total annexed employment is:

$$\text{TAE} = \text{TCE} \times (\text{ANNEX}/\text{CCPOP})$$

where TAE is total annexed employment; TCE is total central-city employment (obtained by summing employees who live in the city and work in the city with employees who live in the suburban ring and work in the city); ANNEX is the population residing in the area annexed by the central city from 1960 to 1970; CCPOP is the central-city population in 1970. The adjusted central-city and suburban-ring employment figures are obtained by *subtracting* total annexed employment from the reported 1970 total central-city employment, and *adding* the same figure to the reported 1970 suburban-ring employment total.

2. To estimate the amount of annexed employment by occupational categories, we must take consideration of the fact that not all occupations are likely to be annexed in equal numbers. Here, the best estimate of annexed employment by occupation is the proportion that each occupation constitutes of total central-city employment. For example, if professional and technical employees constitute 9 per cent of a city's work force, then they are assumed to constitute the same proportion of total annexed employment. Annexed employment by occupation is calculated by simply multiplying total annexed city employment by each occupational category's proportion of total city employment. Or, to take professional and technical workers as an example, the estimation formula is:

$$\text{APTW} = (\text{TPTWCC}/\text{TCE}) \times \text{TAE}$$

where APTW is the estimated number of annexed professional and technical workers; TPTWCC is the total number of professional and technical employees working in the central city, and TCE and TAE are the same as above. The "adjustment" for professional and technical job annexation subtracts annexed professional and technical employment from the reported 1970 total for that occupational category in the central city and adds it to that occupational category in the ring. Similar adjustments are made for the remaining occupational categories.

Adjustments become more complex when correcting commuting patterns for annexation, because residences as well as jobs (workplaces) may be annexed. To adjust commuting data for possible residence and job annexation the following basic assumptions are made:

1. Central-city resident labor-force is increased by annexation in the same proportion as the city's population is increased by annexation.
2. Central-city employment is increased by annexation in the same proportion as the city's population is increased by annexation.

To calculate the adjusted journey-to-work flows we first correct for resident labor-force annexation, and second, for job (workplace) annexation. In the first step, the central-city resident labor-force is decreased and the suburban-ring resident labor-force is increased by the estimate of annexed central-city resident labor-force. However, as in the case of occupational annexation, the estimated annexed resident labor-force must be distributed among the three subcategories of reported place of work for central-city and ring residents (that is, the central city, the ring, and outside the SMSA). For example, the formula for allocating the annexed resident labor-force to the subcategory "live in the city and work in the city" is:

$$ANRLFCC = (ECC/TLC) \times TARLF$$

where ANRLFCC is the annexed resident labor-force that lives in the city and works in the city; ECC is the reported number of employees who live in the city *and* work in the city; TLC is the total number of employees who live in the city, regardless of place of work; TARLF is the total estimated annexed resident labor force computed by multiplying the total central-city resident labor-force (TLC) by the ratio of annexed central-city population to total 1970 central-city population (ANNEX/CCPOP). ANRLFCC is then subtracted from the reported number of central-city residents who work in the central city in 1970.

The other workplace allocations for city residents annexed are computed by substituting ECR (employees living in the city and working in the ring) and ECO (employees living in the city and working outside the SMSA) for ECC in the above formula. These allocations are likewise subtracted from the appropriate place-of-work data reported for central-city residents. Next, the estimated total annexed resident labor-force is *added* to the three suburban-ring *resident* place-of-work subcategories in proportions equivalent to the reported 1970 distribution of suburban-ring residents' place of work. With these computations completed, the resident labor-forces of the central city and suburban ring have been adjusted for *residence* annexation.

The second step involves further adjusting the residence-workplace commuting streams, Ring-City, City-City, Ring-Ring, and City-Ring, for job annexation. (Note that the City-Outside SMSA and Ring-Outside SMSA commuting streams have been corrected for residence annexation but are not affected by job annexation.) To make this adjustment, we first estimate total city employment that has been annexed, by multiplying the number of all SMSA residents *employed* in the central city (TCE) by the ratio of annexed central-city population to total 1970 central-city population. The estimated total annexed city employment is then split into separate proportions of (a) annexed city employees who live in the city and (b) annexed city employees who live in the ring. The estimated total annexed city employment is, at the same time, distributed to ring employment in proportion to the reported residence distribution of ring employees. Finally, each estimate of annexed employment is either subtracted from the respective central city workplace commuting stream (City-City or Ring-City) or added to the respective suburban workplace commuting stream (either City-Ring or Ring-Ring).

Considering again employees living in the central city and working in the central city, the complete equation which adjusts for both residence and job (workplace) annexation is as follows:

$$\text{ADRJCC} = \text{ADRCC} - (\text{ECC}/\text{TCE}) \times [(\text{ANNEX}/\text{CCPOP}) \times \text{TCE}]$$

where ADRJCC is the labor-force living and working in the city adjusted for residence *and* job annexation; ADRCC is the labor-force living and working in the city adjusted for residence annexation; ECC, TCE, ANNEX, and CCPOP are the same as described above. The adjustment formula for those living in the ring and working in the city is the same, except that ADRCC and ECC are replaced by ADRRC and ERC, representing, respectively, those living in the ring and working in the city adjusted for residence annexation, and the reported number living in the ring and working in the city. For the residence-workplace commuting categories City-Ring and Ring-Ring, the total estimated employment annexed must be *added* to each residence adjusted category in proportion to the residence distribution of total ring employment. Thus, if published place-of-work data indicate that 15 per cent of ring employees commute from residences in the central city, then 15 per cent of the estimated annexed jobs are added to that category.

The same two-step procedure may be extended to adjust commuting streams for annexation of employees of particular occupations or occupational categories, such as blue-collar or white-collar workers. This is done by applying the residence and workplace (job) adjustment formulas to place-of-work data subcategorized by residence and occupation. For example, the adjustment formula for residence annexation of white-collar workers living in the city and working in the city is:

$$\text{ADRWHTCC} = \text{WHTCC} - (\text{WHTCC}/\text{TLC}) \\ \times (\text{ANNEX}/\text{CCPOP}) \times \text{TLC}$$

where ADRWHTCC is white-collar workers living in the city and working in the city adjusted for residence annexation; WHTCC is the reported number of white-collar workers living and working in the central city; TLC, ANNEX, and CCPOP are the same as described above.

In our final adjustments of commuting streams by occupational groupings, two sets of annexation adjustment were calculated: First, a two-step adjustment as described above, correcting for both residence and job annexation, second, an adjustment for job annexation only. When adjustments are computed for jobs only and commuting patterns are summed by work location, the workplace totals are equivalent to the city and ring totals presented in Tables 12.5 and 12.6. The two-step adjustment procedure on which the residence-workplace results in Table 12.7 are based does not yield totals equivalent to Tables 12.5 and 12.6, although the results are very close. The differences may be attributed to rounding error and the fact that in the two-step procedure the second set of equations, adjusting for job annexation, contain a term that had been previously altered by the adjustment for residence annexation. It was believed, however, that the two-step procedure provides a more valid adjustment when comparing changes in intrametropolitan commuting streams that may have been altered by residence as well as job annexation.

Black residence-workplace annexation adjustments

Because blacks constitute only a small percentage of the suburban resident population and suburban employment in many SMSAs, it is necessary to consider this fact when adjusting their residence and workplace data for possible annexation. To accomplish this, the total estimated resident labor force annexed as computed above is multiplied by the proportion of the SMSA black labor force *living* in the ring. Similarly, the total estimated number of jobs annexed is multiplied by the proportion of SMSA black labor force *working* in the ring. These two additional adjustments provide us with estimates of (1) the number of black residences annexed and (2) the number of black workplaces annexed. The estimated annexed black residences and workplaces are then used in the two-step procedure described above for adjusting place-of-work data for both residence and job annexation.

Chapter 13
Decentralization and the restructuring of metropolitan America

The centrifugal drift of manufacturing employment and the other trends reported in Chapter 12 may in fact be producing as radical and revolutionary a shift in urban ecology as occurred in the late nineteenth century, when the concentrated core-oriented industrial metropolis emerged as a consequence of the technological changes of the industrial revolution. As a result, the conventional wisdom about the nature of urbanization processes that has been widely accepted throughout the social sciences as a guide for understanding the logic of metropolitan growth and change may be in need of substantial rethinking, re-evaluation, and reshaping. This is the issue that we shall develop in this chapter.

The classic idea was never expressed more straightforwardly than by Hope Tisdale when, in 1942, she wrote that

> urbanization is a process of population concentration. It proceeds in two ways: the multiplication of the points of concentration and the increasing in size of the individual concentrations. . . . Just as long as cities grow in size or multiply in number, urbanization is taking place. . . . Urbanization is a process of becoming. It implies a movement . . . from a state of less concentration to a state of more concentration.

From this widely accepted view, Louis Wirth derived a whole theory of the human consequences of urbanization, using as primary causal variables the size, density, and heterogeneity of the large city. And, consistent with the conventional wisdom, geographers, human ecologists, and land economists in the interwar years produced a classic image of the North American city.

Much of this image involved models of commercial and industrial location within the city. Every city was thought to require at its heart a strong, viable, growing central business district, with adjacent central industrial and commercial zones. At strategic locations within the main sectors of the city, and in a hierarchy of successively smaller communities and neighborhoods, there ought, the model said, to be a hierarchy of business districts providing for the shopping needs and service requirements of the population. Major traffic arteries were seen to support ribbons of highway-oriented businesses, heavy commercial uses, and specialized functional areas such as the "automobile row." Heavy industry was found next to major transportation facilities—ports, and railroad lines and spurs. Light industry was seen as gravitation to the suburbs.

Although a variety of factors were said to be operating in creating this spatial structure (the locational advantages of ports, strategic positions on transportation arteries, availability of fuels, and raw materials all figure prominently in the writings of the period) the principal force was said to be that of urbanization economies.

Urbanization economies involve externalities: the consequences of the close association of many kinds of activity in cities, producing mutual-scale advantages that lower the costs and improve the competitive position of individual firms. Such cost reductions were demonstrated, in a variety of studies, to arise from several sources. The argument is as follows:

1. *Transport costs.* Large cities have superior transport facilities offering significantly lower transport costs to regional and national markets. A city located at a focal point on transport networks is especially suited for easy assembly of raw materials and for ready distribution of products. There are thus market advantages for speedy and cheap distribution, which may be accentuated by the advantages arising out of the size of the local consuming market. In the biggest cities, this local market will account for a very substantial share of service and residentiary activities and of the regionalized activities that need to minimize transport costs, since the local population may form a large part of the total national market. Greater London, for example, has one fifth of the British consumer market, and the New York metropolitan area has been called "one tenth of a nation." In both these cases, these markets account for a significantly larger proportion of the higher quality and fashion demands of the respective nations than the percentages indicate. Thus it may be that, because of its market dominance, the larger city is the ideal place in which to locate a plant even if it is serving a national market, because substantial demands are local and the balance of the output is more easily distributed. A New York location, for example, will obviate the necessity of providing extra facilities in New York, such as showrooms and warehouses, that would otherwise be needed if the firm were located outside New York.

2. *Labor costs.* The large-city labor market is extensive, diverse, and dynamic. In this circumstance, the labor demands of single firms are only a small part of the total demands for labor. Thus, recruitment is easier. This becomes especially important for firms that have seasonally varying

labor needs. The big-city labor market also offers a whole range of skills; consequently, the advantages of the city over the small town are obvious, even though the wage rates of the large city may be somewhat higher than those of the small town. Labor facilities that are readily available in the large city may often also have to be provided in the small town, which raises the real labor costs to the firm.

3. *Advantages of scale.* The larger the city, the higher the scale of services it can supply. Thus, fire and police, gas, electricity, water, waste disposal, education, housing, and roads are all in general better in the larger city than in the small town. Firms developed in otherwise unindustrialized areas generally have to spend much more on social capital than those that have been developed in large cities. These kinds of scale advantages came into play particularly in the industrialization of the largest metropolitan areas in the period following World War I, although interrupted by World War II. Urbanization economies are those that, in particular, have a magnetic effect upon those kinds of industries in which coal was replaced as motive power by electricity, and in which rail and water were replaced as a means of transportation by trucking over the roads. This new industry is "light" industry, concerned with manufacturing consumers' goods. These goods are often branded and serve sheltered markets because of imperfect competition deliberately created by advertising mechanisms. For these industries, transport costs are a low proportion of total costs, because the materials used tend to be semiprocessed and because the products are compact and have high value in relation to weight. Labor requirements are often unskilled or semiskilled, and the firms therefore are more likely to be attracted by large cities, especially by those cities' markets and marketing facilities. Thus, there is a "snowballing" which tends to be accumulative, for, as light industry is attracted to large cities, the magnetic power of large cities for this kind of industry is increased, creating conditions favorable to further growth. This kind of an effect has been called "circular and cumulative causation."

But if these kinds of beneficial externalities were seen to be the forces drawing economic activities into cities, negative externalities were invoked as a means of explaining the spatial distribution of these activities within the city. Traffic congestion, parking problems, and difficulties of loading and unloading were all seen as militating against the most crowded parts of the urban region. Increased competition for labor, combined with the higher costs of urban living (especially rent and transportation) were seen as factors driving up money wages and impairing the competitive capabilities of activities with high wage bills. And, most significantly, the high land values resulting from intense competition for centrally located space in the growing city were seen to be instrumental in "sorting out" activities in the urban area, both within and beyond the central business district.

Industries remaining in or close to the central business district (CBD) were observed to be those that find a great advantage in the accessibility of the CBD, the point of convergence for all city and regional transport routes, offering unusual convenience for buyers and for the assembly of workers. But because industry in central business districts was seen to face severe competition for sites, a hierarchy of land usage was said to result. The "highest" uses in the hierarchy were seen to be those that derive the greatest advantage from centrality and that could therefore pay most by turning the advantage into profit and rents. Housing was low on the list, except for high-class apartment houses. Most types of industry were also seen to be driven out because, for them, the advantages of centrality are relatively small, and the costs of production are lower on the periphery of the city. Many shopping functions likewise were seen to move to outlying suburban districts closer to local community and neighborhood markets.

Therefore, as the process of land use competition took place, the higher uses occupied the central area, and the only economic activities that remained localized within the boundaries of the central business districts were of the following kinds:

1. The *financial district* of banks, insurance companies, lending institutions, brokers, and the like, relying upon speedy personal contact within a closely intertwined set of activities and relationships.
2. *Specialized retail functions*, including department stores that require large population support at the point of convergence of population; and firms, offering high-quality and rare goods, that can satisfy their need for great population support only at the point of maximum convergence of the population of the urban region.
3. *Social and professional functions*, including the *headquarters and main-office function.*

These activities tended to squeeze out housing and industry from the most accessible points. The manufacturing and service industry surviving in the inner city thus had very special features and tended to be found on the edges of the central area in the back streets because land values there were lower, in the classical view. For this kind of industry, central location was imperative; there was no alternative. The industry had to be in the inner city because of its marketing needs, which could not be separated from its manufacturing functions. A simple showroom in the central area would not do. Also, many central manufacturing industries were observed not to have moved because the inner areas supported relatively immobile skilled labor forces. Some also were said to be localized there because of the extreme subdivision of processes within the cluster of interdependent firms. Such firms use ground very extensively; their space requirements in relation to their level of output are very low. In this sense, they can use relatively obsolete multistory buildings in a variety of kinds of adapted space. This kind of space is, of course, available at the lowest rents until it is cleared and rebuilt for some other, more intensive, higher-paying kind of use. Industry in central areas thus tends to move from building to building, being displaced as those buildings are cleared for revenue and reuse. The scale of operation is generally small. Two good examples are printing and clothing.

For job printing in particular, the marketing factor is important, since rush jobs must be completed in a very short space of time. Firms are generally located on the margins of the professional and financial districts. Newspapers, too, tend to be produced in central locations, usually in very large, modern buildings near to the zone of highest land values. Here, centrality is related to the receipt of news. There has to be close contact among the editorial staff, the printers, and the reporters who gather the news, and speedy distribution to the market is, of course, also imperative. On the other hand, periodicals, books, and standardized and regular kinds of job printing are not found in cities, but tend to decentralize into the suburban areas and rural and small-town locations.

Central-area clothing and related industries involve those branches where fashion and style are important—for example, ladies' dresses, gowns, furs, handbags, and fashion jewelry. Styles involve rapid changes in design, which means that output for stock is impossible; large varieties and small quantities are essential. Thus, small scale of activity is appropriate. There is no division of processes or line methods of production, and only a few persons work on a single garment. The work is highly skilled. Thus, in central London there are more than two thousand firms making women's clothing. This kind of industry has no special space requirements. It can use converted obsolescent buildings, as the space required is small; but on the other hand, skilled labor is essential. The trade is thus highly localized, with interdependence arising out of the provision of accessories—embroidery, belts, buttons, and so on—by specialized and separate firms. Reliance on common specialists arises out of the fashion nature of the trade. The crucial point binding the industry to central locations is marketing, the essentiality of immediate displays to buyers. The industry has to be close to a place where buyers can cluster.

Manufacturing and selling are close together because the scale of firms is so small that they cannot afford to run separate premises, and because selling and manufacturing are concerns of separate firms. Subcontracting is common, and people to whom the work is subcontracted need to be located close together. On the other hand, other branches of the clothing industry have quite different spatial distributions. Standardized products made by line methods, where there is subdivision of processes and where there are no requirements for a skilled labor force, have decentralized.

The printing and clothing industries highlight the economies of urbanization and localization that held industry to the central areas of cities, surrounding the central business district. In the classic models, as the CBD grew and land values escalated, other activities would decentralize in a regular sorting-out process. Land values were considered the most important centrifugal force because they imposed increased costs upon industry. Moreover, in inner-city areas, they resulted in cramped and restricted sites, multistory buildings, a shortage of storage space, parking difficulties, deficiency of light and air, and environmental pollution. Therefore, the pressure of space pushed out industry seeking space, light, and air to the periphery of the big city, to smaller towns, or to nonurban areas.

Decentralization: a continuous process

The selective redistribution of population and economic activities in metropolitan areas that we documented in previous chapters attests to the validity of the classical models. Note also the evidence in Figure 13.1, prepared using a table appearing in the work of urban economist Edwin S. Mills (1972:99). In this diagram, the average exponential density gradient of population and several types of employment for a sample of cities is plotted against time on semilogarithmic graph paper, to show whether decentralization was continuous, or evidenced marked changes over the year. For each item plotted (except manufacturing during the Great Depression), continuous decentralization is seen to have been taking place as cities have grown and transportation has improved.

Findings such as these lead some commentators—for example, Banfield (1968)—to argue that all that is happening to cities, even today, is very logical, a straightforward extension of past trends to which the classic models still apply:

> Much of what has happened—as well as of what is happening—in the typical city or metropolitan area can be understood in terms of three imperatives. The first is demographic: if the population of a city increases, the city must expand in one direction or another—up, down, or from the center outward. The second is technological: if it is feasible to transport large numbers of people outward (by train, bus, and automobile) but not upward or downward (by elevator), the city must expand outward. The third is economic: if the distribution of wealth and income is such that some can afford new housing and the time and money to commute considerable distances to work while others cannot, the expanding periphery of the city must be occupied by the first group (the "well-off") while the older, inner parts of the city, where most of the jobs are, must be occupied by the second group (the "not well-off"). The word "imperatives" is used to emphasize the inexorable, constraining character of the three factors that together comprise the logic of metropolitan growth.

One indeed might conclude from inspection of Figure 13.1 that Banfield's argument is true—that since the last burst of inventions producing the

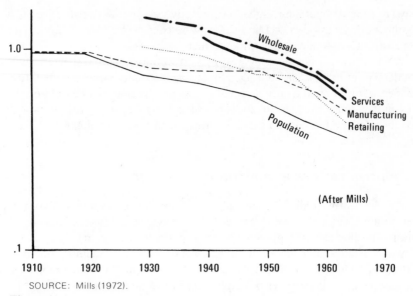

SOURCE: Mills (1972).

Figure 13.1
Decentralization trends since 1910.

upward and downward growth of the city (the skyscraper in 1880, the subway in 1886, the elevator in 1889; Eberhard, 1966), an outward urge has prevailed, with the last major contributing change being the onset of the automobile after 1920.

But also notice that there were increases in the rate of decentralization for both manufacturing and retailing in the years following World War II. It is these cases that are the focus of much current discussion, indicating to some that radical restructuring of metropolitan forms may be taking place, and it is to this issue that we now turn. The point we want to discuss was raised perceptively by the historian Oscar Handlin (and Burchard, 1963), who argued that

> Some decades ago ... a significant change appeared. The immediate local causes seemed to be the two wars, the depression, the new shifts in technology and population. However, these may be but manifestations of some larger turning point in the history of the society of which the modern city is part. The differences between city and country have been attenuated almost to the vanishing point. The movement of people, goods and messages has become so rapid and has extended over such a long period as to create a new situation. To put it bluntly, the urbanization of the whole society may be in the process of destroying the distinctive role of the modern city.

Are we indeed seeing such a radical restructuring of society, of which the changes in industrial and retail location are simply symptomatic?

Some statistics of central-city decline

Certainly many statistics appear to support this view. Between 1947 and 1958, the central cities of the New York region lost 6.0 per cent of their manufacturing jobs, whereas the suburbs gained 37.2 per cent. In other central cities of the old industrial heartland, comparable figures were: Chicago, −18.5 and +49.4 per cent; Philadelphia, −10.4 and +16.4; Detroit, −42.9 and +41.5; Boston, −15.3 and +33.5; Pittsburgh, −25.3 and +18.1; St. Louis, −21.1 and +41.7; Cleveland, −22.4 and +98.4. In the period between 1958 and 1963, the central cities of SMSAs exceeding 250,000 population lost 338,000 manufacturing jobs, whereas 433,000 jobs were added to the suburbs. Between 1958 and 1967, net manufacturing employment changes for selected cities are reported in Table 13.1. Most of the older central cities in the nation's "heartland" continued to experience substantial decline. Where there was growth from sources other than annexation, it was in the "rimland" regions of California, Florida, and Texas, where the greater city areas kept much of the "action" within the city limits.

As one observer remarked, it is hard to keep a good record of those companies which have left or are planning to leave. A partial list of those who have moved or plan to move all or part of their facilities out of New York to the surrounding suburbs as of 1974 gives an indication of the dimensions of the problem:

To Connecticut: American Can Co.; Lone Star Cement Corporation; Bangor Punta Corporation; Howmet Corporation; U.S. Tobacco Corporation; Olin Corporation (chemicals division); Hooker Chemical; Chesebrough-Pond's, Incorporated; Technicolor, Incorporated; Christian Dior Perfumes Corporation; General Telephone and Electronics Corporation; Consolidated Oil; and Stauffer Chemical.
To Westchester County: IBM, Incorporated; Pepsico, Incorporated; Dictaphone Corporation; General Foods Corporation; Flintkote; and AMF.
To northern New Jersey: CPC International; Union Camp; and the American Division of BASF. Even the Fantus Company, the relocation firm that helped plan many of these moves, has taken up new offices in Englewood Cliffs, New Jersey (population 5,810)—along with CPC, Thomas Lipton, Scholastic Magazine, and Volkswagen.

Elsewhere, the story is the same. In 1970, St. Louis lost 43 companies to the suburbs. In two recent years, Boston lost about 75. In Cleveland, recent losses include the frozen-foods division of Stouffer Foods; National Screw and Manufacturing; National Copper and Smelting; Fisher-Fazio-Costa (a major food chain); and the headquarters of B. F. Goodrich Chemical Co. Likewise in Detroit, S. S. Kresge, the retail chain, plans to move its corporate offices, as does the Michigan Automobile Club; Delta and

Table 13.1

Percentage increase in manufacturing employment, 1958–1967
(unadjusted for annexation), selected metropolitan areas

Area	Central city	Suburbs
Washington, D.C.	8.5	141.8
Baltimore, Md.	−5.9	22.0
Boston, Mass.	−11.8	17.0
Newark, N.J.	−10.3	12.2
Paterson-Clifton-Passaic, N.J.	−1.3	33.8
Buffalo, N.Y.	−1.9	3.4
New York, N.Y.	−10.3	36.0
Rochester, N.Y.	18.3	55.9
Philadelphia, Pa.–N.J.	−11.6	30.0
Pittsburgh, Pa.	−13.8	3.7
Providence, R.I.	4.5	12.4
Chicago, Ill.	−4.0	51.6
Indianapolis, Ind.	23.1	36.3
Detroit, Mich.	−1.8	47.6
Minneapolis–St. Paul, Minn.	8.9	146.2
Kansas City, Mo.–Kans.	0	68.3
St. Louis, Mo.–Ill.	−14.9	41.4
Cincinnati, Ohio–Ky.–Ind.	10.6	2.7
Cleveland, Ohio	−5.3	42.6
Columbus, Ohio	18.1	20.4
Dayton, Ohio	12.9	99.5
Milwaukee, Wis.	−6.3	55.6
Miami, Fla.	5.7	116.0
Tampa-St. Petersburg, Fla.	3.8	280.0
Atlanta, Ga.	8.9	86.0
Louisville, Ky.–Ind.	16.7	48.4
New Orleans, La.	12.3	29.2
Dallas, Tex.	41.9	131.0
Houston, Tex.	41.6	12.6
San Antonio, Tex.	28.5	62.5
Los Angeles–Long Beach, Cal.	10.3	23.0
San Diego, Cal.	−24.5	27.5
San Francisco–Oakland, Cal.	−23.5	29.3
San Bernardino–Riverside–Ontario, Cal.	57.3	60.0
Denver, Colo.	6.9	112.7
Portland, Ore.–Wash.	16.5	68.0
Seattle–Everett, Wash.	−25.7	244.7

Pan American Airlines; R. L. Polk, publishers; not to mention Circus
World, the toy manufacturers. Things are so glum that former mayor
Jerome Cavanaugh sometimes refers to "Detroit's sister cities—Nagasaki
and Pompeii." In Los Angeles, the financial district is being deserted by
many of the major banks, beginning with Crocker–Citizens National in 1968
(Cassidy, 1972).

On the retail side, the data reveal trends of the same magnitude (see Tables 13.2 and 13.3). Deflated retail sales declined in many central cities between 1958 and 1967, and there were distinct patterns by city-size class. Central business districts in particular lost their retail function, especially in the smaller metropolitan areas.

The story can be repeated for many personal services. In Chicago, for example, in 1950, 62 per cent of all the metropolitan area's medical specialists

Table 13.2

Percentage increase in retail sales deflated by general price increase, 1958–1967, for selected metropolitan areas

Area	Central city	Suburbs
Washington, D.C.	10.5	134.8
Baltimore, Md.	4.9	128.2
Boston, Mass.	−1.4	79.2
Newark, N.J.	−14.1	37.1
Paterson-Clifton-Passaic, N.J.	0.9	74.5
Buffalo, N.Y.	−9.9	54.7
New York, N.Y.	9.7	60.2
Rochester, N.Y.	18.1	91.3
Philadelphia, Pa.–N.J.	6.2	65.4
Pittsburgh, Pa.	7.8	28.7
Providence, R.I.	−36.3	73.1
Chicago, Ill.	5.3	86.6
Indianapolis, Ind.	20.0	160.8
Detroit, Mich.	0.7	86.4
Minneapolis–St. Paul, Minn.	7.9	149.7
Kansas City, Mo.–Kans.	55.2	64.3
St. Louis, Mo.–Ill.	−7.6	76.2
Cincinnati, Ohio–Ky.–Ind.	4.6	129.4
Cleveland, Ohio	−15.2	269.1
Columbus, Ohio	22.8	141.9
Dayton, Ohio	3.6	125.5
Milwaukee, Wis.	7.5	108.3
Miami, Fla.	−2.5	98.2
Tampa–St. Petersburg, Fla.	30.9	108.9
Atlanta, Ga.	37.7	153.9
Louisville, Ky.–Ind.	14.0	101.8
New Orleans, La.	21.0	141.9
Dallas, Tex.	33.6	119.2
Houston, Tex.	55.9	63.3
San Antonio, Tex.	36.4	79.9
Los Angeles–Long Beach, Calif.	22.2	75.4
San Diego, Calif.	25.6	91.8
San Francisco–Oakland, Calif.	16.3	81.6
Denver, Colo.	11.1	132.4
Portland, Ore.–Wash.	28.1	180.3
Seattle, Wash.	18.0	152.5

Table 13.3

Retail change, by city size class, 1954–1967

	Population of metropolitan area (in thousands)			
	3,000+	1,000–3,000	500–1,000	250–500
(a) Percentage change in sales: current dollars				
CBD	12.1	8.3	−3.7	−6.7
Central city	34.3	26.8	58.4	61.4
Suburbs	132.2	175.0	209.0	193.1
(b) Percentage change in numbers of establishments				
CBD	−26.0	−26.9	−38.2	−37.6
Central city	−26.3	−23.7	−8.4	−8.4
Suburbs	29.9	30.3	51.3	48.0

had their offices in the Loop, another 23 per cent were to be found elsewhere in the central city, and only 15 per cent were in the suburbs. By 1970, the Loop's share had dropped to 25 per cent, the city had another 25 per cent, and fully 50 per cent were to be found in the suburbs. Altogether, Chicago lost 400 firms and 70,000 jobs between 1950 and 1965 in the inner third of the city. This exodus continues with little abatement in the dingy, crime-ridden, and decaying West Side neighborhoods. Forty-five factories employing 3,875 persons left Chicago for the suburbs in 1969. In 1970, 23 firms employing 1,455 persons left. On the West Side, loss of 7,512 industrial jobs meant that 210 retail businesses folded, 4,550 persons employed in non-manufacturing businesses lost their jobs, and retail sales there declined $23.1 million (Young, 1972).

Table 13.4 provides yet another insight into the changes. It shows that a very large share of new private nonresidential construction in the period 1954–1967 was taking place in the suburbs—for example, over 80 per cent of new industrial building in Boston, or 98 per cent of all new community facilities in Dayton, Ohio. And if the statistics are adjusted to control for annexation, the story is even worse. For example, correcting for annexation, the central cities of our 33 largest SMSAs suffered a net loss of 1,146,845 manufacturing jobs between 1947 and 1972. For wholesaling, these same 33 cities lost 302,016 jobs during that period; and for retailing, the cities lost 565,125 jobs. These statistics suggest a field day for the prophets of doom.

Prophecies of disaster

To the reader of the daily press and of the news magazines, loss of jobs in manufacturing and the decline of the retail function of the central business district presages disaster. Consider, for example, the following random clippings from stories appearing in 1972:

the loss of the Giants was an almost comic sidelight to a more serious problem: the exodus of dozens of businesses, including many of the "Fortune 500" from New York to the surrounding suburbs or other cities. And since conditions in Gotham often forecast what may happen elsewhere, America's older cities are finding that they, too, are losing some of their long-time corporate occupants to the burgeoning suburbs. The growing industrialization of the overwhelmingly white suburbs, coupled with the entrapment of minorities in the central cities, paints a picture with the word CRISIS splashed across it in boldface .... If business (and suburban) communities continue in their present mindlessly selfish way we will Los Angelesize our land, Balkanize our region's finances, and South Africanize our economy. (Cassidy, 1972)

The second great migration to the suburbs—the exodus of families and offices—has profoundly altered the economic life of cities, just as the great migration of middle-class families altered the social patterns. Four out of five new jobs created in major metropolitan areas in the past ten years have been outside the cities. This disproportion has been a major cause of the "urban crisis," damaging the ability of cities to pay their way and to provide employment for the people who live in them. (Gooding, 1972)

While concern has mounted in the last decade over the plight of the inner cities in vast metropolitan areas, the aging downtowns of hundreds of smaller cities have been quietly deteriorating with little national notice ... losing businesses and office-users just as surely as their metropolitan cousins.... "In the old days, the farmers all came to town on Saturday night," ... a long-time resident recalled sadly. "Not any more, it's just dead." And a teenage girl driving what is locally known as "the Circuit" said blithely: "This town is for nothing but old fogeys. It's a bummer." (Kneeland, 1972)

Reasons for the exodus

The press is also full of speculations about the reasons for the exodus of manufacturing and retailing, although there is far from any consistency in journalistic perceptions. Some see the shift from the pull of low land values to the push of urban problems:

Business is moving out for a variety of reasons. The explanation immediately after World War II was the attraction of cheap land in the suburbs, permitting single-story factories that were convenient for truck loading. In recent years, the motive is more push than pull. Executives complain about the abominable phone service in many cities, horrendous commuting conditions, rapidly rising crime. Even bomb threats are mentioned, when GT&E actually had a bomb go off in the building, the bosses lost no time in making the final decision to get the hell out. Then there are problems with the work force. Many young women seem to be avoiding the big cities, while young execs no longer consider a move to the New York office a promotion; indeed, they demand differential pay to cover the increased cost of living. There is also the desire to get away from it all, which was one of the big reasons why Xerox moved its top men from Rochester to pastoral Connecticut; the company president felt that they would get a better perspective of the whole company from the

Table 13.4

Percentages of private nonresidential construction in the suburbs, 1960–1965 and 1954–1965

Percentage of valuation of permits authorized for new nonresidential building in—

	Atlanta	Boston	Chicago	Cleveland	Dayton	Detroit	Indianapolis	Los Angeles	New Orleans	New York	Philadelphia	St. Louis	San Francisco
1960–1965													
All types	47	64	65	56	62	69	41	59	42	38	65	41	60
Business	44	68	64	60	66	69	49	60	49	39	70	39	63
Industrial	71	81	77	61	56	70	52	85	58	61	75	67	84
Stores and other mercantile buildings	44	74	67	74	78	80	55	63	66	64	75	75	72
Office buildings	25	52	58	38	53	55	21	41	10	21	52	32	38
Gasoline and service stations	63	91	54	57	98	58	54	60	60	51	66	55	72
Community	50	61	64	44	49	71	33	61	37	31	60	37	58
Educational	59	63	64	51	28	63	24	61	35	29	67	67	57
Hospital and institutional	59	38	56	15	56	61	14	72	44	25	38	35	52
Religious	69	92	73	84	56	81	56	69	35	55	77	86	62
Amusement	31	59	80	60	99	86	58	35	41	19	59	85	74
1954–1965													
All types	43	68	63	58	a	71	44	62	a	44	67	a	63
Business	41	70	61	59	a	73	50	63	a	44	69	a	64
Industrial	66	82	73	60	a	75	61	86	a	75	76	a	84
Stores and other mercantile buildings	40	74	67	73	a	77	52	66	a	71	72	a	72
Office buildings	21	51	39	37	a	58	21	41	a	18	51	a	37
Gasoline and service stations	60	82	59	62	a	65	56	62	a	65	73	a	73
Community	48	67	66	44	a	70	40	63	a	38	68	a	64
Educational	57	72	69	61	a	79	46	59	a	34	72	a	53
Hospital and institutional	32	41	58	33	a	62	10	70	a	32	43	a	53
Religious	59	86	68	81	a	74	59	70	a	61	80	a	65
Amusement	30	64	75	57	a	43	52	50	a	33	72	a	55

Source: Newman (1967).

a. Not available.

new, more isolated locale. But the biggest appeal of the suburbs, of course, is that much of the population, housing and development is there, or headed there. (Cassidy, 1972)

Others see the lure of new kinds of prestige locations, closer to the preferred places of residence of the businessmen and decision makers:

Around the perimeter of O'Hare International Airport are any number of businesses that have O'Hare stuck in their name. The O'Hare area, which had no office space 10 years ago, now accounts for nearly 30 per cent of all suburban office space in the Chicago area. "Using O'Hare in the company name has the psychological aspect of giving the firm prestige," says Loren Trimble, an expert on the area's economic development. "O'Hare carries the connotation of being near the world's busiest airport, of being modern and in the jet age." Proximity to the airport explains part of the phenomenal growth of the northwest suburbs in the last decade. O'Hare has had a very direct effect on the building of fancy hotels, convention facilities and office buildings in the area.

But it is only one of many reasons for the rapid industrial and commercial growth and, most experts agree, not the main one. Most maintain that things would be booming in the area without the world's busiest jetport.

"I say that the principal reason we have 500 new companies in Elk Grove Village is that it is close to where the boss lives," said Marshall Bennett, a partner in Bennett & Kahnweiler, industrial real estate brokers. His firm developed Centex Industrial Park in Elk Grove Village, the largest of its kind in the nation. The northwest suburbs are now Chicago's biggest competitor for the dollar of the industrialist, conventioner and nightclubber. The area is one of the four fastest growing areas in the country. Why this once rural economy grew so rapidly to an industrial and commercial giant is laid mostly to transportation. First came the people. They started the trek to the suburbs in the early 1950's. Later, the bosses and decision-makers in industry started bringing their factories closer to home. Old and dreary sawtoothed-roof factories were abandoned in favor of horizontal layouts in spacious suburban buildings. They were mostly light industries that produced little noise, smoke, odor or other irritants to suburban residents. Many were warehouses. However, employees in these factories and warehouses have not been as keen as their bosses at being around O'Hare. Many of the workers, particularly the blue collar ones, still live in Chicago, while many of the northwest suburbanites still work in the city. This situation has led to considerable reverse commuting. For example, about 75 per cent of the 25,000 employees at Centex Industrial Park in Elk Grove commute from Chicago, while the majority of Elk Grove Village residents work in Chicago. (Young, 1972)

Yet others look at identical situations and see only polarization centering on race:

A series of downtown events have highlighted the problem of the Chicago downtown area and raised fears that the ghostlike character of downtowns in most major U.S. cities will come to pass in Chicago as well. The abandoned appearance of downtowns throughout the country are a product of two factors: race and market and the interplay between them.

The society that is in the majority judges an area's decline or renaissance in this country (whether it is a downtown area or a residential community) by the behavior of white residents and consumers—i.e., by an area's attraction and appeal to white people. This is attributable in part to our social and economic stratification and most particularly to the income distribution in our population. Therefore, since whites possess the wealth, the range of choices available to them, and the ways in which they act on such choices, tend to predestine an area's future. When whites decide to leave an area, a self-destructive process is perceived in white eyes to have been set in motion. This process feeds on itself and results in an inevitable reduction in the money necessary to support commercial facilities, pay for housing and building maintenance and make possible new construction. This in turn further accelerates the exodus of white populations and increases black populations, vacancies and abandonment. (Meltzer, 1972)

Some alternative statistics

What is actually happening and why ? Before an answer can be provided, it must be said that all is not told by the statistics in the previous section. Indeed, there have been some dramatic changes of a directly contrary kind, and the counterpoint they play to the loss of manufacturing jobs and retail sales must be orchestrated into a balanced interpretation of contemporary urban dynamics. For example, who would acknowledge, reading the interpretations and data on Chicago already presented, that between 1953 and 1964 $750 million was invested in construction and improvement of manufacturing plants in Chicago (Hartnett, 1971)?

The problem is that whenever a negative statistic is observed in the city today, someone hastens to cry wolf. Thus, when the usual annual summertime peak in new industrial investment commitments vanished in metropolitan Chicago in 1971 and 1972, and industrial vacancies in the central city started to mount, the Chicago Association of Commerce and Industry produced illustrations asserting that flight to the suburbs was the cause, as Figure 13.2 reveals. But is this a valid interpretation of the statistics ? All the numbers so far have represented *net* change. This may be the source of major errors in interpretation, masking the fact that net change, plus or minus, is made up of many elements.

Consider the data on the sources of employment change in various parts of New York in the years 1967–1969 reported in Table 13.5 (Leone, 1972). Net change in each area is seen to represent the resolution of interacting birth, death, immigration, emigration, and local growth processes. Clearly, given such complexity, any simple explanation is likely to be not only suspect, but almost undoubtedly faulty. Although Manhattan's CBD lost 10,625 jobs because of moveouts and 17,890 jobs because of businesses closing, 24,602 new jobs were created in new businesses, and local nonmovers expanded their employment by 38,804. Much of this was associated with the

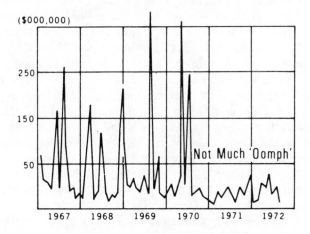

INDUSTRIAL DEVELOPMENT INVESTMENTS
Metropolitan Chicago

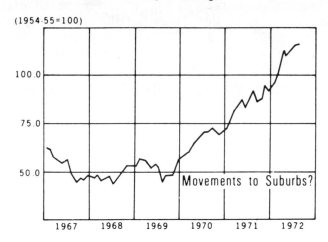

INDEX OF VACANT INDUSTRIAL BUILDINGS
City of Chicago

SOURCE: Chicago Association of Commerce and Industry.
Figure 13.2
Recent industrial statistics for the Chicago region.

office boom, to which we referred in the last chapter and which we will further develop shortly. The decline in jobs in Brooklyn was due to an excess of deaths over births, because that area experienced net immigration. In Queens, too, there was net immigration, and the decline in that case was due to local nonmovers reducing their employment by 10,702.

The dynamic was different in each case. No wonder, then, that many observers have difficulty in squaring the gloom-and-doom perceptions with the most massive downtown office boom in the nation's history, which has transformed central business districts of the nation's major metropolises. Gross floor space in private office buildings in the Manhattan CBD increased only from 126 million square feet in 1936 to 128 million in 1950. But by 1960, the total was 160 million; by 1963, 184 (with an additional 18.8 in public buildings); and by 1970, 226 (with 20.8 in public buildings). In the Dallas CBD, comparable figures were 4.7 million in 1936; 6.0 in 1950; 15.6 in 1960; 17.3 in 1963; and 22.5 in 1970 (Armstrong, 1972). In Chicago, 8.2 million square feet of office space was added to the Loop in the period 1967–1972. In the ten years between 1960 and 1969, the valuation of new office space authorized for construction in the nation's major office centers was, in millions of dollars: New York SMSA, 1,659; Los Angeles SMSA, 1,220; Washington SMSA, 812; Chicago SMSA, 710; San Francisco SMSA, 601; Boston SMSA, 393; Detroit SMSA, 327; Atlanta SMSA, 279; Philadelphia SMSA, 250; Cleveland SMSA, 172; Seattle SMSA, 169; Milwaukee SMSA, 101. This investment boosted private and public office space by 44 per cent in these SMSAs, which, along with Pittsburgh and St. Louis, account for over 70 per cent of the headquarters and headquarters employment of the nation's top five hundred industrials, and over 60 per cent of all of the nation's central administrative office employment (another 7 per cent is in Minneapolis–St. Paul, Houston, Dallas, Cincinnati, and Kansas City; Armstrong, 1972).

Table 13.5

Employment change in New York, 1967–1969

	Manhattan CBD	Brooklyn	Queens	New York SMSA
Births	+24,602	+5,882	+3,713	+45,146
Deaths	−17,890	−8,083	−3,901	−40,574
Immigrants	+5,851	+5,777	+6,892	+9,018
Emigrants	−10,625	−4,960	−3,078	−5,579
Local relocators	+8,136	+1,098	+961	+14,184
Nonmovers	+38,804	−1,338	−10,702	+11,696
Net change	+48,878	−1,624	−6,115	+33,891

Source: Leone (1972), using data from the Dun & Bradstreet Corporation's "Dun's Market Identifier" files, prepared for analysis at the National Bureau of Economic Research.

Interpretation

These apparently contradictory trends, which evidently have defied the abilities of most popular commentators to comprehend their basic importance, are in fact a product of a revolutionary transformation of American metropolitan areas that renders conventional wisdom inoperable.

The initial development of road transport and highway improvements did indeed breed a continuous and gradual decentralization of people and jobs in the core-oriented metropolis through the first half of the twentieth century. But then came a series of technological changes in rates of decentralization, as is observable in Figure 13.1.

On the manufacturing side, superhighways, large-scale trucks, and piggyback combination road and rail transport reversed the pull of rail terminals and docks in central areas on industrial location. Previously, extremely efficient interregional transportation conspired with slow and costly local road transport to constrain industry to central terminal locations. Suddenly this constraint was removed, and peripheral locations became as (or even more) accessible to the metropolitan region than was the traditional city center. With the removal of traditional constraints, the negative externalities of the central city loomed large in choice, both perceptually and in fact. The locational choices of the subsequent massive decentralization of industry were dictated more by social factors and prestige locations than by traditional dollars and cents. Low-status locations—poor, polluted, and black neighborhoods—were left and avoided by new concerns.

For retailing, the significant development was the invention of the planned shopping center, a dramatic and drastic technological change. Although the first planned centers were built in the 1920s, it was not until after 1950 that the rapid diffusion of the new technology transformed retail relationships. In the early years, developers *followed* the growing suburban market with a hierarchy of neighborhood, community, and regional centers. The first centers were oriented to the more affluent sectors of the biggest cities and diffused from them into smaller places and lower-income suburbs. It is in the smaller cities that the change has been most devastating.

As one commentator (Kneeland, 1972) remarked, Portland, Maine

had a population of 75,000 in 1960, a figure that dropped to 63,000 in 1970 as residents left the aging seaport for the rapidly developing suburbs.

In 1958, the year before the first shopping center opened in the area, gross sales of downtown retail merchants were $140-million. Now there are 10 shopping centers in Greater Portland, including the recently opened Maine Mall, an enclosed center with about 50 shops and two huge department stores. Downtown retail sales have fallen to about $40-million annually.

Like a number of other cities, Portland seems to be somewhat resigned to downtown's loss of retail dominance to the shopping center.

In the 1960s in the bigger metropolitan areas, a fundamental change in shopping-center developers' concepts became apparent. A report in *Fortune* magazine (Breckenfeld, 1972) captured some of the flavor of the change:

The shopping center has become the piazza of America. In big metropolitan areas and smaller cities alike. . .indoor piazzas are reshaping much of American life. Giant regional shopping centers have risen by the hundreds across the nation and are still going up by the score. To an amazing degree, they are seizing the role once held by the central business district, not only in retailing but as the social, cultural, and recreational focal point of the entire community. . . . Grim, formal unwelcoming old cities scarcely fit the more relaxed life style of the newly affluent middle class with considerable leisure. . . . In many cities middle-class white shoppers are beginning to abandon downtown to the poor and blacks except from nine to five, Monday through Friday (office working hours).

Again, the suggestion of the social dynamic is there, replacing much of the economic dynamic of conventional wisdom. This social dynamic involves not only the avoidance of the central city by the majority of white residents, but also the creative, style-setting leadership envisaged by major shopping center developers who are busy building on the metropolitan periphery ahead of residential growth. By leading rather than following, the new developers seek to establish the direction of metropolitan growth, set the style and tone for the suburbs to follow, and capitalize on the new opportunities thereby created.

Although part of the social dynamic involves this creative development of new opportunities on the expanding periphery, this is possible only because of the eagerness of most Americans to find a safe haven in the new forms of suburbia thereby created.

What all of this means, of course, is that the black resident of the metropolis finds himself in a central-city ghetto abandoned by both whites and, increasingly, employment. The flight of white citydwellers into the expanding peripheries of metropolitan regions is an accelerating phenomenon as minorities move toward majority status in the city center. The exurban fringes of many of the nation's urban regions have now pushed one hundred miles and more from the traditional city centers. More important, the core orientation implied in use of the terms *central city* and *central business district* is fast on the wane. No longer is it necessary to have a single, viable, growing heart. Today's urban systems appear to be multinodal, multiconnected social systems in action, in which the traditional centralization of the population into metropolitan areas has been counterbalanced by a multifaceted reverse thrust of decentralization. The situation is very different from the period at the end of the nineteenth century, from which we derive the concept of urbanization. Decentralization and an outward urge have replaced centralization and core orientation; differentiation and segregation substitute for the integrative role of the melting pot.

The essence of the new urban system is its linkages and interactions, as shaped by changing modes of transportation and communication. Both places of residence and places of work are responding more and more to social rather than to traditional economic dynamics. At the same time, new communications media, notably television, have contributed to cognitive changes by providing the universal perception of decaying central cities; of old neighborhoods now the domicile of low-income minorities; the immediate on-the-spot coverage of violent crime; the careful documentation of the frustrations of city life; and acute awareness of emerging separatist feelings. It is no accident that the suburbanization of white city-dwellers has increased, supported by rising real incomes and increased leisure time. Similarly, decentralizing commerce and industry avoids the low-status central-city location, and the new office complex seals itself in by defensive techniques. Gradients of distance-accretion are now beginning to replace those of core-centered distance-decay within the larger megalopolitan complexes, as persons of greater wealth and leisure seek home and work among the more remote environments of hills, water, and forest; many aspire to such settings as an ideal. As we noted, core-dominated concentration is on the wane; the multinode, multiconnection system is the rule, with the traditional multifunctional core simply a specialized one among many. It is the spontaneous creation of new communities, the flows that respond to new transportation arteries, the waves emanating from new industrial and retail growth centers, the mutually repulsive interactions of antagonistic social groups, the reverse commuting that results as employment decentralizes, and a variety of other facets of social dynamics that today combine to constitute the new urban systems in America.

These trends and changes should, of course, be put back into the context of changing technology. Concentrated industrial metropolises developed *only* because proximity meant lower transportation and communication costs for those interdependent specialists who had to interact with each other frequently or intensively and could only do so on a face-to-face basis. But shortened distances also meant higher densities and costs of congestion, high rent, loss of privacy, and the like. As soon as technological change permitted, the metropolis was transformed to minimize these negative externalities. The decline of downtown retailing and of central-city manufacturing, alongside the office boom within the major metropolitan areas, are but manifestations of this fundamental transformation of American urbanism, for which a new definitive theory to match the conventional wisdom of the first half of the twentieth century has yet to be written.

PART V
REGIONAL GROWTH
AND URBAN SYSTEMS

Chapter 14
Changing relationships between urban and regional growth

That a fundamental transformation of American urbanism is now taking place should be no surprise. Other such transformations have taken place in the past, and yet others can be expected in the future. The causes are to be found in the relationships between urban and regional growth within the changing environment of national and international economic organization. It is to these relationships that we now turn.

Recall what was said in Chapter 5—that it was impossible to study urban ecology in isolation because cities are the central elements in the spatial organization of regional, national, and supranational socioeconomies, by virtue of the interregional organization in a total "ecological field" of the functions they perform. Throughout the first half of the twentieth century this ecological field has had a particular form.

The central driving force has been

> a great heartland nucleation of industry and the national market, the focus of large-scale national-serving industry, the seedbed of new industry responding to the dynamic structure of national final demand, and the center of high levels of per capita income....

Standing in a dependent relationship to the heartland,

> radiating out across the national landscape are... resource dominant regional hinterlands specializing in the production of resource and intermediate outputs for which the heartland reaches out to satisfy the input requirements of its great manufacturing plants. Here in the hinterlands, resource-endowment is a critical determinant of the particular cumulative advantage of the region and hence its growth potential. (Perloff and Wingo, 1963)

Figure 14.1 portrays the geographic configuration of national market access in the United States in 1960; and Figure 14.2, the related spatial pattern of industrial concentrations. The geography of this relationship—the

271

juxtaposition that constitutes the nation's heartland—was explored at length over two decades ago by Harris (1954). Subsequent workers have demonstrated the theoretical significance of general access within the country, thus cementing our understanding of industrial location processes (Tidemann, 1968).

To appreciate why the nation's hinterlands developed their dependent relationship to this industrial heartland, one must understand the sequence of stages through which the national economy has evolved, and the different urban-regional relationships represented by these stages:

> throughout the evolution...two factors, great migrations and major changes in technology, have particularly influenced the location of relative growth and decline. Both factors have repeatedly been given specific geographical expression through their relationship to resource patterns. Major changes in technology have resulted in critically important changes in the evaluation or definition of particular resources on which the growth of certain urban regions had previously been based. Great migrations have sought to exploit resources —ranging from climate or coal to water to zinc—that were newly appreciated or newly accessible within the national market. Usually...the new appreciation or accessibility has come about, in turn, through some major technological innovation. (Borchert, 1967)

Several stages of "natural resources that count" can be cited in the history of the American economy. In the agricultural period, the natural endowment most valued was arable land with environmental components of climate and water. During industrialization a new set of mineral resources became important. In the twentieth century, service activities and amenity resources have exerted an increasingly strong pull on industry and on people, along with regional differences in governmental expenditures. In each of these stages, urban growth and regional development have been closely interdependent.

America's oldest cities were mercantile outposts of a resource area whose exploitation was organized by the developing metropolitan system of Western Europe. The initial impulses for independent urban growth came at the end of the eighteenth century, when towns were becoming the outlets for capital accumulated in commercial agriculture and the centers of colonial development of the continental interior. Arable land was the resource that counted in regional growth. Regional economies developed a certain archetype: a good deepwater port as the nucleus of an agricultural hinterland well adapted for the production of a stable commodity in demand on the world market. Growth potentials of regions depended on the extent and richness of the hinterlands accessible to the ports. The prototypic American metropolis thus was a port at a strategic location on long-distance oceanic or riverine trade routes, providing a range of mercantile services and determining the terms of trade. This agricultural resource-dominated (but city-centered) expansion of the economy set the stage for subsequent developments by establishing a geography of markets, transport routes, and labor forces that conditioned the nature of succeeding growth.

New resources became important from 1840–1850 onwards, and new locational forces came into play. Foremost was a growing demand for iron, and later steel, and along with it rapid elaboration of productive technologies. Juxtaposition of coal, iron ore, and markets afforded the impetus for manufacturing growth in the northeastern United States, localized by both factors in the physical environment (minerals) and environmental components created by prior growth of the urban system (linkages to succeeding stages of production, in turn located closer to markets). The "heartland" of the American manufacturing belt developed westwards from New York in the area bounded by Lake Superior iron ores, the Pennsylvania coalfields, and the capital, entrepreneurial experience, and engineering trades of the northeast, while at the same time New York cemented its dominance by accentuation of its financial, entrepreneurial, and specialized manufacturing roles. This heartland became not only the heavy industrial center of the country, but has remained the center of national demand, determining patterns of market accessibility ever since (Figures 14.1 and 14.2).

The heartland had initial advantages of both excellent agricultural resources and a key location in the minerals economy. With development, it grew into the urbanized center of the national market. Subsequent metropolitan growth took place in a pattern organized around this national core region until after the Second World War. From 1869 to 1950 there was, for example, a stable pattern of growth in manufacturing employment among the states (Borts, 1967). Continued spread of population and agriculture over the continent pulled processing and servicing activities and new urban growth with them. However, the developmentally-dominant effects still came from growth of the minerals economy until well into the twentieth century, so that a process of "circular and cumulative causation" (Pred, 1966) strengthened and maintained the relations of the national heartland and hinterlands—of core and periphery—and the new metropolitan centers that did emerge did so in sequence with the overall growth, outward spread, and spatial integration of the economy.

In each case, the basic conditions of regional growth were set by the heartland. It served as the lever for successive development of newer peripheral regions by reaching out to them as its input requirements expanded, and it thereby fostered specialization of regional roles in the national economy. The heartland experienced cumulative urban-industrial specialization, whereas each of the hinterlands found its comparative advantage based on narrow and intensive specialization in a few resource subsectors, diversifying only when the extent of specialization enabled the hinterland region to pass through that threshold scale of market necessary to support profitable local enterprise. Flows of raw materials inward and finished products outward articulated the whole (Ullman, 1957).

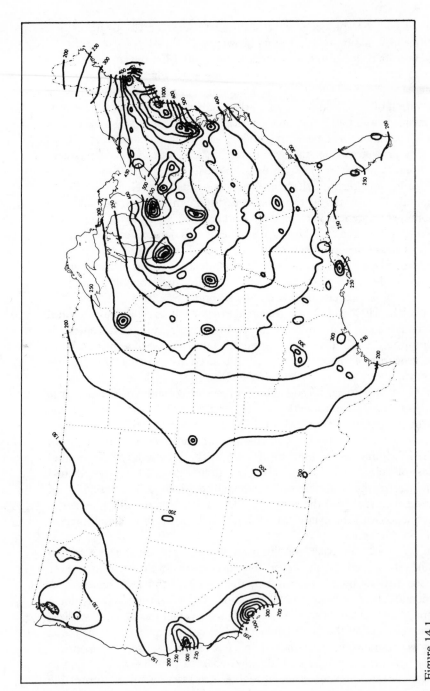

Figure 14.1
National Market Access in 1960. The map contours population potentials.

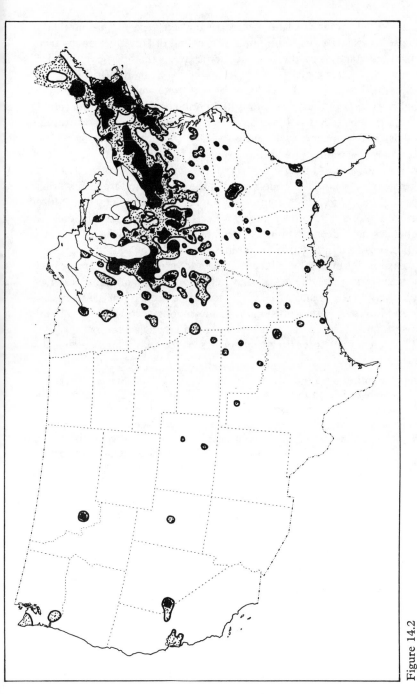

Figure 14.2
Manufacturing areas in 1962. Shades indicate intensity of manufacturing concentrations.

Since 1950, growth of the service sector, increase in the number of "footloose" industries (including final processing of consumer goods using manufactured parts, and the aircraft, aerospace, and defense industries), rapid emergence of a "quaternary" sector of the economy (involving, for example, the research and development industry), expansion and inter-regional migration of the nonjob-oriented population (for example, retirees to Florida, Arizona, and California), rising governmental expenditures and overall rising real incomes all have served to produce yet another trans-formation of the economy and the urban system—and of the traditional export-base concept of hinterland growth. Advantages for economic growth have been found around the "outer rim" of the country, in former hinterland regions (Ullman, 1954), as changing communications technology has reduced the time and costs involved in previous heartland-hinterland relationships. The changes have been cumulative: regional growth within the context of the national pattern of heartland and hinterland had brought these regions to threshold sizes for internal production of a wide variety of goods and services at the very time that changes in the definition of urban resources made possible their rapid advance, free of the traditional constraints of heartland-hinterland leverage. Hence, the explosive metropolitan growth of the South, Southwest, and West, led by the tertiary and quaternary sectors.

The outcome of this latter stage of urban-regional growth is that it is now

> possible to interpret the spatial structure of the United States in ways that will emphasize a pattern consisting of one, metropolitan areas and two, the inter-metropolitan periphery. Except for thinly populated parts of the American interior, the inter-metropolitan periphery includes all the areas that intervene among metropolitan regions and that are, as it were, the reverse image of the trend towards large scale concentrated settlement that has persisted in this country for over half a century. Like a devil's mirror, much of the periphery has developed a socio-economic profile that perversely reflects the very opposite of metropolitan virility. Economically, the inter-metropolitan periphery includes most of the areas that have been declared eligible for federal area redevelopment assistance.... They have a dispro-portionately large share of low-growth and declining industries and a corre-spondingly antiquated economic structure. Nevertheless, one-fifth of the American people are living in these regions of economic distress.

Friedmann and Miller (1965), the authors of the above quotation, go on to say:

> Demographically, the inter-metropolitan periphery has been subjected to a long-term, continuous decline. This trend reflects the movement of people to cities, especially to the large metropolitan concentrations. Although the smaller cities on the periphery have to some extent benefited from migration, their gains have been less, on the average, than for all urban areas. In addition, migration from economically depressed regions has been highly selective, so that the age distribution of the remaining population has become polarized

around the very young and the very old.... Socially, the standard indices of education and health are substantially lower along the periphery than in metropolitan areas. The quality of public services has deteriorated (though their per capita cost has increased), the housing stock is older, and the level of educational attainment is significantly below the average for metropolitan America. Rapid and selective outmigration, a declining economic base, the burden of an aging population, and low incomes have rendered many peripheral communities helpless in their desire to adapt to changing circumstances in the outside world. The remaining population is frequently short both on civic leadership and hope. They can neither grasp the scope of events that have overtaken them nor are they capable of responding creatively to the new situations.

It is in the peripheries that "growth-center" strategies are being proposed as instruments for alleviating problems of unemployment and low incomes, while meeting the need for institution building, community organization, and leadership development. Yet the strategies take as their theoretical base the earlier driving mechanism of the heartland-hinterland relationship, failing to include the self-sustaining growth of metropolitan America that is the new feature of metropolitan dynamics. It is to this discordance that we should now turn.

Traditional regional growth concepts

The traditional theory of regional growth and classical growth-center concepts both stem from the heartland-hinterland dependency relationship. The problem for public policies formulated on the basis of these ideas is, however, that the service-related transformation of the American economic scene in recent decades, along with the attendant emergence of metropolitan areas and the intermetropolitan periphery as the basic regional distinction in the nation, casts serious doubt on this classical basis of growth-center theory. Thus, it is more critical today than ever before that we examine in detail the growth processes operating in the American urban system in the 1970s, and on that basis explore alternative theoretical formulations and new approaches to the uses of growth centers as instruments for inducing regional development. It is in the framework of such alternative approaches that our analysis proceeds.

Following the example of Perloff (1960), Friedmann (1966), and others, the classical theory of regional growth can be set in propositional form, as follows:

1. *A regional economy is an interdependent part of a national economy.* That differential regional economic growth has been until recently an expression of an open, highly dynamic economy, made up of an industrial heartland and resource-dominant hinterlands, is unquestionable. As long as demand and supply conditions change and regions have differing advantages for production, differences in regional growth must be seen as part of the

total economic system, just as are economic specialization and the division of labor. All the elements essential to national economic growth, therefore, are also essential to the economic growth of regions. Chief among these elements are development of natural resources, development of manpower skills, aggressive entrepreneurship, an elaborate infrastructure, and associated external economies. In such circumstances, the degree to which local decisions can shape the future of a regional economy depends upon the degree of self-sustaining local autonomy ("closure") of that economy. Greater closure implies greater autonomy of choice; greater openness implies greater dependency upon changes and choices in other regions and the nation. Clearly, the degree of closure is small in resource-oriented hinterland regions, and this in turn implies that sensitivity to external change centered in the national heartland remains substantial in such cases. The significant change noted later in this chapter is that all of the nation's major urban regions achieved substantial closure in the short run by 1970, with the result that, whatever their economic specialization, their growth rates converged on the national growth rate.

2. *Regional growth is externally induced.* Basic to the growth of regions, according to the classical theory, is their capacity for attracting industries that produce goods for export to other regions of the country. Growth impulses in regional economies come from outside, according to this view, in the form of demands for regional specialties. The nature of these special-ties, alternative sources of them, and changes in the structure of demand therefore determine in large measure the nature and extent of regional growth. Expanding national demands spark heartland growth, which in turn directs differential hinterland growth through growing demands for hinter-land specialties. What more recent evidence suggests, however, is that the growth of major metropolitan regions is based upon the exercise of innovative leadership as new specialties are invented, producing rapid growth based upon a favorable mix of new industries. Further, there is accumulating evidence that the heartland-hinterland "lever" no longer works by means of heartland demands for regional resource-related specialties, as traditionally defined, but is rapidly being replaced by the filtering of older slower-growth industries from the metropolitan regions into the hinterlands as the principal source of hinterland growth.

For this reason, not all of the regions of a country can expect their economic activities and populations to increase at the same rate of speed. Yet every region can hope to enjoy a high and rising per capita income as long as the nation's output and productivity continue to increase, filtering works smoothly, and its people are willing to face up to the need for a degree of out-migration when the overall situation calls for it. In most such in-stances, significantly higher income levels within a lagging hinterland region can be achieved only by combining effective economic development with substantial out-migration.

3. *Export sector growth translates into residentiary sector growth.* Export industries need secondary support in the form of housing, public facilities, retail establishments, service facilities, and many other elements of local infrastructure. In the classical theory, export income supports these local facilities in a basic-nonbasic relationship that involves a simple multiplier linkage between export-industry fortunes and total regional growth.

The essence of the postwar transformation of regional economic relationships is, however, the changing role of the service industries in the traditional tertiary and modern quarternary sectors. Whereas the export-base formulation relegates these industries to a "nonbasic" role, with the magnitude of their growth determined strictly by the multiplier relationship to the export base, more recent evidence conclusively demonstrates that the tertiary and quaternary sectors have become the new propulsive sources of internally generated growth, as we shall see later.

4. *Economic growth takes place in a matrix of urban regions through which the space economy is organized.* The crux of the link between regional growth and modern growth center concepts is that it is cities within the urban system, linked by filtering mechanisms—not the heartland-hinterland lever in the regional system, linked by export-base multipliers—that today organize the economy spatially. The cities are centers of activity and of innovation, focal points of the transport and communications networks, locations of superior accessibility at which firms can most easily reap scale economies and at which industrial complexes can obtain the economies of localization and urbanization. They encourage labor specialization, areal specialization in productive activities, and efficiency in the provision of services. Agricultural enterprise is more efficient in the vicinity of cities. The more prosperous commercialized agricultures encircle the major cities, whereas the peripheries of the great urban regions are characterized by backward lower-income economic systems.

There are two major elements in this city-centered organization of economic activities in space (Berry, 1967b):

(a) A system of cities, arranged in a hierarchy according to the functions performed by each.
(b) Corresponding areas of urban influence, or urban fields surrounding each of the cities in the system.

Impulses of economic change have been shown to be transmitted in such a system simultaneously along three planes:

(a) Outward from heartland metropolises to those of the regional hinterlands.
(b) From higher-level to lower-level centers in the hierarchy, in a pattern of "hierarchical diffusion."
(c) Outward from urban centers into their surrounding urban fields, in radiating "spread effects."

Part of the diffusion mechanism has been shown to reside in the operation of urban labor markets. When growth is sustained over long periods, regional income inequality should be reduced, for any general expansion in a high-income area will result in some industries being priced out of the high-income labor market. There will be a shift of those industries to lower-income regions, especially to smaller urban centers in more peripheral areas. The significance of this "filtering" or "trickle down" process lies not only in its direct effects but also in its indirect ones. If the boom originates in the high-income region, as is very likely, the multiplier effects will be larger in the initiating region, although the relative rise in income may be greater in the underdeveloped region. But the induced effects on real income and employment may be considerably greater in the low-income region if prices there are likely to rise less, and/or if the increase in output per worker is greater. Both contingencies are likely, because of decreasing cost stemming from external economies with the urbanization of the labor force. If the boom can be maintained, industries of higher labor productivity will shift units into lower-income areas, and the low-wage industries will be forced to move into even smaller towns and more isolated areas.

The net result is that the following properties should characterize such urban-regional hierarchical systems:

(a) The size and functions of a central city, the size of its urban field, and the spatial extent of developmental "spread effects" radiating outward from it are proportional.

(b) Because impulses of economic change are transmitted in order from higher to lower centers in the urban hierarchy, continued innovation in large cities remains critical for extension of growth over the complete economic system.

(c) The resulting spatial incidence of economic growth will be a function of distance from the central city. Troughs of economic backwardness will lie in the most inaccessible areas along the peripheries between the least accessible lower-level centers in the hierarchy.

(d) Further, the growth potential of an area situated along an axis between two cities is a function of the intensity of interaction between them, which is in turn a function of their relative location and the quality of transportation arteries connecting them.

It is easy to see how a logical progression of urban growth flows from such events, leading to the following conclusion:

5. *When economic growth is sustained over long periods, it results in progressive integration of the space economy.* If metropolitan development is sustained at high levels, rural-urban differences will be eliminated, and the space economy should be integrated by outward flows of growth impulses through the urban hierarchy and inward migration of labor to cities in a reverse stepwise or ratchet fashion. Troughs of economic backwardness at

the intermetropolitan periphery are thereby eroded, and each area should then find itself within the spheres of influence of a variety of urban centers of a variety of sizes. Concentric bands of agricultural organization and efficiency around metropolitan centers should be eliminated or reduced in importance, and agricultures should also introduce new specialties, taking full advantage of differences in local resource endowments.

It is this full integration of the national space economy that constitutes the objective of regional development strategy. Essentially, what modern growth theory suggests is that continued urban-industrial expansion in major metropolitan regions should lead to catalytic impacts on surrounding areas. Growth impulses and economic advancement should filter and spread to smaller places and ultimately infuse dynamism into even the most tradition-bound peripheries. Growth-center concepts enter the scene if filtering mechanisms are perceived not to be operating quickly enough, if "cumulative causation" leads to growing regional differentials rather than their reduction (as we shall see), or if institutional or historical barriers block diffusion processes. The purpose of spatially selective public investments in growth centers, it is held, is to hasten the focused extension of growth to lower echelons of the hierarchy in outlying regions, and to link the growth centers more closely into the national system by way of higher-echelon centers in the urban hierarchy.

The changing nature of regional growth: the United States in the 1970s

There is an increasing realization that significant changes have occurred in the nature of regional development processes in the United States. These changes indicate the pressing need for new and imaginative approaches to regional policy. We have already noted the rapid rise of the service and quaternary sectors of the economy and the increasing significance of both governmental expenditures and amenity resources in the new growth nexus. Under these conditions, it is now clear that large urban regions (whether located in the heartland or the hinterland) can free themselves of narrow export dependency, as traditionally defined.

The long-run viability of any large metropolitan area, in these new conditions, resides ultimately on its capacity to invent and innovate, i.e., to create new rapid-growth, high-wage activities. What are such activities? Wilbur Thompson (1968) argues that

> the economic base of the larger metropolis is... the creativity of its universities and research parks, the sophistication of its engineering firms and financial institutions, the persuasiveness of its public relations and advertising agencies, the flexibility of its transportation networks and utility systems, and all the dimensions of "infrastructure" that facilitate the quick and orderly transfer from old dying bases to new growing ones.

The link between self-sustaining innovation in metropolitan regions and growth in regions not able to attain the threshold conditions for self-sustaining growth is, in such circumstances, clear:

> Larger urban regions combine a favorable industry mix for growth with a steadily declining share of the various growth industries.... High wage rates of the innovating area, quite consonant with the high skills needed at the beginning of the learning process, become excessive when skill requirements decline, and the industry (or parts of it) "filters down" to the smaller, less industrially sophisticated regions where the cheaper labor can meet the declining skill demands of the filtering industry, thus creating the phenomenon of small towns with low-wage, slow-growth filtered-down industry [at the time when the metropolis has moved on to new bases]. (Thompson, 1968)

The heartland-hinterland lever for growth is thus replaced by the mechanisms of filtering on a regional scale. This possibly can produce convergence and integration of the space economy; however, far more likely is a circular and cumulative process that leads to the innovating area's spiraling away in terms of aggregate growth and incomes, leaving a continuing lag in the peripheral region that only new growth centers can begin to bridge.

At sufficient scale, local markets, infrastructure, and residentiary development thus make the large metropolitan region self-generative of growth. When threshold conditions are reached, however, the metropolis grows at about the rate of the nation. This threshold was argued by Thompson to be a metropolitan area (SMSA) of at least 250,000 population, although we shall show later that a regional population of at least 1,000,000 is probably more accurate. Beneath this scale, growth in the peripheries is related to capturing a share of downward-filtering industry, slow in growth and yielding lower returns. The basic regional distinction is therefore the distinction between self-generative metropolitan America, and the hand-me-down intermetropolitan periphery that is condemned to progress characterized at best by lagged emulation and secondhand growth.

A unit for analysis: the daily urban system

The validity of this alternative view of regional growth processes is to be seen quite clearly in the evidence of the 1970s.

At the outset, however, it is necessary to have a precise functional definition of the nature of the "urban regions" themselves, so that there is clarity in the discussion relative to the units of analysis about which generalizations are being made. Only then will we be able to focus clearly on the problems of inducing growth in those areas not participating fully in the strengths and benefits of a dynamic market economy.

Let us pose the definitional problem in the following way. Suppose we had information available about the likely growth of economic activity in certain urban centers during the 1980s. How might we go about forecasting the growth in population that would result? Or suppose we have earnings figures for workers at their place of employment, and want to study expenditure patterns at their places of residence. How might we go about linking earnings and expenditures in a consistent accounting framework?

The problem that arises is the same in both cases. People no longer live next to their workplace; indeed, the separation of residence and work continues to increase year by year, as both residential neighborhoods and new industrial estates decentralize further afield from historic concentrations in the central cities. Maps showing the commuting "fields" of the nation's major central cities in 1960 (Berry, 1968) reveal the extensive areas of the country in daily commuting contact with legally defined central cities of various size categories in that year (Figure 14.3).

The commuting fields have surely become more extensive since 1960. For population projections, this means that the residential areas (housing markets) related to particular employment clusters (job markets) cover zones extending far beyond both legal city limits and the boundaries of, for example, the Bureau of Budget's Standard Metropolitan Statistical Areas, which are explicitly constructed to exclude all counties for which the number of resident workers commuting to work in the central city drops beneath 15 per cent. Population projections based upon assumptions about economic activity thus need a different unit of accounting if changes in employment are to be related properly to changes in the population supported by the jobs. Such an accounting unit is the Daily Urban System (DUS).

The term *Daily Urban System* was coined by C. A. Doxiadis in 1967, when he suggested in testimony before a committee of the U.S. House of Representatives the delineation of functional boundaries for the nation's emerging urban areas through use of daily commuting fields. He argued that sixty DUSs were forming in the United States, each with an average radius of ninety miles "within which people will move the way they now move within well-organized metropolitan areas."

The DUSs used in what follows differ from those proposed by Doxiadis in that they are based upon the actual evidence about commuting patterns around existing economic centers.

The delineation procedure was as follows, as developed by the Office of Business Economics (OBE), U.S. Department of Commerce (now the Bureau of Economic Analysis, Economic and Social Statistics Administration), on the basis of our earlier studies of commuting (Berry, 1968). First, economic centers were identified; Standard Metropolitan Statistical Areas were chosen where possible. Each SMSA has a large city at its center which serves both as a wholesale and retail trade center and as a labor market center. However, not all SMSAs were made centers of economic areas because some are integral parts of larger metropolitan complexes and evidence a high degree of cross-commuting with other metropolises. The

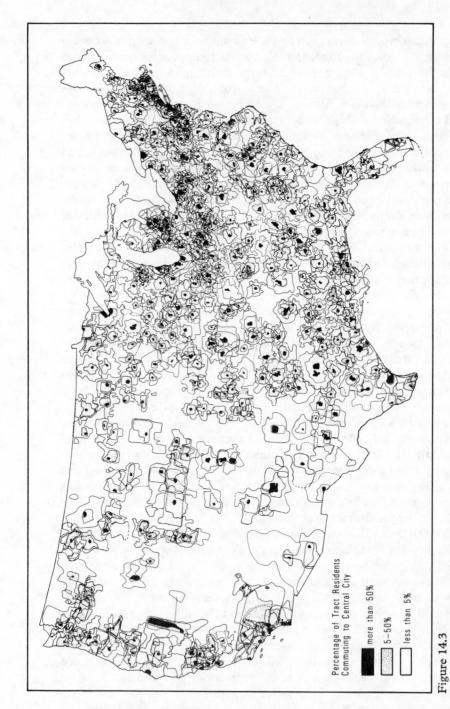

Percentage of Tract Residents
Commuting to Central City

■ more than 50%

▨ 5–50%

□ less than 5%

Figure 14.3
The national pattern of urban commuting fields in 1960.

284

New York City area, for instance, encompasses not only the New York City SMSA, but also Jersey City, Newark, Patterson-Clifton-Passaic, Stamford, Norwalk, and Bridgeport SMSAs. The Seattle economic area includes the Seattle-Everett and the Tacoma SMSAs. In rural parts of the country, where there were no SMSAs, cities of from 25,000 to 50,000 population were utilized as economic centers provided that two other criteria were met. These other criteria were: (1) that the city form a wholesale trade center for the area, and (2) that the area as a whole have a population minimum of about 200,000 people. After economic centers were identified, intervening counties were allocated to the centers. This assignment was made primarily on the basis of the journey-to-work pattern around the economic centers. Comparative time and distance of travel to the economic centers, the interconnection between outlying counties because of the journey-to-work pattern, the road network, and the linkages of counties by such other economic ties as telephone traffic, bank deposits, television viewing, newspaper circulation, and topography were also used to determine placement of peripheral counties into the appropriate economic area. In places where the commuting pattern of adjacent economic sectors overlapped, counties were included in the economic area containing the center with which there was the greatest commuting connection. In the case of cities where the commuting pattern overlapped to a great degree, no attempt was made to separate the two cities; instead, both were included in the same economic area.

A map of this preliminary division of the United States into OBE Economic Areas was distributed in September 1967 to members of state bureaus of business research, state planning agencies, and to field representatives of federal agencies involved in water-resources planning. At the same time, OBE made a series of population forecasts using the preliminary economic area definitions. Reviewing the maps, with reference to the criteria used, the states and federal agencies made several suggestions for revision based on additional data; the areas were revised accordingly. The revised set of 173 OBE Economic Areas is shown in Figure 14.4, dated January 1969.

It is this revised set of OBE areas that we call "Daily Urban Systems" in what follows. The areas have the following statistical and conceptual advantages for urban and regional analysis:

1. They completely disaggregate the United States into subregions, using county units as convenient building blocks.
2. They have a high degree of "closure" with respect to the job and housing markets, and in the tertiary sector of their economies. This means (a) that forecasts based on assumptions about the economic base (job market) can be translated into population forecasts (commuting area = housing market); (b) that income earned within each area can be equated with income spent, plus saving; (c) that a tertiary or "residentiary" economic sector can be identified whose growth, according to traditional economic

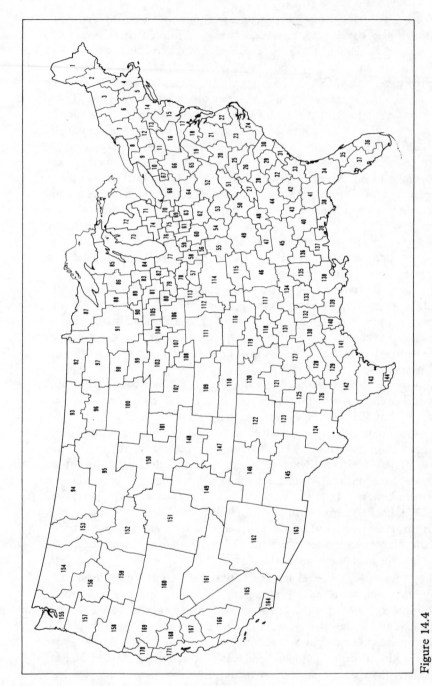

Figure 14.4
The OBE economic areas. Names corresponding to the numbers on the map will be found in Table 14.1.

theory, ought to be related to total sales made within the areas, in contrast to the primary and secondary sectors whose growth is in part related to sales made outside the economic area in larger markets, if traditional theory holds.

3. Because they extend beyond the existing commuting areas in many parts of the country, including counties whose orientation is defined by other forms of communication and marketing, the DUSs have the advantage of including the most likely prospective commuting areas within their limits, enabling the definition to remain stable over the time period of the population projections. The exceptions are to be found in those places where increasing cross-commuting violates the "closure" conditions of section 2 above and suggests that, in the future, adjacent DUSs should be merged as their job and housing markets coalesce.

Each DUS specializes, of course, in the production of certain goods and services in order to take advantage of lower costs of production resulting from the availability of natural resources, and from benefits to be derived through location and through internal and external economies. Each also approaches self-sufficiency in its residentiary industries; that is, most of the services and some of the goods required by the household sector and by local business as intermediate products are produced within the area. Thus the DUSs correspond to the closed trade areas of central place theory, in which the number and type of establishments and their size and trade areas are bounded by the relative transportation costs from hinterland to competing centers. In other words, the closure with respect to residentiary industries includes general and convenience retail and wholesale trade activities and those other services which are difficult or impossible to transport and are most efficiently consumed in the vicinity of their production. On the other hand, the economic areas are open for the most part to the movement of transportable commodities and nontransportable special services, such as education at Cambridge and recreation at Miami, as well as to the effects of federal investment. Hence, if tertiary or quaternary activity is, as we have argued, becoming a source of new growth, the regional definition in no way inhibits recognition of this fact.

Growth characteristics of the DUSs, 1960–1970

Table 14.1 presents the population of each DUS in 1960 and 1970, and shows the decadal percentage changes for that interval. There are wide variations in the growth rates. Why is this so?

First, there are clear relationships between growth and size. Figure 14.5 presents the evidence graphically, plotting the growth rate on the ordinate and the 1960 population on the abscissa (using a logarithmic scale in the latter case). Each dot represents one of the 173 DUSs. Horizontal lines show

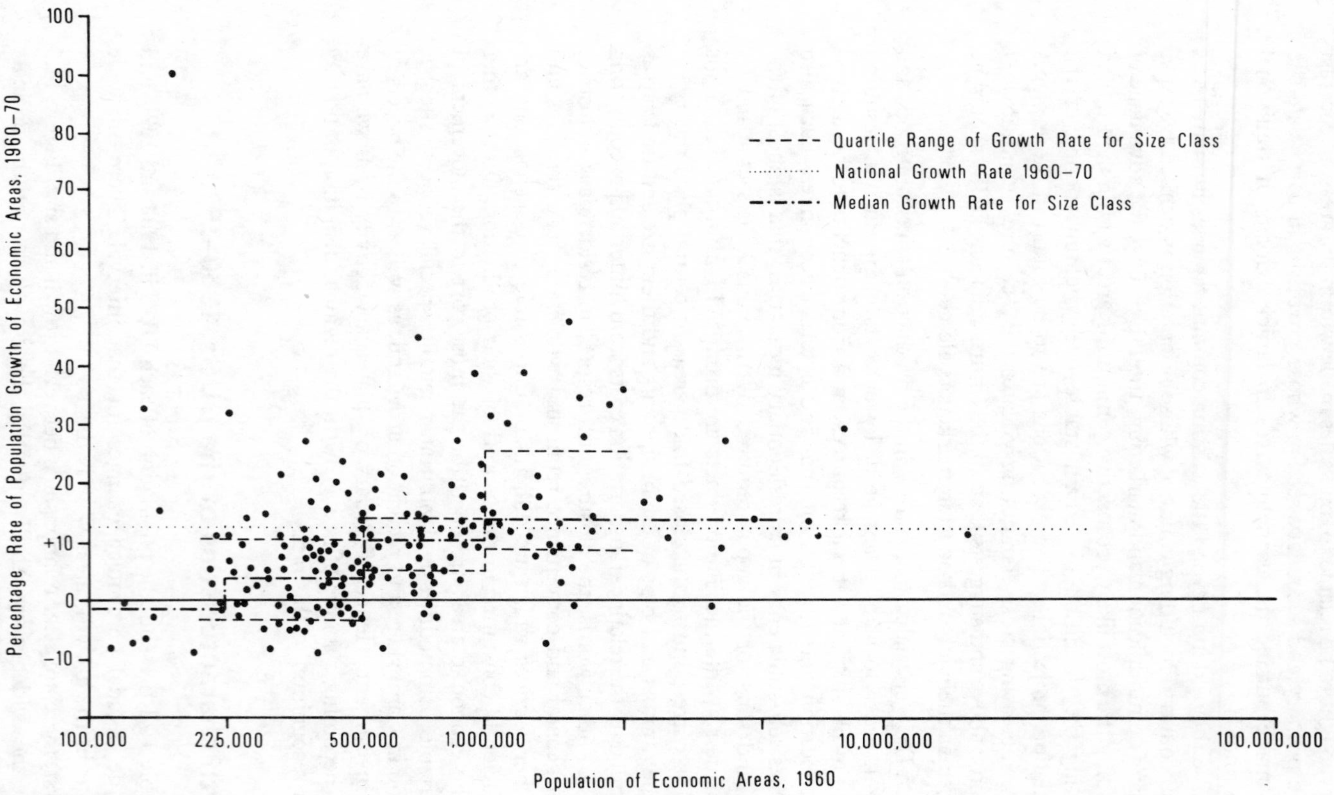

Figure 14.5
Growth rate of the DUSs 1960–1970 related to their population in 1960.

the national 1960–1970 growth rate, and zero-change. The observations have been grouped into roughly equal logarithmic intervals (100,000–225,000; 225,000–500,000; 500,000–1,000,000, etc.,) and for each class the graph also shows the median growth rate for the observations within the class, and the upper and lower quartiles for all except the extreme classes with the fewest observations. Several things are immediately apparent:

1. The median growth rate increases progressively with size of DUS to a population of 1,000,000, and stabilizes thereafter at about the national growth rate. This is consistent with the ideas on self-generative development above a threshold.
2. The interquartile range is stable for size classes of less than 1,000,000, and above that point also for the lower quartile. However, the upper quartile is markedly greater than elsewhere in the size class 1,000,000–2,250,000, indicating accelerated growth of many centers in this size range in particular, as they pass over the threshold of self-sustained growth, and, thereafter, converge again on the national average rate.
3. The median growth rate is negative in the smallest size class, as is the lower quartile in the size range 225,000–500,000, indicating that declining DUSs are disproportionately the smaller ones.

What accounts for some of these relationships? Do certain economic bases result in more ebullient growth, whereas others produce decline? Comparisons of data from Tables 14.1 and 14.2 provide information on shares of regional income derived from employment in different sectors; this information produces relevant insights.

1. The greater the dependence of a DUS on earnings derived from primary economic activities such as agriculture, the lower the growth rate. Indeed, when agricultural earnings exceed 25 per cent of local earnings, there is a declining population.
2. As earnings from secondary sources in the manufacturing industries increase, the population growth rate of the DUS stabilizes around the national average, consistent with Thompson's (1965, 1968) notion of the urban size–ratchet effect.
3. The greater the share of local earnings derived from the tertiary sector, involving residentiary activities, the greater the population growth rate. One reason is the interregional migration of the more elderly nonjob-oriented population, supporting a tertiary sector via transfer payments; another is the differential regional growth of R and D, education, and other "brain power" activities.
4. The greater the earnings derived from governmental sources, the greater the local growth rate.

The growth sectors of the decade were the tertiary (traditionally the non-basic or secondary sectors) and the governmental. Indeed, no DUS with more than 12.5 per cent of its earnings derived from federal sources lost

Table 14.1

Population of daily urban systems, 1960 and 1970, with percentage rate of change, 1960–1970

| | Population (in thousands) | | Percentage change in population |
Daily urban system	1960	1970	1960–1970
1. Bangor, Maine	337	321	−4.7
2. Portland, Maine	691	740	7.1
3. Burlington, Vt.	449	502	11.8
4. Boston, Mass.	5,668	6,338	11.8
5. Hartford, Conn.	2,542	2,966	16.7
6. Albany–Schenectady–Troy, N.Y.	1,192	1,331	11.7
7. Syracuse, N.Y.	1,342	1,444	7.6
8. Rochester, N.Y.	851	1,016	19.4
9. Buffalo, N.Y.	1,590	1,642	3.2
10. Erie, Pa.	444	459	3.6
11. Williamsport, Pa.	405	419	3.5
12. Binghamton, N.Y.–Pa.	695	765	10.1
13. Wilkes-Barre–Hazelton, Pa.	689	692	0.4
74. New York, N.Y.	16,406	18,228	11.1
15. Philadelphia, Pa.–N.J.	6,418	7,281	13.5
16. Harrisburg, Pa.	1,581	1,723	9.0
17. Baltimore, Md.	2,348	2,670	13.7
18. Washington, D.C.–Md.–Va.	2,273	3,090	35.9
19. Staunton, Va.	361	387	7.1
20. Roanoke, Va.	768	803	4.6
21. Richmond, Va.	898	1,004	11.7
22. Norfolk-Portsmouth, Va.	1,002	1,232	23.0
23. Raleigh, N.C.	1,479	1,621	9.6
24. Wilmington, N.C.	448	482	7.5
25. Greensboro–Winston-Salem–High Point, N.C.	1,005	1,142	13.6
26. Charlotte, N.C.	1,285	1,489	15.9
27. Asheville, N.C.	350	391	11.6
28. Greenville, S.C.	741	817	10.3
29. Columbia, S.C.	551	610	10.7
30. Florence, S.C.	406	400	−1.3
31. Charleston, S.C.	368	430	16.9
32. Augusta, Ga.	422	461	9.3
33. Savannah, Ga.	403	417	3.5
34. Jacksonville, Fla.	808	946	7.1
35. Orlando, Fla.	648	941	45.3
36. Miami, Fla.	1,644	2,430	47.9
37. Tampa–St. Petersburg, Fla.	1,299	1,797	38.4
38. Tallahassee, Fla.	310	344	10.9
39. Pensacola, Fla.	313	382	21.9
40. Montgomery, Ala.	669	686	2.6
41. Albany, Ga.	453	460	1.6
42. Macon, Ga.	469	496	5.7

Table 14.1 (*continued*)

Daily urban system	Population (*in thousands*)		Percentage change in population 1960–1970
	1960	*1970*	
43. Columbus, Ga.–Ala.	462	488	5.6
44. Atlanta, Ga.	1,793	2,296	28.0
45. Birmingham, Ala.	1,680	1,660	−1.2
46. Memphis, Tenn.–Ark.	1,613	1,700	5.4
47. Huntsville, Ala.	552	671	21.4
48. Chattanooga, Tenn.–Ga.	650	718	10.4
49. Nashville, Tenn.	1,280	1,426	11.4
50. Knoxville, Tenn.	876	904	3.3
51. Bristol, Va.–Tenn.	718	698	−2.7
52. Huntington–Ashland, W. Va.–Ky.–Ohio	1,422	1,309	−7.9
53. Lexington, Ky.	708	805	13.8
54. Louisville, Ky.–Ind.	1,070	1,204	12.5
55. Evansville, Ind.	747	771	3.1
56. Terre Haute, Ind.	250	252	0.9
57. Springfield, Ill.	471	490	4.0
58. Champaign–Urbana, Ill.	354	390	10.1
59. Lafayette–West Lafayette, Ind.	227	250	10.1
60. Indianapolis, Ind.	1,384	1,613	16.5
61. Muncie, Ind.	501	551	10.1
62. Cincinnati, Ohio–Ky.–Ind.	1,744	1,889	8.3
63. Dayton, Ohio	1,002	1,159	15.7
64. Columbus, Ohio	1,552	1,763	13.6
65. Clarksburg, W. Va.	333	326	−2.1
66. Pittsburgh, Pa.	3,749	3,716	−0.9
67. Youngstown–Warren, Ohio	749	770	2.8
68. Cleveland, Ohio	3,898	4,255	9.2
69. Lima, Ohio	259	276	6.5
70. Toledo, Ohio	967	1,054	8.9
71. Detroit, Mich.	4,582	5,207	13.7
72. Saginaw, Mich.	691	787	14.0
73. Grand Rapids, Mich.	990	1,166	17.8
74. Lansing, Mich.	889	986	10.9
75. Fort Wayne, Ind.	517	597	15.5
76. South Bend, Ind.	681	747	9.6
77. Chicago, Ill.	7,323	8,193	11.9
78. Peoria, Ill.	572	628	9.8
79. Davenport–Rock Island–Moline, Iowa–Ill.	552	605	9.5
80. Cedar Rapids, Iowa	288	330	14.5
81. Dubuque, Iowa	292	301	3.3
82. Rockford, Ill.	492	560	13.7
83. Madison, Wis.	377	455	20.8
84. Milwaukee, Wis.	1,848	2,066	11.8
85. Green Bay, Wis.	831	926	11.4

Table 14.1 (*continued*)

| | Population (*in thousands*) | | Percentage change in population |
Daily urban system	1960	1970	1960–1970
86. Wausau, Wis.	322	350	8.7
87. Duluth–Superior, Minn.–Wis.	449	429	−4.6
88. Eau Claire, Wis.	205	219	6.9
89. La Crosse, Wis.	257	269	4.8
90. Rochester, Minn.	230	245	6.1
91. Minneapolis–St. Paul, Minn.	2,528	2,935	16.1
92. Grand Forks, N.D.	223	220	−1.3
93. Minot, N.D.	189	170	−9.8
94. Great Falls, Mont.	226	222	−1.7
95. Billings, Mont.	245	246	0.7
96. Bismark, N.D.	149	144	−3.2
97. Fargo–Moorhead, N.D.–Minn.	342	335	−2.0
98. Aberdeen, S.D.	142	132	−7.0
99. Sioux Falls, S.D.	372	365	−2.0
100. Rapid City, S.D.	237	231	−2.5
101. Scotts Bluff, Nebr.	116	105	−9.1
102. Grand Island, Nebr.	322	309	−4.0
103. Sioux City, Iowa–Nebr.	467	454	−2.8
104. Fort Dodge, Iowa	280	266	−5.3
105. Waterloo, Iowa	427	426	−0.3
106. Des Moines, Iowa	759	741	−2.3
107. Omaha, Nebr.–Iowa	720	794	10.3
108. Lincoln, Nebr.	320	323	1.0
109. Salina, Kans.	379	342	−9.8
110. Witchita, Kans.	731	723	−1.1
111. Kansas City, Mo.–Kans.	1,538	1,674	8.9
112. Columbia, Mo.	367	396	7.9
113. Quincy, Ill.	301	299	−0.7
114. St. Louis, Mo.–Ill.	2,945	3,248	10.3
115. Paducah, Ky.	580	558	3.9
116. Springfield, Mo.	777	816	5.1
117. Little Rock–No. Little Rock, Ark.	771	864	12.0
118. Fort Smith, Ark.–Okla.	252	289	14.4
119. Tulsa, Okla.	891	1,014	13.9
120. Oklahoma City, Okla.	1,040	1,156	11.2
121. Wichita Falls, Tex.	460	455	−1.1
122. Amarillo, Tex.	451	437	−3.0
123. Lubbock, Tex.	326	328	0.6
124. Odessa, Tex.	337	319	−5.4
125. Abilene, Tex.	290	264	−9.0
126. San Angelo, Tex.	126	124	−0.9
127. Dallas, Tex.	2,063	2,736	32.6
128. Waco, Tex.	374	403	7.8
129. Austin, Tex.	452	559	23.7
130. Tyler, Tex.	518	545	5.2

Table 14.1. (*continued*)

Daily urban system	Population (in thousands)		Percentage change in population 1960–1970
	1960	*1970*	
131. Texarkana, Tex.–Ark.	315	329	4.6
132. Shreveport, La.	445	453	1.7
133. Monroe, La.	515	532	3.8
134. Greenville, Miss.	556	506	−8.9
135. Jackson, Miss.	489	510	4.4
136. Meridian, Miss.	402	465	15.8
137. Mobile, Ala.	664	724	8.9
138. New Orleans, La.	1,884	2,148	14.0
139. Lake Charles, La.	655	748	14.1
140. Beaumont–Port Arthur– Orange, Tex.	373	394	5.6
141. Houston, Tex.	1,758	2,362	34.4
142. San Antonio, Tex.	1,065	1,229	15.4
143. Corpus Christi, Tex.	465	516	11.0
144. Brownsville–Harlingen–San Benito, Tex.	369	355	−3.8
145. El Paso, Tex.	646	681	5.4
146. Albuquerque, N.M.	500	572	14.4
147. Pueblo, Colo.	424	509	20.1
148. Denver, Colo.	1,169	1,523	30.3
149. Grand Junction, Colo.	239	251	5.1
150. Cheyenne, Wyo.	221	229	3.5
151. Salt Lake City, Utah	898	1,061	18.2
152. Idaho Falls, Idaho	286	300	4.9
153. Butte, Mont.	213	234	10.3
154. Spokane, Wash.	659	687	4.1
155. Seattle–Everett, Wash.	1,879	2,363	25.8
156. Yakima, Wash.	398	406	2.0
157. Portland, Ore.–Wash.	1,348	1,637	21.4
158. Eugene, Ore.	458	541	18.0
159. Boise City, Idaho	241	265	9.8
160. Reno, Nev.	142	191	33.9
161. Las Vegas, Nev.	166	317	91.1
162. Phoenix, Ariz.	945	1,316	39.3
163. Tucson, Ariz.	357	454	27.3
164. San Diego, Calif.	1,033	1,357	31.4
165. Los Angeles–Long Beach, Calif.	8,087	10,436	29.0
166. Fresno, Calif.	916	1,036	13.1
167. Stockton, Calif.	537	643	19.6
168. Sacramento, Calif.	854	1,089	27.6
169. Redding, Calif.	153	176	15.3
170. Eureka, Calif.	132	121	−7.9
171. San Francisco–Oakland, Calif.	4,001	5,090	27.2
172. Anchorage, Alaska	226	300	32.8
173. Honolulu, Hawaii	632	768	21.5

population in the decade, and the greater the federal contribution to local earnings, the greater the decadal growth rate. This finding is of particular importance, for it emphasizes that among the unplanned consequences of the concentration of governmental expenditures in particular places are systematic impacts on the growth rates of those places. Thus, when one views the issue of inducing growth in lagging areas, he must consider what alternative pattern might have been produced by some other pattern of regional expenditures of the federal dollar.

Putting the data together, we find that the joint relationships between DUS size and economic base are as reported in Table 14.3. In general, both increased diversification and increasing size produce greater stabilization of the growth rate at the national average. Small size combined with a regional economy specialized in the primary sectors usually means a declining population. The accelerated growth of the upper-quartile DUSs in the million-plus size range is associated with massive reliance upon governmental expenditures. And because these expenditures produce higher-than-average incomes, the decade was characterized by (1) a positive association between per capita incomes and population growth rates, and (2) a stability relationship between growth rates of income and of population. The dynamic in these latter two relationships involves positive relations between population growth and growth rates of earnings in retail trade, service activities, and from federal civilian employment, on the one hand, alongside similar positive relationships with the growth rate of manufacturing earnings, on the other hand.

Such, then, are the bases of the broad regional differences in population growth rates depicted in Figure 14.6. But this map also reveals interesting intraregional growth differences—for example, the negative growth rates of the traditional core cities that we documented earlier, as well as accelerated new growth directions penetrating the former interurban peripheries. The exurban fringes of many of the nation's daily urban systems have now pushed more than one hundred miles from the central cities. Many previously sparsely settled areas are quickly becoming more populated as people are moving in increasing numbers to rural towns and villages and other remote environments. In fact, Bureau of the Census data on growth and migration patterns between 1970 and 1973 indicate that nonmetropolitan counties for the first time in recent history grew at a faster *rate* than metropolitan counties (see Beale, 1975; Beale and Fuguitt, 1975).

Concentrative migration from industrial urbanization has clearly ended. Migration now takes place largely between metropolitan areas on an interregional scale, and intraregionally through an accelerating dispersion of people and jobs outward into widening peripheries. As a result there has been a dramatic shift of new growth—residential, industrial, commercial—to the expanding suburban, exurban, and even rural segments of daily urban systems on an intraregional scale, as well as to the nation's "rimland" on an interregional scale.

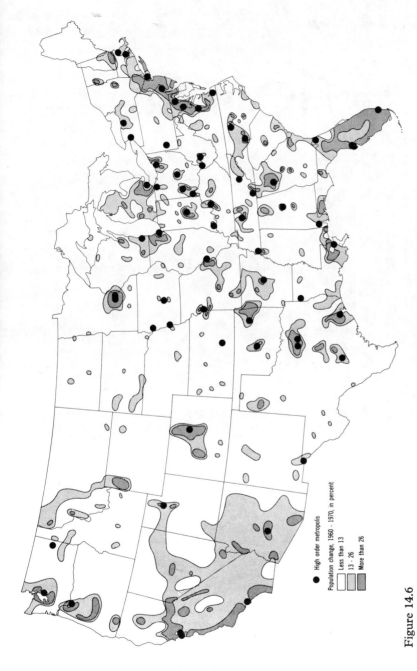

Figure 14.6
Major metropolitan centers and regions of above-national population growth, 1960–1970.

● High order metropolis

Population change, 1960 - 1970, in percent

Less than 13
13 - 26
More than 26

Table 14.2
Basic and residentiary earnings of the daily urban systems in 1967, expressed as a percentage of total earnings

Economic area	Basic						Residentiary
	Agriculture	Mining	Manufacturing	Federal government	Other	Total	
1. Bangor, Maine	7.49	0.09	25.52	10.36	1.26	44.73	55.27
2. Portland, Maine	2.76	0.06	33.58	2.32		38.72	61.28
3. Burlington, Vt.	5.09	0.69	29.04	0.99	3.82	39.64	60.36
4. Boston, Mass.	0.72	0.06	29.32	2.39	4.34	36.83	63.17
5. Hartford, Conn.	1.20	0.10	39.92	1.71	0.93	43.86	56.14
6. Albany–Schenectady–Troy, N.Y.	1.96	0.45	27.60	8.01	1.85	39.87	60.13
7. Syracuse, N.Y.	3.38	0.52	29.47	5.68	1.58	40.63	59.37
8. Rochester, N.Y.	2.88	0.23	44.14	0.30		47.56	52.44
9. Buffalo, N.Y.	2.10	0.35	37.36	1.76	0.45	42.02	57.98
10. Erie, Pa.	2.87	0.33	45.03	0.43	0.54	49.21	50.79
11. Williamsport, Pa.	2.16	2.08	38.16	5.03	1.58	49.01	50.99
12. Binghampton, N.Y.–Pa.	4.32	0.12	39.77	3.17	0.51	47.90	52.10
13. Wilkes-Barre–Hazelton, Pa.	1.60	1.91	33.77	0.53	1.44	39.25	60.75
14. New York, N.Y.	0.38	0.12	24.65	0.66	11.38	37.20	62.80
15. Philadelphia, Pa.–N.J.	1.12	0.26	32.61	2.20		36.19	63.81
16. Harrisburg, Pa.	3.97	0.73	33.81	1.20	1.81	41.52	58.48
17. Baltimore, Md.	2.49	0.05	24.04	9.64		36.22	63.78
18. Washington, D.C.–Md.–Va.	0.77	0.14	2.05	32.08	6.13	41.17	58.83
19. Staunton, Va.	5.74	1.74	30.38	3.20	2.00	43.06	56.94
20. Roanoke, Va.	3.88	0.36	37.48	1.03	2.64	45.39	54.61
21. Richmond, Va.	3.33	0.20	21.23	7.93	2.94	35.64	64.36
22. Norfolk–Portsmouth, Va.	2.71	0.02	12.84	35.72		51.29	48.71
23. Raleigh, N.C.	12.99	0.31	18.07	13.72		45.09	54.91
24. Wilmington, N.C.	10.45	0.05	13.56	29.44		53.50	46.50

25. Greensboro–Winston-Salem–High Point, N.C.	3.49	0.17	42.12	0.75	1.94	48.47	51.53
26. Charlotte, N.C.	2.06	0.17	40.56	0.63	1.75	45.17	54.83
27. Asheville, N.C.	4.37	0.40	36.85	0.81	1.05	43.47	56.53
28. Greenville, S.C.	1.45	0.14	46.84	0.83		49.26	50.74
29. Columbia, S.C.	5.63	0.24	18.36	16.29		40.52	59.48
30. Florence, S.C.	14.63	0.22	29.85	3.38	0.48	48.57	51.43
31. Charleston, S.C.	3.69	0.03	12.98	34.16	0.30	51.16	48.84
32. Augusta, Ga.	6.20	0.41	26.54	22.77		55.93	44.07
33. Savannah, Ga.	9.99	0.06	21.80	6.32	3.39	41.56	58.44
34. Jacksonville, Fla.	3.72	0.55	12.49	13.46	6.58	36.80	63.20
35. Orlando, Fla.	5.29	0.06	19.03	4.24	8.80	37.42	62.58
36. Miami, Fla.	4.49	0.24	10.86	2.43	17.98	36.00	64.00
37. Tampa–St. Petersburg, Fla.	4.69	1.28	13.60	2.38	11.91	33.86	66.14
38. Tallahassee, Fla.	6.17	0.11	11.56	27.26		45.11	54.89
39. Pensacola, Fla.	1.52	0.15	15.72	33.17		50.55	49.45
40. Montgomery, Ala.	6.96	0.19	19.00	16.27	0.04	42.47	57.53
41. Albany, Ga.	20.87	0.35	17.88	6.57		45.67	54.33
42. Macon, Ga.	11.90	1.98	18.20	16.63		48.71	51.29
43. Columbus, Ga.–Ala.	3.17	0.12	22.45	27.20		52.93	47.07
44. Atlanta, Ga.	1.55	0.29	24.05	1.89	8.42	36.20	63.80
45. Birmingham, Ala.	4.27	1.51	30.29	3.34	1.12	40.52	59.48
46. Memphis, Tenn.–Ark.	13.86	0.20	19.55	2.97	3.04	39.62	60.38
47. Huntsville, Ala.	6.64	0.31	25.56	11.40	4.99	48.89	51.11
48. Chattanooga, Tenn.	3.37	0.64	41.95	0.67		46.62	53.38
49. Nashville, Tenn.	5.88	0.38	23.53	6.10		35.90	64.10
50. Knoxville, Tenn.	1.95	2.80	31.04	4.29	0.43	40.50	59.50
51. Bristol, Va.–Tenn.	3.00	10.15	28.51	0.96	1.75	44.37	55.63
52. Huntington–Ashland, W.Va.–Ky.	1.02	11.77	24.01	3.14	3.61	43.55	56.45
53. Lexington, Ky.	11.84	3.19	18.59	4.37	0.48	38.47	61.53
54. Louisville, Ky.–Ind.	3.25	0.26	28.44	8.46	1.17	41.58	58.42

Table 14.2 (*continued*)

Economic area	Basic					Total	Residentiary
	Agriculture	Mining	Manufacturing	Federal government	Other		
55. Evansville, Ind.	8.08	6.24	26.38	0.53	0.93	42.17	57.83
56. Terre Haute, Ind.	8.75	3.32	19.90	4.31	4.99	41.27	58.73
57. Springfield, Ill.	10.48	0.93	19.70	5.46	4.02	40.59	59.41
58. Champaign–Urbana, Ill.	11.25	0.40	16.79	16.05	0.48	44.96	55.04
59. Lafayette–West Lafayette, Ind.	13.80	0.25	26.46	8.65	0.94	50.09	49.91
60. Indianapolis, Ind.	3.17	0.32	33.02	1.57	1.17	39.26	60.74
61. Munice, Ind.	3.67	0.17	50.71	0.35		54.90	45.10
62. Cincinnati, Ohio–Ky.–Ind.	1.83	0.16	34.50	0.54	0.97	38.00	62.00
63. Dayton, Ohio	2.25	0.19	40.71	5.94		49.10	50.90
64. Columbus, Ohio	1.69	0.80	28.91	3.94	0.50	35.84	64.16
65. Clarksburg, W.Va.	1.22	14.30	19.66	4.45	4.91	45.54	55.46
66. Pittsburgh, Pa.	0.79	2.87	36.53	0.45	2.87	43.50	56.50
67. Youngstown–Warren, Ohio	0.96	0.42	50.46	0.41	0.60	52.85	47.15
68. Cleveland, Ohio	1.12	0.39	41.70	0.37	0.67	44.24	55.76
69. Lima, Ohio	8.43	0.21	38.69	0.36	0.50	48.20	51.80
70. Toledo, Ohio	4.22	1.00	36.14	0.40	1.04	42.80	57.20
71. Detroit, Mich.	0.48	0.09	42.94	0.39	0.17	44.07	55.93
72. Saginaw, Mich.	4.42	1.04	37.66	4.78	0.31	48.21	51.79
73. Grand Rapids, Mich.	3.33	0.17	38.85	0.42	0.44	43.21	56.79
74. Lansing, Mich.	2.95	0.15	38.15	4.95	0.42	46.62	53.38
75. Fort Wayne, Ind.	5.26	0.23	40.52	0.35	0.73	47.08	52.92
76. South Bend, Ind.	3.86	0.09	45.11	0.36		49.42	50.58
77. Chicago, Ill.	1.09	0.27	32.12	0.99	4.04	38.52	61.48
78. Peoria, Ill.	9.52	1.22	30.31	0.32	0.63	42.00	58.00
79. Davenport–Rock Island–Moline, Iowa–Ill.	8.74	0.20	34.12	0.47		43.52	56.48

80. Cedar Rapids, Iowa	11.70	0.43	29.68	0.36		42.17	57.83
81. Dubuque, Iowa	22.86	0.91	17.75	0.50		42.02	57.98
82. Rockford, Ill.	7.06	0.24	43.84	0.27		51.42	48.58
83. Madison, Wis.	9.05	0.26	15.85	12.43	1.70	39.30	60.70
84. Milwaukee, Wis.	2.28	0.15	39.58	0.47	0.23	42.71	57.29
85. Green Bay, Wis.	5.76	2.94	32.04	3.68		44.43	55.57
86. Wausau, Wis.	9.79	0.23	28.37	3.25	0.37	42.01	57.99
87. Duluth–Superior, Minn.–Wis.	0.42	13.28	15.54	6.75	8.31	44.30	55.70
88. Eau Claire, Wis.	10.15	0.26	23.66	5.74	2.64	42.46	57.54
89. La Crosse, Wis.	12.69	0.60	19.95	5.31	2.72	41.27	58.73
90. Rochester, Minn.	12.64	0.26	25.50	0.38	7.53	46.31	53.69
91. Minneapolis–St. Paul, Minn.	5.41	0.16	23.46	0.63	5.42	35.08	64.92
92. Grand Forks, N.D.	30.49	0.18	3.56	10.62	1.40	46.26	53.74
93. Minot, N.D.	27.04	2.89	0.91	11.25	1.61	43.71	56.29
94. Great Falls, Mont.	21.90	0.87	3.11	12.58	2.64	41.09	58.91
95. Billings, Mont.	15.80	3.42	4.77	4.94	6.04	34.97	65.03
96. Bismarck, N.D.	18.62	1.63	1.42	7.24	8.50	37.42	62.58
97. Fargo–Moorhead, N.D.–Minn.	19.12	0.16	2.33	5.69	11.71	39.00	61.00
98. Aberdeen, S.D.	33.79	0.18	2.96	3.53	2.08	42.54	57.46
99. Sioux Falls, S.D.	26.86	0.25	8.21	2.13	2.59	40.03	59.97
100. Rapid City, S.D.	28.41	3.69	2.34	13.13	0.39	47.96	52.04
101. Scotts Bluff, Nebr.	30.67	1.88	2.74	2.43	4.75	42.47	57.53
102. Grand Island, Nebr.	31.46	0.25	9.61	0.54	2.26	44.12	55.88
103. Sioux City, Iowa—Nebr.	26.21	0.14	11.52	1.36	1.81	41.03	58.97
104. Fort Dodge, Iowa	34.43	0.24	9.14	0.41	1.27	45.50	54.50
105. Waterloo, Iowa	17.80	0.51	27.35	0.60		46.26	53.74
106. Des Moines, Iowa	11.65	0.38	16.03	1.66	5.28	35.00	65.00
107. Omaha, Nebr.–Iowa	9.58	0.40	14.84	4.64	8.01	37.46	62.54
108. Lincoln, Nebr.	16.57	0.10	10.06	5.50	3.28	35.51	64.49
109. Salina, Kans.	26.83	4.05	5.58	3.45	1.96	41.87	58.13
110. Witchita, Kans.	10.23	2.17	25.89	2.56	0.86	41.72	58.28
111. Kansas City, Mo.–Kans.	5.90	0.21	19.30	3.86	5.83	35.10	64.90

Table 14.2 (continued).

| | Basic | | | | | | |
Economic area	Agriculture	Mining	Manufacturing	Federal government	Other	Total	Residentiary
112. Columbia, Mo.	17.68	0.42	10.23	11.27	1.67	41.27	58.73
113. Quincy, Ill.	15.41	0.37	29.75	0.42	0.61	46.56	53.44
114. St. Louis, Mo.–Ill.	1.86	1.57	28.86	2.90	1.23	36.41	63.59
115. Paducah, Ky.	20.08	1.08	17.38	3.57	1.21	43.31	56.69
116. Springfield, Mo.	7.02	1.14	21.53	3.87	5.16	38.71	61.29
117. Little Rock–No. Little Rock, Ark.	8.53	0.88	17.56	3.92	5.35	36.24	63.76
118. Fort Smith, Ark.–Okla.	3.38	1.60	23.57	4.52	0.70	33.77	66.23
119. Tulsa, Okla.	2.62	10.42	20.48	0.80	2.39	36.72	63.28
120. Oklahoma City, Okla.	7.62	4.89	9.00	12.07	2.01	35.59	64.41
121. Wichita Falls, Tex.	11.11	4.36	5.76	30.52		51.75	48.25
122. Amarillo, Tex.	19.91	5.01	7.84	8.44	2.22	43.42	56.58
123. Lubbock, Tex.	33.91	2.20	3.75	2.57	0.23	42.66	57.34
124. Odessa, Tex.	7.74	24.28	5.70	2.94	1.43	42.09	57.91
125. Abilene, Tex.	12.20	6.23	7.04	9.74	4.24	39.45	60.55
126. San Angelo, Tex.	16.21	3.20	4.90	12.24	1.62	38.15	61.85
127. Dallas, Tex.	1.22	1.81	25.35	1.80	7.21	37.40	62.60
128. Waco, Tex.	5.06	0.31	10.99	33.96		50.32	49.68
129. Austin, Tex.	5.56	0.93	6.88	22.35	0.95	36.67	63.33
130. Tyler, Tex.	1.52	5.18	25.39	5.49	1.97	39.55	60.45
131. Texarkana, Tex.–Ark.	3.92	1.28	27.72	8.61	1.27	42.80	57.20
132. Shreveport, La.	3.63	5.52	18.73	6.21	2.98	37.06	62.94
133. Monroe, La.	15.80	1.93	13.61	8.85	1.12	41.32	58.68
134. Greenville, Miss.	24.39	1.79	23.42	0.78		50.38	49.62
135. Jackson, Miss.	8.97	1.54	17.14	1.29	4.70	33.64	66.36
136. Meridian, Miss.	6.32	2.31	24.76	5.52	1.32	40.24	59.76

137. Mobile, Ala.	2.34	0.22	23.48	14.95	1.13	42.13	57.87
138. New Orleans, La.	2.29	5.72	16.88	0.70	9.99	35.58	64.42
139. Lake Charles, La.	8.97	11.13	9.91	12.71	3.57	46.29	53.71
140. Beaumont–Port Arthur–Orange, Tex.	0.88	2.58	36.18	0.78	8.81	49.23	50.77
141. Houston, Tex.	1.65	5.87	19.66	0.70	11.59	39.47	60.53
142. San Antonio, Tex.	3.98	1.47	7.38	24.56	0.51	37.90	62.10
143. Corpus Christi, Tex.	8.31	8.32	7.78	13.28	0.72	38.41	61.59
144. Brownsville–Harlingen–San Benito, Tex.	21.06	2.09	5.01	7.85	4.17	40.18	59.82
145. El Paso, Tex.	7.68	6.39	7.63	19.97	2.55	44.23	55.77
146. Albuquerque, N.M.	1.56	2.49	5.24	16.92	3.83	30.05	69.95
147. Pueblo, Col.	5.55	0.79	10.44	25.65	1.71	44.14	55.86
148. Denver, Col.	4.00	1.98	13.83	5.80	7.99	33.60	66.40
149. Grand Junction, Col.	7.87	12.21	4.53	6.62	8.09	39.32	60.68
150. Cheyenne, Wyo.	8.10	9.76	5.91	10.82	7.15	41.74	58.26
151. Salt Lake City, Utah	3.74	3.93	12.75	13.74	2.53	36.68	63.32
152. Idaho Falls, Idaho	21.59	0.97	9.26	1.11	6.65	39.58	60.42
153. Butte, Mont.	4.49	6.84	17.90	4.73	6.85	40.81	59.19
154. Spokane, Wash.	14.04	1.62	12.60	6.88	1.30	36.43	63.57
155. Seattle–Everett, Wash.	1.25	0.15	27.23	6.12	1.37	36.13	62.87
156. Yakima, Wash.	17.57	0.14	13.52	4.78	1.15	37.17	63.83
157. Portland, Ore.–Wash.	3.44	0.18	22.48	2.44	3.97	32.50	67.50
158. Eugene, Ore.	2.52	0.75	31.19	3.80	1.42	39.68	60.32
159. Boise City, Idaho	14.52	0.14	11.04	5.34	5.40	36.45	63.55
160. Reno, Nev.	2.89	3.88	2.50	8.94	23.44	41.65	58.35
161. Las Vegas, Nev.	1.32	1.26	3.39	4.74	35.64	46.35	53.65
162. Phoenix, Ariz.	6.77	2.18	17.63	5.99	2.39	34.97	65.03
163. Tucson, Ariz.	1.20	7.61	7.43	18.69	3.30	38.23	61.77
164. San Diego, Calif.	1.56	0.12	14.97	26.59	0.96	44.21	55.79
165. Los Angeles–Long Beach, Calif.	1.54	0.58	27.91	1.99	4.34	36.37	63.63

Table 14.2 (continued).

Economic area	Basic					Total	Residentiary
	Agriculture	Mining	Manufacturing	Federal government	Other		
166. Fresno, Calif.	20.76	3.61	7.29	7.91	0.68	39.57	60.43
167. Stockton, Calif.	13.22	0.27	14.55	11.01	3.01	39.72	60.28
168. Sacramento, Calif.	4.57	0.17	8.23	22.36	3.01	38.33	61.67
169. Redding, Calif.	1.38	0.37	21.83	14.46	5.54	43.58	56.42
170. Eureka, Calif.	4.43	0.19	28.81	9.95	0.96	44.34	55.66
171. San Francisco—Oakland Calif.	1.96	0.23	17.19	7.97	5.42	32.77	67.23
172. Anchorage, Alaska	2.47	3.13	3.73	32.44	8.10	49.86	50.14
173. Honolulu, Hawaii	5.12	0.01	4.55	22.86	6.09	38.63	61.37

302

Table 14.3

Median growth rates of various types of economic areas, 1960–1970[a]

Population size class of economic area in 1970 (in thousands)	Median growth rate if focus of area is a center of metropolitan status (per cent)				Median growth rate if focus of area is a center of less than metropolitan status (per cent)						
					Source of earnings in 1967						
	>12.5% Federal government	>60% Residentiary	Federal and residentiary	Diversified sources	>12.5% Federal government	>60% Residentiary	Federal and residentiary	Diversified sources	>25% agriculture	Federal and agriculture	>12.5% mining
100–225	b	b	b	b	15.3	−2.05	c	8.6	−8.05	c	c
225–500	c	1.7	23.7	−3.0	9.7	4.9	c	4.0	−4.0	−2.5	−4.6
500–1,000	7.2	11.0	18.2	15.5	10.7	12.0	c	9.6	c	c	c
1,000–2,250	27.2	18.9	15.4	11.8	9.6	c	11.7	9.0	b	b	c
2,250–5,000	35.9	14.9	c	11.5	b	b	b	b	b	b	b
5,000–10,000	c	12.7	c	c	b	b	b	b	b	b	b
10,000 and over	c	11.1	c	c	b	b	b	b	b	b	b

[a] National Growth Rate 1960–1970 was 13.3 per cent.
[b] No centers of this level.
[c] No centers of this type.

303

To reiterate a point we developed in the previous chapter: a multinode, multiconnective system has replaced the core dominated metropolis as the basic urban unit. Many suburban and exurban areas now provide all essential services and numerous specialized services formerly concentrated in the city core; new outlying locations provide for shopping needs, jobs, entertainment, medical care, and the like. It is these and other facets of ecological change such as the spontaneous growth of the exurban periphery that today combine to constitute the daily urban systems that far transcend the Bureau of the Budget's more narrowly defined SMSAs.

Chapter 15
Latent structure of the American urban system: an exercise in factorial ecology

The urban–regional growth dynamics discussed in Chapter 14 have many consequences, one of which is that today there are markedly different types of cities and suburbs with equally marked differences in their spatial distributions across the country. Those differences are important, because it has been shown many times that different kinds of cities have different mixtures and intensities of social, economic, and environmental problems, different political cultures, and the like.

The literature is replete with efforts to reduce the complexity of such differences by classifying urban centers into relatively uniform types, in the belief that a typology aids in sampling and generalization, and in prediction from sample evidence. Best known and most frequently used are the so-called "functional" classifications, developed from data on the economic specialties of cities (Harris, 1943; Pownall, 1953; Nelson, 1955; Hart, 1955; Steigenga, 1955; Jones and Forstall, 1963; Forstall, 1967). Not unexpectedly, each functional classification has ended up by differentiating between mining towns, manufacturing cities, service centers, college towns, and the like.

The cynic is inclined to say, "So what?" As Smith points out, the classifiers all too often fail to answer the question: "Classification for what purpose?" (R. Smith, 1965a, 1965b). To compound the brute empiricism of classification for classification's sake, consumers often have used the resulting classifications uncritically. They have frequently sought explanation of some phenomenon of interest in a priori class differences, without asking whether similarities and differences between classes of cities have any relevance to the questions they want answered.

This chapter was prepared in the belief that some rethinking of the city-classification problem would provide a framework within which the consumer might be induced to address the issue of theoretical relevance more directly. As such, the intent is to challenge, to provide an alternative point of departure, to try applying the methods of factorial ecology to the problem, rather than to assert that there now exists some superior scheme for typifying the diverse elements in the American urban landscape. However, the results indicate clearly enough that any such classification must comprehend the system growth dynamics outlined in Chapter 14.

Methodology

The classical taxonomic approach

The decision to place cities in the same class has usually been based on observed or "manifest" similarities. To take an example, avoiding for the moment the most fundamental questions of *what* similarities for *which* units of observation, any one or a combination of the 97 variables listed in the appendix to this chapter could qualify as providing this manifest evidence when measured for such units of observation as the 1762 urban places in the United States that had ten thousand or more inhabitants within their legal city limits in both 1950 and 1960.

The traditional city classifier would approach the resulting 1762×97 data matrix in a special way, developing a series of criteria for class membership and allocating cities to classes on their basis. Most frequently he would work with the variables one at a time, splitting the cities into relatively uniform subsets in a series of steps.

Consider a classification based on the percentages of the city's labor force in various sectors of the economy. After studying the frequency distributions of cities on each of the variables, the taxonomist might decide that it is important to separate towns with more than 20 per cent of their workers employed in mining from those with less. Similarly he finds a "break" of 40 per cent in manufacturing and 25 per cent in retailing employment "significant." On reflection, he also decides that manufacturing is "more important" than mining, and mining is "more important" than retailing. Thus he splits towns into subsets based first on manufacturing, then on mining, and finally on retailing, so that his taxonomic "tree" takes the form illustrated in Figure 15.1. A relatively straightforward set of mutually exclusive types of town results.

The search for latent dimensions

Clearly the traditional approach will not suffice when a large number of variables are available. If 97 variables are each used to split cities into two groups, as in Figure 15.1, the number of classes possible is 2^{97}. In practice, of course, many of the combinations will not be realized because many of the variables differentiate towns in the same ways. For example, high-

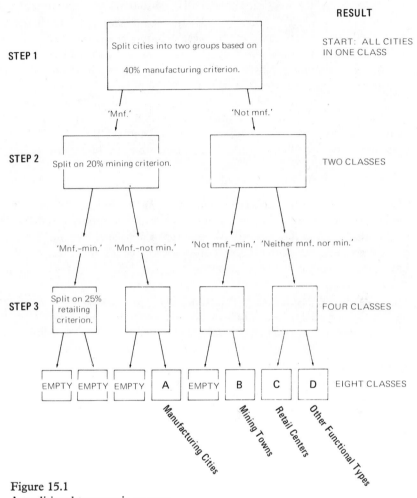

Figure 15.1
A traditional taxonomic process.

income communities also have populations with high educational levels, large proportions of white-collar and professional workers, and, more fundamentally, all of the outward symptoms of high socioeconomic status. We might therefore postulate that the manifest similarities of cities are due to certain fundamental "latent" traits, tendencies, or progenitors like socioeconomic status, resulting from basic cultural traits and processes, such as the aggressive pursuit of economic achievement and related "success." A particular method of analysis, factor analysis, provides a means of searching for these causal factors by separating and identifying clusters of closely interdependent variables whose interpretation is keyed to latent structure and processes.

Take the 1762×97 city-data matrix D and its 97 separate column vectors, $d_1, d_2, d_3, \ldots, d_{97}$ (one for each variable). The fundamental hypothesis of factor analysis is that each of these columns is a product of different combinations of the same underlying factors or common causes and that these latent factors are substantially fewer than the manifest variables. To change the example for a moment, if a column of the d terms were to represent the scores of 1762 students on a test, one would be able to predict these scores on the basis of the underlying verbal and quantitative abilities of the students and the "mixture" of these latent abilities tested by the particular test, also taking into account a random-error term:

$$d_i = \Lambda\theta_i + \Sigma_i$$

where d_i is the manifest (test) vector (1762×1), Λ contains the coefficients measuring each student's latent verbal-ability and quantitative-ability scores (1762×2), θ_i identifies the verbal-quantitative "mixture" of test $_i$ (2×1), and Σ_i is a vector of random errors and other "disturbance" terms for the test (1762×1). Obviously Λ is unchanged for different tests, but θ_i and Σ_i vary.

Educational psychologists determine the Λ for students and the θ_i for tests by factor analysis. In the same way, therefore, factor analysis of the 1762×97 city-data matrix should produce *factor scores for cities* (coefficients measuring each city's rating on the latent urban dimensions) and *factor loadings for manifest variables* (coefficients measuring the mixture of latent dimensions present in each manifest variable). This output should in turn enable the fundamental dimensions of the American urban system to be interpreted and new city classifications to be built *on the basis of those latent traits that bear usefully on the purposes for which the classifications are to be used*. The purposive orientation is critical.

Likewise the full output of scores of the 1762 urban places on each of the latent dimensions should enable users (T. Clark, 1968; Wood, 1964) to focus more clearly on the central issues in city classifications as they bear on their subsequent analytic use:

1. What dimensions of variation are relevant to the analysis; that is, which structural features are of explanatory significance?
2. Are they of equal relevance, or should they receive different weights; that is, what is the relative explanatory power of the dimensions?
3. How many classes, or types, of cities are needed?

When these questions have been satisfactorily answered, it is possible to prepare a classification that minimizes within-group differences, producing classes containing cities that are as much alike as possible (Berry, 1961b). The frame is then set for sampling, for cross-tabulation, or for covariance analysis and other forms of hypothesis-testing or estimation.

American urban dimensions, 1960

To provide factorial output of the type described above, principal-axis factor analysis was applied to the 1762 × 97 data matrix after the 97 variables had been normalized. Only factors whose eigenvalues were greater than unity were extracted. Factor scores were produced after rotating the principal axes to orthogonal simple structure by using the normal varimax criterion.

The complete output of a factor analysis usually includes a correlation matrix for all variables, unrotated and varimax-rotated factor loadings indicating correlations of the primary variables with the latent dimensions, eigenvalues telling the amount of original variance accounted for by the underlying dimensions, and factor scores for each city on each dimension. Obviously not all of these can be presented here. Nor is it appropriate to divert attention from the substantive features of the urban dimensions by reviewing the method, which has been effectively described elsewhere (Ray and Berry, 1965). Instead we focus on the 14 latent dimensions of variation of American cities in 1960 that resulted from the analysis, presenting only those elements of the output that are necessary for purposes of illustration and identification. Together the dimensions account for 77 per cent of the original variance of the 97 primary variables. Table 15.1 summarizes the dimensions that were identified.

An important feature of factor analysis is that the latent dimensions are uncorrelated. They thus contribute additively to the character of each urban place. A first finding is therefore very significant. The elements of the urban economic base traditionally built into functional classifications are largely

Table 15.1

Latent dimensions of the American urban system in 1960

Factor number	Factor description
1	Functional size of cities in an urban hierarchy
2	Socioeconomic status of the city residents
3	Stage in family cycle of the city residents
4	Nonwhite population and home ownership
5	Recent population growth experience
6	Economic base: college towns
7	Population proportion that is foreign-born or of foreign stock
8	Recent employment expansion
9	Economic base: manufacturing
10	Extent of female participation in the labor force
11	Economic base: specialized service centers
12	Economic base: military
13	Economic base: mining
14	Extent to which elderly males participate in labor force

independent of the position of centers in the urban hierarchy, socioeconomic dimensions, and growth behavior. There is much more to urban character than can be explained on the basis of simple economic differences. This being so, we divide what follows into two parts, the first dealing with the systemwide sociocultural factors and the second more briefly with the better-known differences in economic base. Some recent evidence on the declining significance of the traditional categories of economic functions is then introduced, examples of composite community profiles and the multivariate classification problem are discussed, and the question of the universality of the latent structure of urban systems is addressed by comparing our results with those of other studies.

Systemwide factors

Functional size of centers in the urban hierarchy

A substantial number of variables—such as the number of inhabitants, size of the urban labor force, and employment levels—are highly correlated reflections of the total "functional size" or aggregate economic power of cities, or more generally of the status of towns within the nation's urban hierarchy. It was this dimension that was identified first by the factor analysis (see Table 15.2). That increasing numbers and types of economic functions are now distributed according to size of center has been observed by other research workers (C. Clark, 1945; Stanford Research Institute, 1968; Stanback and Knight, 1970). For retail and service activities this is of course perfectly consistent with central-place theory (Berry, 1967a).

Not unexpectedly therefore the highest-scoring cities on the dimension are almost all "national business centers," according to Rand McNally's hierarchical ratings (City Rating Guide, 1964). There is also, however, a close relationship of total manufacturing employment to size of city, consistent with the increasing significance of market orientation for industry and location of various types of manufacturing according to the relative scale economies and externalities afforded by different-sized cities (Berry and Horton, 1970). Not surprisingly, then, 1960–1963 economic classification by the International City Managers' Association (ICMA; now the International City Management Association) of cities with the highest scores on this factor was either diversified manufacturing (MR) or diversified retailing (RM) (Forstall, 1967).

The only other variable to correlate with this hierarchical pattern was the date that a city passed 100,000 population, indicating that the largest cities were those places that were fortunate enough to achieve earliest eminence, a position maintained by self-sustaining growth as they progressed to higher levels of the urban size ratchet (Thompson, 1965).

Table 15.2

Functional size of centers in the urban hierarchy

	Loadings on constituent variables: factor one	
Number	*Name*	*Loadings*
79	Population 1960	0.968
78	Population 1965	0.958
74	Total labor force 1960	0.963
65	Retail labor force 1960	0.974
66	Service labor force 1960	0.943
64	Wholesale labor force 1960	0.909
68	Transportation labor force 1960	0.911
69	FIRE labor force 1960	0.912
70	Public Administration labor force 1960	0.888
71	HEWO labor force 1960	0.889
73	Miscellaneous labor force 1960	0.930
63	Manufacturing labor force 1960	0.777
44	Service employment 1963	0.853
43	Retail employment 1963	0.826
42	Wholesale employment 1963	0.817
41	Manufacturing employment 1963	0.616
48	Service employment 1958	0.704
47	Retail employment 1958	0.643
46	Wholesale employment 1958	0.730
45	Manufacturing employment 1958	0.571
21	Area of city	0.493
16	Date city passed 100,000 population	− 0.433

(continued)

Socioeconomic status

Communities also vary in socioeconomic status (see Table 15.3). The notion that a social mobility process affects individuals during their lifetime is well known. People are born into particular families, from which they achieve initial status, and they are first socialized in the physical and social environment provided by the neighborhood within which the family's residence is located. Education affords the opportunity for upward (or, for lack of it, downward) mobility from that initial status through access to different occupations and therefore different incomes. In turn, financial resources provide access to residential environments of varying quality and to different levels and standards of consumption. These levels and standards represent the base any individual provides for the next generation.

Because developers prefer to build homes in particular price ranges for a market made up of families of similar resources, and because individual purchasers prefer to buy homes in areas that will be occupied by others of similar social status and life style, small communities tend to be relatively homogeneous in terms of the type and quality of housing they provide and the families residing in them (Berry and Horton, 1970). Thus size and quality

Table 15.2 (*continued*)

| | | | | | *ICMA* |
| | | | | *RANALLY* | *economic* |
Rank	*Code*	*City*	*Scores*	*city rating*	*class*
1	189	New York, N.Y.	6.24	1AAAA	MR
2	659	Chicago, Ill.	5.40	1AAA	MR
3	1663	Los Angeles, Calif.	5.21	1AA	MR
4	437	Philadelphia, Pa.	4.69	1AA	MM
5	769	Detroit, Mich.	4.45	1AA	MR
6	1416	Houston, Tex.	4.32	1A	MR
7	504	Cleveland, Ohio	4.17	1AA	MM
8	1396	Dallas, Tex.	3.92	1A	MR
9	1028	Baltimore, Md.	3.86	1A	MR
10	1719	San Francisco, Calif.	3.76	1AA	RM
11	851	Milwaukee, Wis.	3.65	1A	MM
12	1455	San Antonio, Tex.	3.65	2AAA	RM
13	1342	New Orleans, La.	3.64	1A	RM
14	965	St. Louis, Mo.	3.62	1AA	MR
15	1570	Seattle, Wash.	3.61	1A	MR
16	1038	Washington, D.C.	3.58	1AA	G
17	30	Boston, Mass.	3.57	1AA	MR
18	951	Kansas City, Mo.	3.50	1A	MR
19	1504	Denver, Colo.	3.48	1A	RM
20	1249	Memphis, Tenn.	3.40	2AAA	MR
21	439	Pittsburgh, Pa.	3.39	1AA	MR
22	1135	Atlanta, Ga.	3.39	1A	MR
23	890	Minneapolis, Minn.	3.35	1AA	MR
24	1533	Phoenix, Ariz.	3.31	2AAA	RM
25	1717	San Diego, Calif.	3.31	2AA	MM
26	613	Indianapolis, Ind.	3.29	1A	MR
27	1366	Oklahoma City, Okla.	3.29	2AAA	RR
28	506	Columbus, Ohio	3.20	2AAA	MR
29	1586	Portland, Ore.	3.18	1A	RM
30	502	Cincinnati, Ohio	3.17	1A	MR
31	1404	Fort Worth, Tex.	3.17	2AA	MR
32	145	Buffalo, N.Y.	3.11	1A	MR

Notes: The names of variables contain the following abbreviations: FIRE—finance, insurance, and real estate: HEWO—health, education and welfare.

The list of cities contains all Rand McNally (RANALLY) class 1 "national business centers" except Miami, Fla. (1A) and Newark, N.J. (1B). Their factor scores are 2.78 and 2.84, respectively.

of homes, educational levels, occupations, and incomes are highly inter-correlated, each indexing one facet of the differentiation of communities according to socioeconomic status.

The complete range of community types according to social status is, however, clouded by the use of city data, for such statistics are affected by

Table 15.3

Socioeconomic status of community residents

Loadings on constituent variables: factor two

Number	Name	Loadings
59	Median income	0.876
56	Per cent incomes exceeding $10,000	0.897
55	Per cent incomes below $3,000	−0.747
28	Per cent with high-school education	0.834
3	Median school years	0.805
38	Value occupied housing units	0.676
58	Median rent	0.534
11	Per cent housing units sound	0.755
10	Per cent housing units owner-occupied	0.413
35	Median rooms/housing unit	0.557
61	Per cent white collar	0.783
32	Unemployment rate	−0.514

The highest-status communities

Rank	Code	City	Scores	Sample median incomes
1	208	Scarsdale, N.Y.	4.05	22,177
2	752	Winnetka, Ill.	4.03	20,166
3	683	Glencoe, Ill.	3.60	20,136
4	1415	Highland Park, Tex.	3.53	13,707
5	1276	Mountain Brook, Ala.	3.26	14,689
6	157	Garden City, N.Y.	3.21	13,875
7	282	Millburn, N.J.	3.00	
8	1724	San Marino, Calif.	2.98	
9	1698	Piedmont, Calif.	2.97	
10	734	River Forest, Ill.	2.93	
11	751	Wilmette, Ill.	2.91	
12	1020	Prairie Village, Kan.	2.85	
13	578	Upper Arlington, Ohio	2.81	
14	868	Whitefish Bay, Wis.	2.80	
15	413	Lower Merion Twp., Pa.	2.77	
16	566	Shaker Heights, Ohio	2.73	

(continued)

the fragmented overlay of political units on urban systems. The nation's highest-status politically independent communities are therefore suburbs of the largest and richest metropolitan areas. On the other hand, the lowest-status communities lie in the nation's isolated poverty regions. The greatest cluster occurs in Texas along the Rio Grande border with Mexico.

The lowest-status politically independent communities within the more affluent larger metropolitan areas are somewhat better off than small towns in areas of rural poverty, although they too suffer from the same deficiencies of unsound small housing units, low levels of education, inferior job

Table 15.3 (*continued*)

| | | | | Sample |
Rank	Code	City	Scores	median incomes
		The lowest-status communities		
1	1434	Mercedes, Tex.	−4.42	2,395
2	1400	Eagle Pass, Tex.	−4.25	2,436
3	1473	Weslaco, Tex.	−3.98	2,604
4	1426	Laredo, Tex.	−3.50	
5	1456	San Benito, Tex.	−3.45	
6	1401	Edinburg, Tex.	−3.37	
7	1449	Pharr, Tex.	−3.33	
8	1389	Brownsville, Tex.	−3.21	
9	1438	Mission, Tex.	−2.98	
10	1453	Robstown, Tex.	−2.92	
11	1166	Belle Glade, Fla.	−2.82	
12	1397	Del Rio, Tex.	−2.62	
13	1468	Uvalde, Tex.	−2.59	
14	1311	Helena, Ark.	−2.43	
15	1457	San Marcos, Tex.	−2.38	
16	1279	Prichard, Ala.	−2.28	

opportunity, high unemployment, and low incomes. On the other hand, major central cities all have average status scores because they are very large polities spanning a wide range of community areas of all status levels within them, and because they have a diversified economic structure with a broad range of job types.

Stage-in-life cycle

There is variability not simply according to social status but also according to the stage in the life cycle of community residents (Table 15.4). If there were no geographic mobility, a community would age naturally along with its residents. Initially, perhaps, it would house new, young families actively engaged in childrearing. Later it would reach a stage in which family sizes were at their maximum. Thereafter it would experience increasing median ages, diminishing family sizes, and a declining population as children leave home. Final dissolution of the original families with death of the partners would lead to regeneration of the community. Another cycle would begin as new families moved in.

But there is another reason for life-cycle differentiation. If a community specializes in providing the kinds of housing units and residential environments that are appropriate to families at a particular stage of the life cycle, and if there is mobility, with people moving in and out as their housing needs change, the specialized character of the community can be maintained over a long period of time.

Table 15.4

Stage in the life cycle of community residents

Loadings on constituent variables: factor three		
Number	Name	Loadings
12	Median age	−0.806
13	Per cent population under 18	0.869
62	Per cent population over 65	−0.838
22	Fertility rate	0.775
23	Population per household	0.906
36	Persons per dwelling unit	0.869
54	Per cent homes built 1950–1960	0.627
57	Rate of growth 1950–1960	0.484

Youthful child-rearing communities			
Rank	Code	City	Scores
1	654	Carpentersville, Ill.	3.59
2	1514	Thornton, Colo.	3.47
3	1656	La Puente, Calif.	3.19
4	738	Rolling Meadows, Ill.	3.12
5	1463	Weslaco, Tex.	3.09
6	875	Brooklyn Center, Minn.	3.02
7	873	Bloomington, Minn.	2.91
8	1401	Edinburg, Tex.	2.91
9	1449	Pharr, Tex.	2.91
10	879	Crystal, Minn.	2.90
11	494	Brook Park, Ohio	2.89
12	777	Garden City, Mich.	2.87
13	876	Brooklyn Park, Minn.	2.81
14	705	Markham, Ill.	2.77
15	714	Mundelein, Ill.	2.76
16	943	Florissant, Mo.	2.74

Communities with many elderly residents			
Rank	Code	City	Scores
1	1192	Miami Beach, Fla.	−4.87
2	1606	Beverly Hills, Calif.	−3.55
3	1187	Lake Worth, Fla.	−3.51
4	1692	Palm Springs, Calif.	−3.36
5	1415	Highland Park, Tex.	−3.23
6	1168	Bradenton, Fla.	−3.02
7	1206	St. Petersburg, Fla.	−2.95
8	1649	Huntington Park, Calif.	−2.91
9	1731	Santa Cruz, Calif.	−2.72
10	1183	Hollywood, Fla.	−2.64
11	1734	Santa Monica, Calif.	−2.54
12	1740	South Pasadena, Calif.	−2.54
13	1169	Clearwater, Fla.	−2.46
14	1208	Sarasota, Fla.	−2.40
15	1173	De Land, Fla.	−2.37
16	1467	University Park, Tex.	−2.35

In actuality any community experiences a mixture of these two patterns, but communities do differ according to the stage in the life cycle of their residents: by age levels and mixes, fertility rates, family sizes, and population-growth characteristics.

Many of the nation's young-family communities in 1960 were bedroom suburbs within large metropolitan areas. At the other extreme were the retirement communities of Florida and California. The older family communities of the large central cities, being specific apartment neighborhoods (e.g., along Chicago's North Shore) cannot, of course, be identified in a study of legal cities (Berry and Horton, 1970).

Nonwhite and foreign-born populations

Apart from size, social status, and stage in life cycle, two other cultural components important in differentiating American cities were revealed by the analysis: patterns of the nonwhite population and of the population that is foreign born or of foreign stock (see Tables 15.5 and 15.6). The regional variations are quite familiar, the one emphasizing the cities of the South and the Manufacturing Belt, and the other the peripheral points of entry.

Recent population growth and recent increase in employment

Population growth and employment expansion operate independently of other structural features and of each other (see Tables 15.7 and 15.8). This of course reflects the fact that the highest growth rates will represent extension of development into new areas. However, the broad regional differentiation of population growth in the nation's southern and western "rimland" contrasts with the largely intrametropolitan differentiation of employment expansion around the peripheries of the largest metropolitan areas.

Female participation in the labor force

Table 15.9 reveals differences in the percentages of women in the labor force, in labor-force participation rates, and inversely in fertility rates. Textile towns in the American South have the highest female-employment levels, whereas retirement communities, mining towns, heavy industrial complexes, and young-family suburbs have the lowest.

Elderly males in the labor force

Table 15.10 separates a group of communities in which a relatively high number of males remain employed past the normal retirement age of 65. Interestingly, most of these communities are metropolitan suburbs housing high proportions of independent businessmen; a few are isolated towns in

Table 15.5

Nonwhite population in communities

Loadings on constituent variables: factor four		
Number	*Name*	*Loadings*
60	Per cent population nonwhite	0.850
37	Per cent housing units nonwhite	0.848
75	Per cent housing units nonwhite with over 1.01 persons per room	0.831
76	Nonwhite housing units owned	0.826
77	Median income nonwhite	0.798
29	Per cent married couples without own household	0.465

Substantially nonwhite communities				
Rank	*Code*	*City*	*Scores*	*Sample per cent nonwhite*
1	1268	Fairfield, Ala.	2.80	53
2	248	Englewood, N.J.	2.60	27
3	745	Urbana, Ill.	2.59	6
4	1174	Delray Beach, Fla.	2.59	44
5	1181	Hallandale, Fla.	2.59	42
6	655	Centerville, Ill.	2.51	56
7	773	Ecorse, Mich.	2.43	
8	1213	Winter Park, Fla.	2.34	
9	1761	Hilo, Hawaii	2.34	
10	1303	Yazoo City, Miss.	2.31	
11	1204	Riviera Beach, Fla.	2.30	
12	284	Montclair, N.J.	2.27	
13	393	Farrell, Pa.	2.25	
14	1117	Aiken, S.C.	2.25	
15	788	Inkster, Mich.	2.24	
16	1261	Bessemer, Ala.	2.22	

peripheral regions from which many individuals in the active younger age groups have emigrated.

Functional town types

American cities are also differentiated according to the nature of their economic bases. Fine distinctions of functional types were not possible in this study because of the somewhat gross input variables used, but five functional types were isolated. In every case the towns displaying high degrees of functional specialization tended to be quite small, as the ensuing tables indicate.

Table 15.6

Foreign-born or foreign-stock population

Loadings on constituent variables: factor seven		
Number	Name	Loadings
25	Per cent population foreign born	0.822
26	Per cent population foreign stock	0.755
1	Per cent population foreign born; mother tongue not English	0.567
53	Per cent elementary school children in private school	0.480
9	Per cent housing units one-unit	−0.463
34	Per cent using public transport	0.390

Communities with greatest foreign component				
Rank	Code	City	Scores	Sample per cent foreign-born
1	1192	Miami Beach, Fla.	3.74	33
2	1400	Eagle Pass, Tex.	3.50	28
3	1473	Weslaco, Tex.	3.38	17
4	1648	Huntington Beach, Calif.	3.17	8
5	1449	Pharr, Tex.	3.01	15
6	1434	Mercedes, Tex.	2.92	
7	252	Garfield, N.J.	2.83	
8	1401	Edinburg, Tex.	2.69	
9	301	Passaic, N.J.	2.68	
10	783	Hamtramck, Mich.	2.67	
11	1181	Hallandale, Fla.	2.66	
12	1438	Mission, Tex.	2.66	
13	352	Ambridge, Pa.	2.58	
14	95	Central Falls, R.I.	2.57	
15	251	Fort Lee, N.J.	2.55	
16	601	East Chicago, Ind.	2.55	

"Native American" communities			
Rank	Code	City	Scores
1	576	Troy, Ohio	−4.01
2	1444	Palestine, Tex.	−2.88
3	1392	Cleburne, Tex.	−2.80
4	489	Bellefontaine, Ohio	−2.46
5	1418	Hurst, Tex.	−2.46
6	580	Van Wert, Ohio	−2.42
7	1409	Garland, Tex.	−2.42
8	1390	Brownwood, Tex.	−2.35
9	706	Mattoon, Ill.	−2.32
10	1411	Greenville, Tex.	−2.30
11	1398	Denison, Tex.	−2.25
12	461	Sunbury, Pa.	−2.19
13	1413	Haltom City, Tex.	−2.19
14	496	Bucyrus, Ohio	−2.17
15	1433	Marshall, Tex.	−2.17
16	508	Coshocton, Ohio	−2.15

Table 15.7

Recent population growth

Loadings on constituent variables: factor five		
Number	Name	Loadings
2	Per cent population residing in state of birth	−0.685
39	Vacancy rate, owner-occupied units	0.674
96	Growth rate 1960–1965	0.609
27	Per cent moved after 1958	0.600
52	Per cent population over 5 years old who are migrants	0.574
40	Vacancy rate, rental units	0.551

New communities: rapid post-1960 growth			
Rank	Code	City	Scores
1	1535	Scottsdale, Ariz.	6.36
2	1648	Huntington Beach, Calif.	5.72
3	1181	Hallandale, Fla.	4.88
4	1201	Pinellas Park, Fla.	4.66
5	1183	Hollywood, Fla.	4.60
6	1192	Miami Beach, Fla.	4.55
7	1175	Eau Gallie, Fla.	4.45
8	1692	Palm Springs, Calif.	4.39
9	1167	Boynton Beach, Fla.	4.30
10	1187	Lake Worth, Fla.	3.95
11	1548	Las Vegas, Nev.	3.80
12	1169	Clearwater, Fla.	3.79
13	1658	Livermore, Calif.	3.57
14	1203	Pompano Beach, Fla.	3.45
15	1208	Sarasota, Fla.	3.45
16	1633	Escondido, Calif.	3.45

Most stable communities since 1960			
Rank	Code	City	Scores
1	290	New Hanover Twp., N.J.	−2.35
2	407	Lansdowne, Pa.	−2.29
3	187	New Hyde Park, N.Y.	−2.21
4	1761	Hilo, Hawaii	−2.21
5	461	Sunbury, Pa.	−2.11
6	277	Maplewood, N.J.	−2.07
7	202	Rensselaer, N.Y.	−2.06
8	56	Medford, Mass.	−1.96
9	40	Fall River, Mass.	−1.91
10	28	Belmont, Mass.	−1.89
11	64	Newton, Mass.	−1.86
12	370	Carnegie, Pa.	−1.86
13	386	Dunmore, Pa.	−1.86
14	217	Watervliet, N.Y.	−1.84
15	433	North Braddock, Pa.	−1.81
16	577	University Heights, Ohio	−1.81

Table 15.8

Recent growth in employment

	Loadings on constituent variables: factor eight	
Number	*Name*	*Loadings*
82	Growth in manufacturing employment, 1958–1963	0.798
83	Growth in wholesale	0.757
81	Growth in retail	0.724
80	Growth in service	0.707

		Growth centers	
Rank	*Code*	*City*	*Scores*
1	689	Hinsdale, Ill.	4.72
2	1147	Forest Park, Ga.	3.91
3	859	South Milwaukee, Wis.	3.74
4	935	Berkeley, Mo.	3.71
5	1755	Westminster, Calif.	3.48
6	712	Mount Prospect, Ill.	3.42
7	586	Whitehall, Ohio	3.35
8	875	Brooklyn Center, Minn.	3.18
9	716	Niles, Ill.	3.02
10	589	Willowick, Ohio	2.98
11	494	Brook Park, Ohio	2.90
12	878	Coon Rapids, Minn.	2.87
13	583	Warrensville Heights, Ohio	2.77
14	478	Yeadon, Pa.	2.75
15	392	Falls Twp., Pa.	2.73
16	879	Crystal, Minn.	2.65

Manufacturing towns

Table 15.11 identifies the variables separating towns according to manufacturing specialization and gives examples. The pattern is a very familiar one in the northeastern manufacturing belt and a few cities on the southern and western peripheries.

Mining towns

Small mining towns also separate clearly, as in previous classifications (Table 15.12).

College towns

A well-dispersed set of relatively small towns is supported by colleges and universities (see Table 15.13).

Table 15.9

Female participation in the labor force

	Loadings on constituent variables: factor ten		
Number	*Name*		*Loadings*
5	Females over 14 in labor force		−0.808
31	Married women in labor force		−0.805
4	Cumulative fertility rate		0.652
30	Nonworker/worker ratio		0.576

		Towns with greatest participation	
Rank	*Code*	*City*	*Scores*
1	1166	Belle Glade, Fla.	−4.97
2	1149	Griffin, Ga.	−3.45
3	1154	Milledgeville, Ga.	−3.32
4	1257	Alexander City, Ala.	−3.17
5	1277	Opelika, Ala.	−2.93
6	1760	Fairbanks, Alaska	−2.88
7	1566	Pullman, Wash.	−2.87
8	137	Willimantic, Conn.	−2.68
9	1292	Greenwood, Miss.	−2.60
10	48	Lawrence, Mass.	−2.59
11	1087	Concord, N.C.	−2.55
12	1151	La Grange, Ga.	−2.52
13	1037	Takoma Park, Md.	−2.44
14	1122	Gaffney, S.C.	−2.39
15	1491	Moscow, Idaho	−2.39
16	1029	Cambridge, Md.	−2.37
17	1125	Greenwood, S.C.	−2.31
18	1084	Burlington, N.C.	−2.30
19	1106	Reidsville, N.C.	−2.30
20	1113	Thomasville, N.C.	−2.30

		Towns with least participation	
Rank	*Code*	*City*	*Scores*
1	422	Monessen, Pa.	3.19
2	1206	St. Petersburg, Fla.	2.99
3	383	Donora, Pa.	2.96
4	1187	Lake Worth, Fla.	2.73
5	387	Duquesne, Pa.	2.58
6	402	Johnstown, Pa.	2.56
7	1724	San Marino, Calif.	2.52
8	780	Grosse Pointe Farms, Mich.	2.52
9	592	Youngstown, Ohio	2.38
10	1183	Hollywood, Fla.	2.38

Table 15.10

Elderly working males (+) and interstate commuting (−)

Loadings: factor fifteen		
Number	Name	Loadings
33	Per cent working outside county of residence	−0.628
6	Males over 65 in labor force	0.406

Positive-scoring communities			
Rank	Code	City	Scores
1	701	Lincolnwood, Ill.	3.90
2	1606	Beverley Hills, Calif.	3.32
3	1449	Pharr, Tex.	2.88
4	115	Groton, Conn.	2.86
5	1623	Culver City, Calif.	2.75
6	1568	Renton, Wash.	2.71
7	758	Alpena, Mich.	2.51
8	577	University Heights, Ohio	2.46
9	788	Inkster, Mich.	2.38
10	1554	Bellevue, Wash.	2.38
11	1607	Brawley, Calif.	2.36
12	683	Glencoe, Ill.	2.35

Negative-scoring communities			
Rank	Code	City	Scores
1	1147	Forest Park, Ga.	−4.79
2	1064	Vienna, Va.	−3.86
3	1039	Alexandria, Va.	−3.82
4	1163	Warner Robins, Ga.	−3.59
5	276	Maple Shade, N.J.	−3.40
6	478	Yeadon, Pa.	−3.37
7	1046	Fairfax, Va.	−3.30
8	1648	Huntington Beach, Calif.	−3.14
9	1043	Chesapeake, Va.	−3.13
10	1047	Falls Church, Va.	−3.12
11	1355	Del City, Okla.	−3.04
12	1037	Takoma Park, Md.	−3.00

Military installations

Military installations also provide a source of support for many small towns, as shown in Table 15.14.

Service centers

Finally, relatively uniformly distributed throughout the West and South are towns functioning primarily as local service centers. Conversely, the

Table 15.11

Manufacturing towns

Loadings on constituent variables: factor nine		
Number	Name	Loadings
8	Manufacturing ratio	0.812
85	Per cent labor force engaged in manufacturing	0.671
51	Employment/residence ratio	0.388

The manufacturing towns				
Rank	Code	City	Scores	Sample ICMA economic classes
1	1721	San Jose, Calif.	6.44	MR
2	1632	El Segundo, Calif.	3.83	MM
3	115	Groton, Conn.	3.13	MM
4	494	Brook Park, Ohio	2.98	MM
5	672	East Peoria, Ill.	2.91	MM
6	1080	Weirton, W.Va.	2.65	
7	587	Wickliffe, Ohio	2.64	
8	1096	Hickory, N.C.	2.59	
9	773	Ecorse, Mich.	2.58	
10	820	Trenton, Mich.	2.54	
11	601	East Chicago, Ind.	2.52	
12	349	Aliquippa, Pa.	2.37	
13	212	Tonawanda, N.Y.	2.35	
14	1568	Renton, Wash.	2.33	
15	473	West Mifflin, Pa.	2.28	
16	275	Manville, N.J.	2.27	

midwestern and northeastern sections of the country, along with the Piedmont area, are characterized by towns of specialized function (see Table 15.15).

General dimensions versus economic types

An increasing number of economic activities are now market-oriented. Locationally these activities tend to be differentiated according to access to national markets and position in the urban hierarchy rather than by classical location factors. Moreover, although there are broad regional patterns according to which the lowest status communities are found in regions of rural poverty and retirement communities are located in the West and South, the more important elements of socioeconomic differentiation are increasingly *intra*metropolitan in a nation that is fully metropolitanized.

The metropolitan centers are multifunctional, and much of their growth

Table 15.12

Mining towns

	Loadings on constituent variables: factor fourteen		
Number	Name	Loadings
89	Per cent labor force engaged in mining	−0.835
67	Total mining employment	−0.741

		The mining towns		
Rank	Code	City	Scores	ICMA economic class
1	1524	Grants, N. Mex.	−9.78	MG
2	1378	Andrews, Tex.	−7.38	MG
3	1420	Kermit, Tex.	−7.34	MG
4	1520	Carlsbad, N. Mex.	−7.14	MG
5	1436	Midland, Tex.	−6.24	MG
6	1525	Hobbs, N. Mex.	−6.12	MG
7	886	Hibbing, Minn.	−5.79	MG
8	904	Virginia, Minn.	−5.57	MG
9	1352	Bartlesville, Okla.	−5.41	MG
10	1442	Odessa, Tex.	−5.18	RR/MG
11	1522	Farmington, N. Mex.	−5.13	RR/MG
12	1013	Liberal, Kans.	−4.77	RR/MG

is self-generative. Hence the traditional approach to classification based on economic functions today tends to separate only relatively small communities in terms of the following:

1. The few remaining economic activities for which traditional non-metropolitan location factors still prevail; among these are cases of raw-materials orientation (generally mining and agricultural processing) and labor orientation (textiles in poverty regions).
2. Activities located by noneconomic determinants (colleges, military bases).

It follows that the traditional economic approach to city classification is of minimal and declining relevance. .

Another question follows: Why mix intermetropolitan and intrametropolitan differences by using legal cities as units of observation, rather than the "real" urban systems discussed in Chapter 14? Might not the structural dimensions be confounded by this mixture? Parallel analyses at the intermetropolitan and intrametropolitan levels do not suggest this (Berry and Neils, 1969). The principal dimensions of intermetropolitan and intrametropolitan differentiation appear to be the same: for example, size (in intrametropolitan studies it is useful to suppress size as a variable, because it is usually measured for fragmented political units that bear a discordant relationship to community areas and neighborhood units within the metropo-

Table 15.13

College towns

	Loadings on constituent variables: factor six	
Number	Name	Loadings
50	College population 1963	0.821
49	College population 1960	0.809
84	Change in college population 1960–1963	0.783
24	Population living in group quarters	0.734
93	Per cent population employed in HEWO	0.652

	The college towns		
Rank	Code	City	Scores
1	458	State College, Pa.	5.09
2	1085	Chapel Hill, N.C.	4.84
3	484	Athens, Ohio	4.71
4	772	East Lansing, Mich.	4.60
5	639	West Lafayette, Ind.	4.20
6	1393	College Station, Tex.	4.11
7	1566	Pullman, Wash.	4.05
8	1260	Auburn, Ala.	3.92
9	597	Bloomington, Ind.	3.84
10	1372	Stillwater, Okla.	3.77
11	530	Kent, Ohio	3.73
12	802	Mount Pleasant, Mich.	3.70
13	309	Princeton, N.J.	3.64
14	492	Bowling Green, Ohio	3.60
15	922	Iowa City, Iowa	3.59

lis), socioeconomic status, stage-in-life cycle, and recent growth behavior, as shown in Chapter 7.

Community profiles

One outcome of the analysis is a set of scores for each town. Together these scores enable "community profiles" to be developed.

For example, if one looks at the sample output in Table 15.16, New York City scores high on size, is slightly below average on status and family structure, and has a high concentration of foreign-born residents. Weslaco, Texas, is a very low-status community with large families, many foreign-born residents, and a low growth rate. Palm Springs, California, which is above average in status, has small families, a substantial foreign-born element, and a high growth rate. East Peoria, Illinois, is a small, low-status manufacturing town. State College, Pennsylvania, is a small college town with above-average growth.

Table 15.14

Military installations

Loadings on constituent variables: factor twelve		
Number	Name	Loadings
94	Per cent labor force in armed forces	0.845
72	Total armed forces employment	0.648

The bases				
Rank	Code	City	Scores	Sample ICMA economic class
1	290	New Hanover Twp., N.J.	13.58	AF
2	732	Rantoul, Ill.	7.63	AF
3	245	Eatontown, N.J.	6.60	AF
4	1313	Jacksonville, Ark.	6.60	AF
5	98	Newport, R.I.	6.44	AF
6	1286	Biloxi, Miss.	6.22	AF
7	1620	Coronado, Calif.	6.14	AF
8	1086	Fayetteville, N.C.	5.58	MR
9	1701	Port Hueneme, Calif.	5.21	AF
10	115	Groton, Conn.	5.10	MM
11	1591	Alameda, Calif.	5.06	AF
12	1422	Killeen, Tex.	4.68	AF
13	1098	Jacksonville, N.C.	4.53	AF
14	1350	Altus, Okla.	4.48	AF
15	1650	Imperial Beach, Calif.	4.24	
16	1684	Novato, Calif.	4.02	
17	1759	Anchorage, Alaska	3.96	
18	1055	Norfolk, Va.	3.94	
19	1676	Monterey, Calif.	3.91	
20	1760	Fairbanks, Alaska	3.78	

Classification of the cities

Any research worker has in such community profiles the basic data needed for analytically meaningful city classification, because the redundancies of overlapping variables have been eliminated by factor analysis. With such materials in hand, the questions to be answered by the user remain. To reiterate:

1. Which of the structural dimensions are relevant to this study? Which afford explanation?
2. Are they of equal relevance, or should they be weighted differentially, according to external criteria?
3. How many classes of cities are needed?

Table 15.15
Service centers

Loadings on constituent variables: factor eleven		
Number	Name	Loadings
87	Per cent labor force in retail occupations	0.746
88	Per cent labor force in service	0.691
90	Per cent labor force in transport	0.620
95	Per cent labor force in miscellaneous	0.610
86	Per cent labor force in wholesale	0.505
91	Per cent labor force in FIRE	0.505
92	Per cent labor force in public administration	0.432

The centers				
Rank	Code	City	Scores	Sample ICMA economic class
1	1548	Las Vegas, Nev.	2.33	X
2	1240	East Ridge, Tenn.	2.19	S
3	222	Atlantic City, N.J.	1.90	RR/X
4	994	North Platte, Neb.	1.89	RR/T
5	1523	Gallup, N. Mex.	1.79	RR
6	1482	Havre, Mont.	1.70	RR/T
7	1538	Yuma, Ariz.	1.70	
8	1392	Cleburne, Tex.	1.67	
9	1551	Sparks, Nev.	1.66	
10	1312	Hot Springs, Ark.	1.61	
11	1549	North Las Vegas, Nev.	1.61	
12	1710	Roseville, Calif.	1.60	
13	861	Superior, Wis.	1.59	
14	956	Moberly, Mo.	1.58	
15	1194	North Miami, Fla.	1.55	

Generality of the latent structure

How general are the dimensions described above? Are they latent in systems of cities elsewhere? Have they been consistent through time? One author (Gerald Hodge, 1968) argues for complete universality:

1. Common structural features underlie the development of all centers within a region.
2. Structural features of centers tend to be the same from region to region regardless of the stage or character of regional development.
3. Urban structure may be defined in terms of a set of "independent" dimensions covering at least (a) size of population, (b) quality of physical development, (c) age structure of population, (d) education level of population, (e) economic base, (f) ethnic and/or religious orientation, (g) welfare, and (h) geographical situation.

Table 15.16
Sample community profiles

	1 Size	2 Status	3 Family cycle	4 Nonwhite population	7 Foreign-born	5 Population growth	9 Manu-facturing	14 Mining	6 College	12 Military
189 New York City	6.24	−0.30	−0.24	−0.61	1.63	−0.03	−0.32	−0.12	−0.73	0.32
208 Scarsdale, N.Y.	−0.65	4.05	−0.35	1.71	0.70	−0.01	1.01	0.28	0.45	0.34
1473 Weslaco, Tex.	−0.34	−3.98	3.09	−0.83	3.38	−1.17	−3.27	0.59	0.05	−0.59
1692 Palm Springs, Cal.	−0.81	1.16	−3.36	0.33	1.22	4.39	−0.46	0.44	−1.03	−0.42
1268 Fairfield, Ala.	−0.94	−0.47	0.48	2.80	−0.48	−0.91	0.34	0.15	−0.43	−0.82
1535 Scottsdale, Ariz.	−0.56	1.17	0.18	−0.48	0.68	6.36	−0.19	1.82	−0.50	−2.10
1761 Hilo, Hawaii	0.25	−0.34	1.45	2.34	1.39	−2.21	−1.06	−1.07	−0.42	0.21
1122 Gaffney, S.C.	−1.00	−0.70	0.39	1.59	−0.09	−0.40	0.15	0.05	0.81	−0.55*
701 Lincolnwood, Ill.	−0.61	2.25	0.77	0.30	0.93	−1.49	0.41	0.75	−0.44	−0.90**
672 East Peoria, Ill.	−0.97	−0.80	0.00	−0.62	−0.41	0.13	2.91	−0.26	−0.19	0.29
1520 Carlsbad, N. Mex.	−0.19	−0.09	0.68	−0.48	0.31	0.94	−0.72	−7.14	−0.64	−0.56
458 State College, Pa.	−1.17	1.34	−0.72	−0.11	0.36	1.17	−1.27	0.16	5.09	0.30
98 Newport, R.I.	−0.13	−0.20	−0.55	0.74	0.46	−0.22	−0.93	0.97	0.91	6.44

Note: The scores have been standardized for each factor to zero arithmetic average and unit standard deviation.

* Female labor force score of −2.39.

** Elderly male labor force score of 3.90.

4. The economic base of urban centers tends to act independently of other urban structural features.

Clearly we need to review other studies that have been completed to see if Hodge's assertions are valid. If they are, considerable economy in urban studies is suggested, enabling the developing science to pass beyond the easily perceived structural entities of city classifications by supplementing and ultimately superseding them in attention by organizational and developmental ideas.

To move beyond morphology, however, one needs to be sure that there are certain time-constant aspects of the system, an enduring architecture whose physiology and organization, reversible changes in time, and adaptive or homeostatic adjustments to environmental pressure can be described and then explained by the development or evolution of the system—the irreversible secular changes that accumulate in time (Berry and Neils, 1969).

In effect we are only at the beginning of what must be a long-term effort to describe the processes giving rise to the structural organization and orderly functioning of urban activities and the innovations giving rise to periodic transformations in the structural arrangements. A first step is to determine whether urban systems have common latent structures.

Other studies of the United States

There has been a long history of multivariate studies of American cities, largely overlooked until recently. These include Price's study of American metropolitan centers in 1930 (Price, 1942), reanalyzed comparatively by Perle in 1960 (Perle, 1964); Hofstaetter's study (based on work by Thorndike) of American cities of 30,000 to 50,000 inhabitants in 1930 (Hofstaetter, 1952); Kaplan's 1950 study of 370 selected cities with populations exceeding 25,000 (Kaplan, 1958); Hadden and Borgatta's equivalent 1960 investigation (Hadden and Borgatta, 1965; Tropman, 1969); and Mayer's analysis of the 1960 Standard Metropolitan Statistical Areas (as presented in Berry and Neils, 1969).

Using 15 variables, Price found four dominant dimensions of metropolitan centers: size, nonservice occupational specialization, socioeconomic status, and trade-center orientation. Perle confirmed these factors for 1960, using the same set of input variables. Hofstaetter, using 23 variables that he thought indexed the quality of urban environments, found the principal dimensions to be socioeconomic status, degree of industrialization, and prevalence of slum conditions. Kaplan's factors in a 47-variable study were size, socioeconomic status, population stability and growth, relative ethnic and racial homogeneity, and age-sex structure (life cycle). Hadden and Borgatta, in a study closely corresponding to the one reported in this chapter, produced 16 factors in all, from data comprising 65 variables for 644 cities: socioeconomic status, nonwhite population, age composition, educational centers, residential mobility, population density, foreign-born concentration, total population, wholesale concentration, retail concentration, manufacturing concentration, durable manufacturing concentration, communication centers, public administration centers, high school education, and transportation centers. Mayer's factors (data 212 SMSAs × 66 variables) were similar: socioeconomic status, age and size of city, family structure, growth 1950–1960, commercial versus manufacturing orientation, foreign population, nonwhite population, unemployment and male labor force (inversely with female employment), institutional or military population, relative isolation, use of public transport, and low-density development.

Clearly Hodge's generalizations hold in the United States for the period 1930–1960. Differences in the output of the various studies simply reflect differences in the subset of variables; the larger-scale Hadden-Borgatta work, and our own, embrace the data subsets and variety of results of the other research workers.

Canada

The first multivariate studies of Canada were completed by King for the years 1951 and 1961 (King, 1966). Subsequently Ray et al. (1968) restudied the Canadian urban scene in a broader framework of variables from the 1961 census and a wider interpretive context.

King (106 cities × 52 variables) found dimensions of socioeconomic status (related also to differences between English and French Canada), relative isolation with primary-industry orientation, smaller specialized manufacturing towns, and so forth. Ray and his associates (113 cities × 95 variables) reiterated the basic socioeconomic significance of English-French contrasts in Canada and identified several functional types of city mining, service centers, manufacturing, and metropolitan growth poles.

A separate postwar growth pattern emerged, as did British Columbian and Prairie city types based on distinctive Asiatic and Slavic cultural components. The similarities in latent structure are substantial.

Britain

One of the best presentations of results of multivariate urban studies is *British Towns*, by Moser and Scott (1961). In a pathbreaking study, the authors examined 157 towns in England and Wales with respect to 60 different variables. The main object of their work was "to classify British towns into a few relatively homogeneous categories, or to see whether such a classification makes sense." They used eight main categories of variable: population size and structure (7 variables), population change (8), households and housing (10), economic functions and employment characteristics (15), social class (4), voting behavior (7), health (7), and education (2).

Prior to classification the authors found it necessary to isolate the basic patterns according to which the towns varied, "because the many series that describe towns are not independent; they overlap in the story they tell.... Towns with a high proportion of heavy industry tend, on the whole, to have low 'social class' proportions, a substantial Labor vote, high infant mortality, and so on" (Moser and Scott, 1961). Four common factors were found to account for the correlations among the primary variables: social class; age of the area, including growth in 1931–1951; recent (1951–1958) growth; and housing conditions, including overcrowding. "The essence [of the analysis was] to investigate how much of the total variability of towns exhibited in the primary variables [could] be accounted for and expressed in a smaller number of new independent variates, the principal components" (Moser and Scott, 1961).

Notable in these results is the correlation in the Moser and Scott "social status" dimension of North American socioeconomic status and age-structure elements. In Britain higher status is accompanied by such characteristics as higher proportions of older persons, smaller families, and lower birth rates. In turn, the highest-status communities represent a combination of resorts and retirement communities, exclusive residential suburbs, and professional administrative centers. The universal validity of factorial results postulated by Hodge is thus called into question.

Because the common factors summarized the essential differences among towns contained in the entire set of original primary variables, Moser and Scott could simplify the classification problem. Each town was given a score on each common factor, and towns were then allocated to groups on the basis of relative scores on the four factors. The 14 groups of towns that were identified fall into three major categories:

1. Resort, administrative, and commercial centers.
   (a) Seaside resorts.
   (b) Spas; professional and administrative centers.
   (c) Commercial centers.
2. Industrial towns.
   (a) Railway centers.
   (b) Ports.
   (c) Textile centers of Yorkshire and Lancashire.
   (d) Industrial centers of the northwest and Welsh mining towns.
   (e) Metal manufacturing centers.
3. Suburbs and suburban-type towns.
   (a) Exclusive residential suburbs.
   (b) Older mixed residential suburbs.
   (c) Newer mixed residential suburbs.
   (d) Light industrial suburbs, national defense centers, and towns within the influence of large metropolitan conurbations.
   (e) Older working-class industrial suburbs.
   (f) Newer industrial suburbs.

In the grouping "the general aim [was] to minimize within-group [differ-ences] and to maximize those between groups" (Moser and Scott, 1961).

Yugoslavia

Moving beyond the North Atlantic context, Fisher (1966), using 1961 data, analyzed 55 selected urban centers in Yugoslavia with respect to 26 variables. He interpreted the most important latent dimension as comprising an index of relative development in which status and proportion of popula-tion in the economically active childrearing age groups are highly related. This factor indicated a broad difference between the "developed" and the "underdeveloped" regions of Yugoslavia.

Fisher also found several functional types (construction and transporta-tion; traditional cultural, commercial, and administrative centers; industrial towns) and a factor identifying recent growth and change, but these, he felt, were secondary to the principal dimension.

Chile

Several analyses of Chilean data have been completed (Berry, 1969b): employment structure of 105 communes with populations exceeding 15,000 in both 1952 and 1960; 59 social, economic, political, and demographic variables for 80 urban communes in 1960; and exploration of data on transportation and traffic flows for 94 urban places in 1962–1965.

The principal factors in the first analysis were the functional sizes of centers in the urban hierarchy and a contrast between traditional towns of the agricultural heartland of the country and mining towns on the periphery. The larger analysis reiterated the factors of size and of traditionalism versus modernism. The latter factor represented, as in Yugoslavia, a combination of socioeconomic status and age structure. In addition, separate factors were identified for recent growth, mineral exploitation, manufacturing, and cer-tain elements of voting behavior.

India

Ahmad's 102-city, 62-variable analysis of the largest Indian cities in 1961 (Ahmad, 1965) identified as factors certain themes which are by now already familiar: size, recent change, and economic specialization (com-mercial, industrial). In addition, certain broad regional differences were also noted—between northern and southern India in the sex composition of cities and the position of women in the labor force, between eastern and western India in migration patterns, and so on—each reflecting broad regional differences in India's cultures. Types of town were shown to be highly differentiated by region within the country.

Nigeria and Ghana

Studies of the urban systems of Nigeria and Ghana have been completed by Mabogunge (1965a, 1965b) and McNulty (1969a, 1969b), respectively.

McNulty found a principal factor differentiating urban populations according to age structures and sex ratios. This factor, he felt, reflected the migration of males in the active age groups to growing commercial and service centers, leaving behind high proportions of poorly-educated young, old, and females in areas of primary occupational specialization. A similar dimension was found by Mabogunge in Nigeria. These dimensions are not unlike Fisher's "modernism-traditionalism" scale for Yugoslavia.

A second factor found by McNulty related to functional type: highly specialized mining towns were distinguished from centers offering diversified employment in commerce, services, and manufacturing. Both McNulty and Mabogunge argued that these dimensions are structural correlates of the development process, particularly as overlaid on the countries by colonial capital, and that urban structural change, in turn, is attendant on development through the interrelated processes of migration and increasing economic diversification.

Conclusions

Extending and modifying Hodge's arguments, several conclusions about the latent structure of urban systems follow:

1. The economic base of urban centers tends to act independently of other urban structural features (with the exception of hierarchical organization of market-oriented activities; see item 2 below), and, to the extent that there is geographic specialization based on locational factors other than market orientation, each broad economic function will lead to its own distinctive economic town type. Public activities—military bases, educational centers, public administration—act as any other specialized economic base.
2. Every urban system is organized systemwide into a hierarchy of centers based on aggregate economic power. The functional size of centers in an urban hierarchy is a universal latent dimension.
3. In every society the principal dimensions of socioeconomic differentiation are those of social status and age structure, or stage-in-life cycle. However, only at the highest levels of development do these factors appear to operate independently. At somewhat lower levels of welfare (Britain) there remains a correlation between income and family structure, and only the rich elderly can segregate themselves in retirement resorts and spas; at lower-income levels there is a great mixture of family types in the same residential areas. Further down the scale still (Yugoslavia, Chile,

West Africa), status and age-structure differences combine in broad regional patterns of development versus underdevelopment or modernism versus traditionalism, often expressed spatially in the differences between the national core region, or heartland, and the periphery, or hinterlands. In India, which lacks a single heartland, the pattern is one of relative accessibility to the national metropolises of Bombay, Calcutta, Delhi, and Madras. In both the United States and Canada the factor of relative accessibility at the national level is independent of status and life-cycle variations but remains correlated with manufacturing as an economic specialization.

4. A culturally heterogeneous society will be characterized by separate ethnic or racial dimensions if the cultural groups are clustered in particular cities. If the groups occupy different status levels and have different family structures, the cultural differences may override other socioeconomic dimensions, as in the case of English-French contrasts in Canada.

5. Generally each new stage of growth will act independently of prior structural features if it is based on innovations giving rise to structural transformations. Thus distinct phases or stages of growth should each result in a separate latent dimension indexing a distinct pattern of variation of urban centers.

Such then, are the latent dimensions of manifest urban differences—the underlying causes of distinctive town types. We believe that their recognition and use can provide a much needed basis for a systematic ecological classification of urban communities.

APPENDIX
List of Variables

Number	Mnemonic	Name	Transformation
1	P/FOR.B/NENG	Per cent foreign-born with mother tongue not English	None
2	P/POP/SOB	Per cent native population residing in state of birth	None
3	MDSYO.25	Median school years for persons over 25	None
4	CUMFERTRATE	Cumulative fertility rate	None
5	FOVR14LF	Females over 14 in labor force	None
6	MOVR65LF	Males over 65 in labor force	None
7	P/MNF	Per cent in manufacturing	None
8	MNF TOTEMP	Manufacturing ratio, 1958: manufacturing employment as per cent of "aggregate employment"	None
9	P/HV 1-UNT	Per cent of all housing units in one-housing-unit structures, 1960	None
10	P/HVOWN	Per cent of occupied housing units which are owner-occupied, 1960	None

Number	Mnemonic	Name	Transformation
11	P/HV-SOUND	Per cent of all housing units sound and with all plumbing facilities, 1960	None
12	M AGE	Median age, 1960	None
13	P/U. 18	Per cent of population under 18, 1960	None
14	1st CENSUS	Year place first appeared in census	None
15	PASS 2-5TH	Year place passed 2–5,000	None
16	PASS 10TH	Year place passed 10,000	None
17	PASS 25TH	Year place passed 25,000	None
18	PASS 50TH	Year place passed 50,000	None
19	PASS 100TH	Year place passed 100,000	None
20	PASS 10TH R	Year place first over 10,000 and in Rand Metro Area	None
21	AREA	Area, 1960	None
22	FERT. RATE	Fertility ratio	Sq. root
23	POP/HH	Population per household	Sq. root
24	P/GP Q	Per cent living in group quarters	Sq. root
25	P/FOR. BORN	Per cent foreign-born	Sq. root
26	P/FOR. ST	Per cent foreign stock	Sq. root
27	P/MVED 58+	Per cent moved in after 1958	Log
28	P/HS 4 YRS	Per cent completing over 4 years high school	Sq. root
29	P/MWHH	Per cent married couples without own household	Sq. root
30	NW/W	Nonworker/worker ratio	Sq. root
31	MWINLF	Married women in the labor force	Sq. root
32	P/UNEM	Per cent unemployed	Log
33	P/W. OUT	Per cent working outside county of residence	Sq. root
34	P/PB. TRANS	Per cent using public transport	Sq. root
35	MD. ROOMS	Median rooms of all housing units	Sq. root
36	PERS/D.U.	Median persons per dwelling unit	Log
37	P/HV NW	Per cent housing units occupied by non-whites	Sq. root
38	VAL. OCC. UN	Median value owner-occupied units	Log
39	VAC. RAT. OCU.	Vacancy rate owner-occupied units	Log
40	VAC. RAT. RNT	Vacancy rate rentals	Sq. root
41	MNF63	Manufacturing employment, 1963	Log
42	WHL 63	Wholesaling employment, 1963	Log
43	RET 63	Retail employment, 1963	Log
44	SERV 63	Services employment, 1963	Log
45	MNF 58	Manufacturing employment, 1958	Log
46	WHL 58	Wholesaling employment, 1958	Log
47	RET 58	Retail employment, 1958	Log
48	SERV 58	Services employment, 1958	Log
49	COLL 60	College enrollment 1960 in 100s	Log
50	COLL 63	College enrollment 1963 in 100s	Log
51	E/R RAT 58	Employment/residence ratio, 1958	Sq. root
52	P/Of MIGR	Per cent of persons 5 and over who are migrants, 1960	Sq. root
53	P/ELMPRIV	Per cent of elementary school children in private school, 1960	Sq. root

Number	Mnemonic	Name	Transformation
54	P/HV 50–60	Per cent of all housing units in structures built 1950–1960, 1960	Sq. root
55	P/V 3,000	Per cent of families with incomes under $3,000, 1959	Sq. root
56	P/O 10,000	Per cent of families with incomes over $10,000, 1959	Log
57	P/50–60	Per cent change in population 1950–1960	Sq. root
58	MDRENT	Median gross rental of renter-occupied housing units, 1960	Sq. root
59	MDINC	Median income of families, 1959	Log
60	P/NW	Per cent of population which is non-white, 1960	Sq. root
61	P/WC	Per cent of employed persons in white-collar occupations, 1960	Sq. root
62	P/O.65	Per cent of population 65 and over, 1960	Sq. root
63	MNF.LF	Manufacturing, 1960	Log
64	WHL.LF	Wholesale trade, 1960	Log
65	RET.LF	Retail trade, 1960	Log
66	SERV.LF	Services, 1960	Log
67	MIN.LF	Mining, 1960	Log
68	Tran.LF	Transport and communications, 1960	Log
69	FIRE.LF	FIRE, 1960	Log
70	PUBAD.LF	Public administration, 1960	Log
71	HEWO.LF	HEWO, 1960	Log
72	ARMED.LF	Armed Forces, 1960	Log
73	MISCELL.	Miscellaneous, 1960	Log
74	TOT.LF	Total employed labor force, 1960	Log
75	NW HU 1.01+	Per cent of nonwhite housing units with over 1.01 persons per room	Log
76	NW HU OWN	Per cent of nonwhite housing units which are owner-occupied	Log
77	MED. INC. NW	Median income of nonwhite families	Log
78	POP 65	1965 estimated population	Log
79	POP 60	1960 population	Log
80	63/58 MANU	Manufacturing employment growth, 1958–1963	Log
81	63/58 WHL	Wholesaling employment growth, 1958–1963	Log
82	63/58 RET	Retail employment growth, 1958–1963	Log
83	63/58 SERV	Services employment growth, 1958–1963	Log
84	63/58 COLL	College enrollment growth, 1958–1963	Log
85	PR MANU	Per cent employed in manufacturing, 1960	Log/Log
86	PR WHL	Per cent employed in wholesaling, 1960	Log/Log
87	PR RET	Per cent employed in retail trade, 1960	Log/Log
88	PR SERV	Per cent employed in services, 1960	Log/Log
89	PR MINING	Per cent employed in mining, 1960	Log/Log
90	PR TRAN	Per cent employed in transport and communications, 1960	Log/Log

Number	Mnemonic	Name	Transformation
91	PR FIRE	Per cent employed in FIRE, 1960	Log/Log
92	PR PUB AD	Per cent employed in public administration, 1960	Log/Log
93	PR HEWO	Per cent employed in HEWO, 1960	Log/Log
94	PR ARMED SRV	Per cent employed in Armed Forces, 1960	Log/Log
95	PR MISC	Miscellaneous, 1960	Log/Log
96	PR GROWTH	Per cent growth 1960–1965	Log/Log
97	DENSITY	Population density	None

Chapter 16
Structural implications of increasing system size: a three-level analysis

The strongest latent dimension of urban systems reported in Chapter 15 was the functional size of cities in the urban hierarchy. What are some of the implications of differences in size for the internal structure of communities? More generally, what are the structural implications of increasing system size? This is the question that is probed in this chapter—one that serves to link structural components of smaller sociospatial systems examined in earlier chapters to broader societal structure, which is the focus of the chapters that follow.

The question of the relationship between the size and the structure of systems is one with which social scientists have exhibited continued interest. Spencer's analysis of ethnographic materials (1877) led him to infer that structural change invariably follows from an increase in the size of communal units. Similarly, Simmel's work with social aggregates (1902, 1903a) convinced him that the size of any group determines its internal structure. Durkheim's comparative examination of the large segmented social systems of nineteenth-century China and Russia (1893) pointed out that mere increase in numbers was not a sufficient condition to bring about structural differentiation. Along with an increase in mass (physical density), Durkheim observed, there must be an increase in interpersonal contacts (dynamic density) for structural change to occur. However, Durkheim viewed "dynamic density" to be largely predicated upon "physical density," so that in the final analysis he also attributed structural change to a society's increase in numbers (Parsons, 1937:322).

Since Durkheim's time relatively little research has been conducted on the influence of size on the internal structure of sociospatial systems such as communities and societies. Contemporary social scientists have, for the most part, neglected this larger issue, preferring instead to study the problem in the context of formal organizations (Hawley, 1967).[1] The objective of this chapter, therefore, is to examine the structural implications of system size in and across institutions, communities, and societies. We shall try to ascertain if similar structural changes occur in these different types of systems as they expand in size. We shall also try to determine if any particular structural component (or function) tends to dominate others in institutions, communities, and societies as they enlarge. Finally, we shall assess the generalizability to communal and societal systems of the current research findings and hypotheses on the structural effects of size in formal organizations.

Our exploratory hypotheses are derived from the principle of nonproportional change, as formulated by Scottish biologist D'Arcy Thompson (1917) and explicated more recently by Kenneth Boulding (1953). In short, this principle states that as any organic-like system grows, the proportions of its significant parts or functions *cannot* remain constant. The principle of nonproportional change rests on the premise that increasing size exacerbates particular system problems, which in turn result in a disproportionate growth in functions responsible for solving these problems.

Perhaps the most critical problem large-scale social systems face is maintaining communication among their parts. To compensate for the increased problems of communication, we would expect larger social systems to divert relatively greater proportions of their human resources to communicative functions. Our initial hypothesis, therefore, is that as institutions, communities, and societies enlarge, roles whose primary function is to aid communication will increase disproportionately.

A related problem that large social systems face is managing and coordinating their parts so that the system continues to operate as an integrated unit. This problem has received a good deal of attention at the institutional level (among others Akers and Campbell, 1970; Blau, 1970, 1972; Blau and Schoenherr, 1971; Hawley et al., 1965; Hendershot and James, 1972; Holdaway and Blowers, 1971; Kasarda, 1973; Klatzky, 1970; Meyer, 1972; Pondy, 1969; and Rushing, 1967), with the bulk of the findings showing that the proportion of managers and supervisors declines with large organization size. The explanation of the negative relationship, as postulated by Blau (1970), is that economies of managerial scale resulting from increased spans of control in large organizations exceed diseconomies resulting from greater

---

[1] The most notable exceptions include Zipf's studies (1941, 1949) of allometric relationships between system size and social-economic diversity, Duncan and Reiss's analysis (1956) of the structural characteristics of urban and rural communities, Clemente and Sturgis's (1972) investigation of community size and division of labor, Kasarda's studies (1972a, 1972b) of ecological expansion, and Mayhew's examination (1973) of Mosca's hypothesis (1939) relating proportion of ruling elite to political system size. Anthropologists, using data on primitive societies, have also examined the effects of system size on degree of sociocultural complexity. Naroll (1956) found an allometric relationship between settlement size and nonlineage associations. Likewise, Carneiro (1967) showed that the number of traits in a single community society varies approximately with the square root of population size.

complexity, so that the total effect of large size is a decline in the proportion of managers and supervisors. If the empirical findings and hypotheses for institutional systems are generalizable to communal and societal systems, then we would expect the proportion of managerial activities to decline in both communities and societies as they expand in size.

The third problem we shall examine is the effect of increasing system size on the proportion of professional and technical specialists in social systems. The professional and technical specialist informs, advises, and supports those who are directly engaged in attaining the goals of the system. We hypothesize that increasing system size will be matched by a disproportionate growth in professional and technical functions in social systems. This hypothesis is based on two beliefs: first, that in modern systems, large size generates greater problems of information gathering, evaluation, and planning which can be handled more efficiently by specialized professional and technical staff; and second, that given the specialized nature of most professionals and technicians, large sociospatial systems are better able to maintain (employ) these supportive activities on a full-time basis. The net result should be an increase in the overall proportion of professional and technical specialists in larger systems.

Taken together, the managerial, communicative, and professional and technical functions form the administrative fabric of modern sociospatial systems. What we are proposing, then, is not only will the overall administrative fabric of these systems change with increasing size, but also that its significant parts (its managerial, communicative, and professional and technical components) will be differently affected. As a consequence, we expect the internal structure of large systems to be quite different from the internal structure of their smaller counterparts.

To examine the relative merit of our hypotheses, data were gathered to measure the managerial, communicative, and professional and technical structure of institutional, communal, and societal systems. Our institutional sociospatial systems are represented by 178 school systems in the state of Colorado. We selected school systems because they each performed the same task, had a comparable technology, distributed themselves on a continuum of size, and had clearly defined data on managerial, professional and technical, and communicative functions. The 1969–1970 annual report of the superintendent of each school system (Colorado Department of Education, 1971) presents a breakdown by function performed of all the system's personnel. We aggregated these data into four categories of employees. Those employees defined as performing *managerial functions* include superintendents, assistant superintendents, principals, assistant principals, directors, assistant directors, administrative assistants, business managers, coordinators, and supervisors. The *communication* category includes all secretaries and clerical personnel. *Professional and technical* staff include guidance counselors, librarians, psychologists, social workers, speech therapists, and school nurses. *Instructional* staff consists of all classroom teachers in the system. The relative size of the managerial, communicative, and professional and technical components of each school system is measured by the percentage of personnel engaged in the respective categories. Our measure of functional size is the total number of persons employed by each school system.

Communal systems are represented by 207 communities in the state of Wisconsin with populations of less than 25,000. The 1970 Census of Population fourth-count summary tapes (U.S. Bureau of the Census, 1972) were used to obtain data on each community's occupational structure. The tapes were used because published reports of the 1970 census provide detailed occupational characteristics only for communities (that is, places) of 2,500 or more inhabitants. This precludes analysis of communities smaller than 2,500, a large and important set for the present analysis, whereas in Wisconsin small incorporated centers in nonmetropolitan areas are treated as minor civil divisions. Detailed occupational data on minor civil divisions are available on the fourth-count tapes; therefore one can extract the occupational characteristics of a large sample of smaller communities.

For each communal unit, the managerial component is measured by the percentage of its employed residents classified as managers, officials, and proprietors. The communicative component is measured by the percentage of its employed residents engaged in clerical and kindred activities, and the professional and technical component is measured by the percentage of employed residents classified as professional, technical, and kindred. The functional size of each communal system is indexed by the total number of economically active residents.

Two methodological problems affected the selection of the sample. First, since the occupational data were secured from a 20 per cent sample of the total population, communities containing fewer than 150 employed residents were eliminated to reduce possible bias caused by a sampling error. Second, because these data were gathered on a de jure (official place of residence) rather than a de facto (where they actually work) basis by the Bureau of the Census, community structure is not accurately depicted by census data for those communities that have a substantial proportion of residents employed in large cities or other outlying communities. To mitigate the potential problem caused by commuting, our sample is further restricted to nonmetropolitan communities that are at least 25 miles from any other community of 25,000 or more. We believe that restricting our sample to smaller, more isolated communities improves the epistemic correlation between the sociological concept of community and its measure, provides units with greater system closure, and lessens the likelihood that the community's various administrative and support functions will be performed primarily by large cities. Nevertheless, we do not wish to convey the impression that our sample of communities is at all self-contained and free of the influence of large metropolitan areas. Numerous studies have documented the impact that mass society can have on small, relatively isolated communities (Stein, 1964; Vidich and Bensman, 1958; Warren, 1972) and our earlier discussions of daily urban systems (Chapter 14) and functional classifications (Chapter 15) reinforces the idea that these communities are linked into larger urban networks.

The third type of sociospatial system, societies, is represented by all nonagriculturally based nations for which complete and comparable data are available on industrial and occupational structure. Our final sample consists of 43 nations which have more than one half their economically active population engaged in either secondary or tertiary activities. We selected Argentina, Australia, Austria, Barbados, Belgium, Canada, Ceylon, Chile, Colombia, Costa Rica, Czechoslovakia, Finland, France, Germany (Republic of), Hungary, Ireland, Italy, Jamaica, Japan, Jordan, Kuwait, Libya, Luxemburg, Mauritius, the Netherlands, New Caledonia, New Zealand, Norway, Panama, Peru, Portugal, Ryukyu, South Africa (Republic of), Spain, Sweden, Switzerland, Syria, Tunisia, Trinidad and Tobago, United Kingdom, United States, Uruguay, and Venezuela.

Occupational and industrial data for the societal units were obtained from the international *Year Book of Labour Statistics* (International Labour Office, 1969, 1970). The managerial component for each society is measured by the percentage of its economically active population classified as administrative, executive, and managerial workers. The communicative component is indexed by the percentage of economically active population performing clerical and related functions; and the professional and technical component is measured by the percentage of economically active population engaged in professional, technical, and related functions. Functional size is measured by the total number of economically active persons in each society.

A total administrative component for the institutional, communal, and societal systems was also computed. This component was obtained by summing the number of managerial, communication, and professional and technical workers in each system and dividing that sum by the total number of workers. Such a measure is analogous to what has been labeled "administrative intensity" in research on formal organizations (Pondy, 1969).

Before we examine the implications of functional size for system structure, it will be useful to present the means and standard deviations of each structural component for our samples of institutions, communities, and societies. These are presented in Table 16.1.

The relative size of the managerial, communicative, and professional and technical components differs in and across system levels. Comparing the mean structural components for the three types of systems, one may observe that communal systems have the largest components for every set of administrative functions. As a whole, administrative functions constitute approximately one third of the functions performed by communal systems,

Table 16.1

Means and standard deviations of selected components of social-system structure, by type of social system

Structural components	Institutions (N = 178)	Communities (N = 207)	Societies (N = 43)
Managerial			
mean	8.09	9.54	2.34
standard deviation	(3.14)	(4.78)	(1.69)
Communicative			
mean	3.98	11.92	7.93
standard deviation	(3.41)	(4.48)	(3.58)
Professional and technical			
mean	3.83	12.13	6.59
standard deviation	(3.03)	(5.74)	(2.90)
Total administrative			
mean	15.90	33.59	16.86
standard deviation	(5.35)	(9.05)	(7.09)

whereas they constitute only about 16 per cent of the total functions performed by institutional and societal systems. Viewing the relative size of the structural components in the three types of systems, we see that the managerial component of institutions clearly exceeds their communicative and professional and technical components. However, the opposite is the case in communal and societal systems. Such discrepancies among institutions, communities, and societies in the mean size of their structural components should not be surprising given the diverse nature of these social systems.

The important issue for analysis is: What happens to these structural components as the fractional size of each type of system expands? Table 16.2 tells us. It presents the total administrative components and the managerial, communicative, and professional and technical components of institutions, communities, and societies by the size of each system.

The influence of functional size on the total administrative component of the systems is positive and consistent across the three levels of the social-system hierarchy. The mean percentage of administrative employees for our sample of institutions expands from 12.88 in the smallest school systems to 20.48 in the largest. Administrative intensity similarly increases from 31.59 per cent in the smallest communities to 40.58 per cent in the largest, as well as from 13.57 per cent in the smallest societies to 25.17 per cent in the largest. These positive gradients between the size of institutional, communal, and societal systems and their overall administrative structure suggest that a ramification of organizational complexity does indeed occur with increases in system size.

It is also important to note how the relationship of managerial, communicative, and professional and technical functions to other system functions changes differentially with increasing system size. For institutions, we observe that the managerial component declines as the system enlarges. This result corroborates recent findings on the negative influence of size on the managerial component of formal organizations. On the other hand, functions whose primary task is to aid communication increase dramatically with system size. Likewise, professional and technical functions expand as the institutions enlarge. Since large organization size generates greater relative increases in communicative and professional and technical functions than it does a relative decline in managerial functions, a monotonic increase occurs in the overall administrative structure of institutions as they expand in size.

Looking next at structural changes in communities, we see that communicative functions again show the largest relative increase with expanding system size. Professional and technical components also tend to increase with larger system size. However, the relationship of the managerial component to community size is not consistent. Medium-size communities have, on the average, greater percentages of residents performing managerial functions than do either small or large communities. It is questionable, therefore, whether findings on managerial scale among institutions can be generalized to larger, more diffuse sociospatial systems.

This difficulty becomes even more apparent when we examine size-structure relationships for societal systems. The relationship between the functional size of societies and the proportion of their economically active population engaged in managerial-related functions is exactly opposite that found in institutions. As nonagricultural societies expand in size, greater

Table 16.2

Mean components of social-system structure, by type and size of social system

Type and size of social system	Total administrative component	Managerial component	Communicative component	Professional and technical component
Institution size				
Under 25 (N = 71)	12.88	9.23	1.19	2.46
25–99 (N = 64)	17.11	8.03	4.55	4.53
100–750 (N = 33)	18.70	6.49	7.29	4.93
Over 750 (N = 10)	20.48	5.66	9.30	5.52
Community size				
150–499 (N = 128)	31.59	9.26	11.01	11.33
500–1999 (N = 60)	36.23	10.03	12.99	13.21
2000–5000 (N = 16)	38.10	10.08	13.95	14.07
Over 5000 (N = 3)	40.58	8.57	18.08	13.94
Society size				
Under 500,000 (N = 11)	13.57	1.72	6.49	5.36
500,000–5 million (N = 20)	16.80	2.18	7.98	6.73
5–25 million (N = 9)	18.26	2.67	8.27	7.31
Over 25 million (N = 3)	25.17	4.64	12.56	8.00

Table 16.3

Product-moment correlation coefficients between size (log) of social systems and their structural components, by type of social system[a]

Structural component	Institutions (N = 178)	Communities (N = 207)	Societies (N = 43)
Managerial	−.431 (.001)	.045 (.258)	.359 (.009)
Communicative	.776 (.001)	.302 (.001)	.371 (.007)
Professional and technical	.358 (.001)	.229 (.001)	.306 (.023)
Total administrative	.438 (.001)	.320 (.001)	.398 (.004)

[a] Levels of statistical significance are in parentheses.

proportions of their functions are devoted to managerial tasks. Positive relationships also exist between the size of societies and both their communicative and professional and technical components. Consistent with findings among institutions and communities, the communicative component of societies exhibits the largest relative increase with expanding system size.

The results in Table 16.2 tentatively support our proposition not only that the overall administrative structures of sociospatial systems change with expanding size, but also that their significant parts are differentially affected. A major shortcoming of the analysis, however, is that we have grouped interval scale structural variables into size categories and simply compared their means. Although presenting results in this manner illustrates the relative magnitude of the structural components at different levels of system size, the cut-points remain somewhat arbitrary and may mask important variation in system structure.

To avoid the problem of cut-points and examine more precisely the relationship between system size and structure, we turned to correlation analysis. Product-moment correlations were computed between the size of each social system and its structural components.[2] Table 16.3 presents the correlation coefficients and their levels of statistical significance.

[2] The sizes of the social systems were transformed via logarithms ($\log_{10}$) to reduce the skewed distribution resulting from the presence in our samples of a few very large institutions, communities, and societies. The mean logarithms and standard deviations (shown in parentheses) of the institution, communal, and societal sizes are 1.70 (0.58), 2.52 (0.41), and 6.31 (0.76), respectively. For a theoretical as well as empirical rationale for transforming system size by means of logarithms in studies of system structure, see Mayhew, et al. (1972).

As our previous data have suggested, large size generates greater degrees of administrative intensity in each of the three types of social systems. The correlations between size and the total administrative components of social systems range from .320 for communities to .438 for institutions. Further examination of Table 16.3 indicates that the positive correlations between the size of the three types of social systems and their total administrative intensity results primarily from the positive influence large system size has on communicative and professional and technical functions. This is particularly so among institutions where a rather strong negative correlation exists between size and the proportion of functions devoted to managerial tasks.

The inconsistency across system levels of the effect of size on the managerial structure of social systems is once more apparent in Table 16.3. These substantial differences in the correlations between system size and the managerial components of institutions, communities, and societies suggests that type of social system exerts a strong interactive influence. Hence, the generalizability to communal and societal systems of findings and hypotheses on managerial scale among institutions becomes problematic.

Our other two hypotheses receive clear support from the data presented in Table 16.3. Positive and consistent relationships exist between the size of institutions, communities, and societies and the magnitude of their communicative and their professional and technical components. The fact that the correlations between system size and communicative functions are the largest in each type of social system further supports the contention that communication is the most critical problem faced by large social systems.

To provide yet another test of this contention, census data were gathered on the proportion of residents in 216 metropolitan communities (SMSAs) in 1970 who performed managerial, communicative, and professional and technical functions in each SMSA. Strong support was in the offing. Correlation analysis of the relationship between SMSA size ($\log_{10}$) and the proportion of total employees devoted to managerial, communicative, and professional and technical functions yielded coefficients of $+.05$, $+.40$, and $+.16$, respectively. Cross-tabular analysis of the structural components by size categories showed communicative functions exhibiting by far the steepest gradients as system size expanded.

Organization size and administrative structure

Having examined the influence of size on the relation of a system's managerial, communicative, and professional and technical functions to the whole, we next wish to determine how the composition of the administrative fabric itself changes with expanding system size. In conducting this final phase of the analysis, we shall focus on school systems.

We chose the institutional units over communities and societies for three reasons: first, our sources of data enable us to obtain more precise measures of managerial, communicative, and professional and technical functions for school systems than for communities or societies. Second, the school systems more closely approximate closed systems; and third, since most studies of administrative structure have used institutional units, not only is comparability improved, but the likelihood of adding to the research literature on organizational structure is increased as well.

What, then, are the implications of system size for the relative magnitude of the functions which consitute the administrative fabric of institutions? This question is answered in Table 16.4, which presents—by size of the school system—the mean percentages of administrative personnel who perform managerial, communicative, and professional and technical functions.

Decomposition of administrative personnel by functional task shows that within small organizations, managerial tasks form the bulk of administrative functions. This should be expected: in smaller organizations management has more direct contact with personnel and therefore less need for large clerical and professional advisory staffs. Indeed, it seems likely that in small organizations the magnitude of facilitative functions such as record keeping, communication, evalution, and reporting is small enough to be handled almost entirely by management. With the expansion of functional size, however, communication, evaluation, and planning become increasingly difficult, resulting in larger proportions of administrative resources being drawn over to clerical and professional staff functions. These structural changes are well illustrated in Table 16.4, which shows that the proportion of administrative functions devoted to managerial tasks declines from 76.90 per cent in the smallest institutions to 27.78 in the largest institutions, whereas the proportion engaged primarily in communication increases from 7.65 per cent to 45.27 per cent, and the proportion performing professional staff functions increases from 15.45 per cent to 26.98 per cent. The remarkable expansion of the communicative (clerical) elements with increasing system size again illustrates how important those elements are in holding large organizations together.

Correlation analysis of the relationship between school system size and

Table 16.4

Mean percentages of administrative personnel performing managerial, communicative, and professional staff functions, by size of school system, 1970

Functional size	Managerial functions	Communicative functions	Professional staff functions
Under 25 (N = 71)	76.90	7.65	15.45
25–99 (N = 64)	47.65	26.80	25.54
100–750 (N = 33)	35.23	38.15	26.61
Over 750 (N = 10)	27.78	45.23	26.98

the proportion of administrative personnel devoted to managerial, communicative, and professional staff functions also yields impressive results. Product-moment correlations between the logarithm of size and the proportion of administrative personnel devoted to such functions are $-.70$, $+.72$, and $+.30$, respectively. In terms of explained variance, increasing functional size accounts for 49 per cent of the decline of the managerial elements and 51 per cent of the concomitant expansion of the communicative elements. Although the professional staff element of the administrative fabric is not as greatly influenced by functional size as are the managerial and communicative elements, size still accounts for 9 per cent of its variance.

In sum, the above results indicate that functional size has substantial implications for the composition of the administrative fabric of institutions. Most affected by large size are those administrative tasks whose primary functions are communication and management. The strong opposing correlations, however, indicate that communicative functions increase with system size at a much faster rate than do managerial functions. The net result is a disproportionate amount of human resources being drawn into communicative functions in larger organizations.

Summary and inferences

Although these findings should be regarded as exploratory, they do support the general proposition that functional size has a pervasive influence on the internal organization of modern sociospatial systems. We observed that large size promotes greater administrative intensity in institutions, communities, and industrialized societies. Large size also differentially affects the managerial, communicative, and professional and technical structure of systems. The most prominent structural changes occur in communication. As institutions, communities, and societies expand, substantially greater proportions of their personnel are devoted to communicative (clerical) functions. It may therefore be inferred that the major role of holding large sociospatial systems together rests with those whose primary function is facilitating communication.

A second inference that may be drawn is that large size promotes an increase in the proportion of professional and technical specialists within modern social systems. Two reasons are suggested for the disproportionate growth of professional and technical functions. First, large functional size generates additional problems of information gathering, evaluation, and planning which usually can be handled more efficiently by specialized professional and technical staff. Second, with larger numbers of system personnel to draw on the guidance and advice of professional and technical experts, relatively more of such experts can be effectively employed on a full-time basis.

The final inference to be drawn is that research findings showing a negative relationship between size and managerial structure of formal organizations cannot be applied to larger and more diffuse sociospatial systems. On the contrary, as we move to ever larger systems the relationship changes from negative to positive. Perhaps the diverse nature of the parts of communities and, more particularly, of societies requires a greater complement of managers and officials to ensure that the larger system operates in an orderly manner. Another possible explanation for the reversal in correlations is that as we move from institutions to communities to societies, the degree of system interdependence (or dominance) increases. Under such circumstances, a working hierarchy emerges in which larger sociospatial systems perform extralocal administrative functions for smaller systems. Since the previous two chapters on urban systems demonstrate that functional size is the single best indicator of hierarchical position, it seems more than plausible that the managerial components of large communities and societies are substantially greater because of the extralocal administrative functions they perform. In the next section, Chapter 19 will further highlight this likelihood.

PART VI
COMPARATIVE URBAN STRUCTURE AND PLANNED CHANGE

Chapter 17
The role of planning in comparative urban change: the United Kingdom, the Soviet Union, and the United States

We now turn to the broadest issues that arise in contemporary urban ecology; that is, the comparative study of the relationships between urban structure and societal change. Our purpose in this chapter is to examine the role of urban policy and planning in three diverse societal contexts. First, we focus on the United Kingdom, where a tradition of public involvement has attempted to engineer urban change to achieve national social goals. Next, we look at another form of planned urban development, the city of socialist man. Finally, we return to the American scene, where public policy and urban planning remain committed to private initiative and corporate growth within a mosaic of cultural pluralism. The two chapters that follow probe a similar range of issues in the context of the Third World.

The questions raised are of fundamental importance if the body of ecological theory is to grow. Recall that one of the basic assumptions of traditional ecological theory was that urban structure evolved largely through the spontaneous, unplanned operation of the market. Conscious design was either absent or trivial in influencing the ultimate nature of urban form. This assumption, of course, reflected the sociopolitical and economic milieu of nineteenth- and early twentieth-century America, where the dynamics of privatism and laissez faire enterprise prevailed. Competition served as the main driving force, sorting individuals, groups, and institutions into spatial and functional niches, relatively unfettered by public intervention. Even today, free-enterprise dynamics predominate in establishing urban order in the United States, although numerous government policies have been initiated in recent years in an effort to counter the social consequences of laissez faire urbanization. Unfortunately, these policies have proven ineffective in most cases and in some have only exacerbated existing social ills. In the United Kingdom and the Soviet Union, on the other hand, variously effective means of regulating urban structure and change have developed. Public counterpoints to private interests have emerged, with the intent of explicitly directing the urbanization process. In these cases, social policy goals have become part of the urbanization process, with directed change contributing to social and structural outcomes that differ markedly from those predicated by traditional ecological theory.

Great Britain

Industrial urbanization came earlier to Britain than it did to North America. As early as 1843, Robert Vaughan proclaimed that "ours is the age of great cities." From 1801 to 1911, Britain's urban areas accounted for 94 per cent of the country's population increase. One third of the urban growth was due to net immigration from rural areas (Lawton, 1972). But there was no planning of urbanization in the birthplace of the industrial metropolis. Victorian urban growth brought new factories and workshops in abundance; to handle the products of the new specialization, port facilities and railway terminals, adjacent to the old city, became the commercial core and the symbol of the new urbanization. In association with the railway terminals and docks were to be found the wholesale warehouses and retail commodity markets; separating the old city from the newer residential developments were the factories and workshops. Horse-drawn city transport, and later the suburban railways and tramways, gave rise to better-quality suburban development in areas previously beyond walking distance from the center. The character of the suburbs owed much to the pattern of communications, the pre-existing types of residential areas within the city, and the nature of land ownership in the areas under development. In Britain, the middle classes were the first to move out of the city to single-family homes in new suburbs, thus breaking the older tradition of the preindustrial city, in which the workers' suburbs were peripheral, often beyond the walls. Interestingly, this move was not paralleled extensively elsewhere in Europe, where to this day "suburban" frequently implies lower-class, and where the historic claims of centrality and the preference for apartment living still conspire to contain many continental cities to more restricted areas. It is this difference, emerging as it did in the mid-nineteenth century, that Choay (1965) has argued is the source of the two dominant ideologies in European urban planning in this century. British New Towns concepts sought a new balance of town and country, reflecting the new middle-class orientation to the suburbs and beyond, whereas the French *grand ensemble* focused on the apartment house and on the higher-density living it represents. To Choay, the work of Ebenezer Howard, Lewis Mumford, and Frank Lloyd Wright on the one hand, and that of Le Corbusier on the other, exemplify these two main currents of thought; in this sense Wright's Broadacre City is the ultimate in decentralization of the family house, whereas Le Corbusier's Ville Radieuse is the archetype of vertical steel and glass apartment towers, cruciform in shape, elevated on stilts, and with wide pedestrian spaces between.

In Victorian Britain, housing for the new factory workers was built in courts, back from the street—confined, poorly lit, badly ventilated, and lacking even rudimentary water supplies and sanitation. Often cellars were used as dwellings to increase the housing capacity. A graphic picture of the urban ecology and social structure of the times was provided by Friedrich Engels in 1844:

> Manchester contains, at its heart, a rather extended commercial district, perhaps half a mile long and about as broad, and consisting almost wholly of offices and warehouses. Nearly the whole district is abandoned by dwellers, and is lonely and deserted at night.... The district is cut through by certain main thoroughfares upon which the vast traffic concentrates, and in which the ground level is lined with brilliant shops. In these streets the upper floors are occupied, here and there, and there is a good deal of life upon them until late at night. With the exception of this commercial district, all Manchester proper, all Salford and Hulme...are all unmixed working people's quarters, stretching like a girdle, averaging a mile and a half in breadth, around the commercial district. Outside, beyond this girdle, lives the upper and middle bourgeoisie in regularly laid out streets in the vicinity of working quarters...the upper bourgeoisie in remoter villas with gardens...in free, wholesome country air, in fine, comfortable homes, passed every half or quarter hour by omnibuses going into the city. And the finest part of the arrangement is this, that the members of the money aristocracy can take the shortest road through the middle of all the labouring districts without ever seeing that they are in the midst of the grimy misery that lurks to the right and left. For the thoroughfares leading from the Exchange in all directions out of the city are lined, on both sides, with an almost unbroken series of shops, and are so kept in the hands of the middle and lower bourgeoisie...[that] they suffice to conceal from the eyes of the wealthy men and women of strong stomachs and weak nerves the misery and grime which form the complement of their wealth.... I know very well that this hypocritical plan is more or less common to all great cities; I know too, that the retail dealers are forced by the nature of their business to take possession of the great highways; I know that there are more good buildings than bad ones upon such streets everywhere, and that the value of land is greater near them than in remote districts; but at the same time, I have never seen so systematic a shutting out of the working class from the thoroughfares, so tender a concealment of everything which might affront the eye and the nerves of the bourgeoisie, as in Manchester. And yet, in other respects, Manchester is less built according to plan after official regulations, is more outgrowth of accident, than any other city; and when I consider in this connection the eager assurances of the middle class, that the working class is doing famously, I cannot help feeling that the liberal manufacturers, the Big Wigs of Manchester, are not so innocent after all, in the matter of this sensitive method of construction.

The spatial organization described by Engels is quite analogous to the one described later for the North American industrial metropolis by R. E. Park and E. W. Burgess; because of its ills, it not only became anathema to the socialist ideologies of the later nineteenth century, but also was the stimulus of planned urban change.

One of these was the New Towns movement. In Ebenezer Howard's mind (1898), the New Town was to achieve three goals: (1) rearrangement of the inhuman mass of the big industrial city on a human scale in new towns of strictly limited size; (2) balance between the quantity of housing and the number of jobs; and (3) public ownership of the land to paralyze speculation and thereby control development. His main concern was the plight of the overcrowded cities and depleted country districts, and he concluded that there were not just two alternatives—town life and country life—but a third, town life in a country setting "in a perfect combination." Working with the distinguished planner Raymond Unwin, he developed and built the first garden city at Letchworth in 1903 as a "city in a garden." Later, with F. J. Osborn and C. B. Purdom involved, the Garden Cities and Town Planning Association led in the construction of Welwyn Garden City in 1920.

But the First World War intervened. After the war Britain was faced with an immense housing shortage, and it was agreed that the private sector was incapable of meeting housing needs at rents people could pay. Housing Acts in 1919 and 1924 enabled local authorities to build and plan public housing estates. From 1919 to the mid-thirties, local-authority housing and other state-aided construction accounted for about one half of all new housing (as they do today), and a major policy aim had emerged: to provide satisfactory housing for all at rents within their capacity to pay. Most of the new public-housing programs lacked broader planning content, however, and the private sector of the housing industry, relatively unconstrained, produced increasing urban sprawl radiating from towns in the form of speculative profit-seeking ribbon developments responding to the demands of Britons for homes in the countryside. In response to such speculation came a succession of Town Planning Acts, the most important being that of 1932, which authorized local governments "to control development; to secure proper statutory conditions–amenity and convenience; to preserve existing buildings...." Local authorities were authorized to zone land and to reserve land from development; both of these ingredients were to assume major postwar importance in controlling urban form.

The social impact of these changes was substantial. Indeed, Robson (1969) has concluded that "in the twentieth century, the development of council housing and of town planning...[has] largely invalidated many of the bases on which the classical models (i.e. Wirth) have been built." He refers to the lower spatial and social mobility in British cities; the critical importance of the local area as a facet of the individual's social world, involving significant face-to-face relationships and kinship networks; the persuasive influence of age structure on social outlook in housing estates where early occupants stayed and aged together; and the continuing importance of more rigid social class roles in the British way of life. Each of these factors was to play its role in postwar Britain, just as they had converged to create new life styles in the council house estates.

But in 1932 Britain was still experiencing the mass unemployment brought on by the Great Depression, although that year saw the beginning of the upturn in trade and finance. At the time that A. Trystan Edwards was circulating proposals for a hundred New Towns widely spread throughout the country, and at the time in 1935 that the London County Council appropriated £2 million for acquisition of land as part of a green belt system, the Royal Commission on the Distribution of the Industrial Population was created under the chairmanship of Sir Montague Barlow. The Barlow Report was delayed by the beginning of the Second World War. When published in 1940, it recommended controlling London's growth, redevelopment of congested urban areas, decentralization or dispersal of industry and population from such areas, and the encouragement of "balanced" regional growth; to achieve the last two, the report recommended the creation of garden cities or garden suburbs, satellite towns, trading estates, and the development of existing small towns and outlying regional centers. Sir Patrick Abercrombie was one of six members of the Barlow Commission signing minority reports urging stronger recommendations and a New Towns Policy, which he subsequently helped fashion.

The postwar plans: new towns

The war resulted in widespread devastation, especially in London. But as early as October 1940, Lord Reith had been appointed Minister of Works and Buildings, and in March 1941 he asked the London County Council to prepare plans for postwar reconstruction. The LCC hired Patrick Abercrombie to work as consultant with the Council's architect, J. H. Forshaw. The result was The County of London Plan (1943). Abercrombie pictured London in concentric-ring terms. The inner ring was to have population and industry reduced. A surrounding suburban ring required no action. This was girdled by a green belt in which urban growth was to be stopped. Finally, in the surrounding country ring, the planned expansion of existing towns and eight totally new towns were to house overspill population and industry— 525,000 people in the planned expansions and 350,000 people in the new towns.

Many of his population assumptions have been shown to be wildly astray, but nevertheless Abercrombie's was the ideal that guided postwar planning, supported by the 1943 and 1944 Town and Country Planning Acts which strengthened the machinery for intervention and control of private development by the system of planning permissions. A new Labor government came to power by a landslide majority after the war, and brought with it enthusiastic commitment to planning, ideological opposition to profit-seeking development, and leftover wartime controls, most notably the complete control of all building licenses.

Planning could proceed affirmatively because most housing was to be publicly built, new construction was to be licensed in accordance with national investment priorities, and physical design was to be an instrument of the new social policy. Plans were made at all levels of government, and powers were provided for their implementation.

In 1945 Lord Reith was appointed to chair an Advisory Committee to the Minister of Town and Country Planning

> to consider the general questions of the establishment, development, organization and administration that will arise in the promotion of New Towns in furtherance of a policy of planned decentralization from congested urban areas; and . . . to suggest guiding principles on which such towns should be established and developed as self-contained and balanced communities for work and living.

Three subsequent reports were prepared, and culminated in the New Towns Act of 1946.

Between 1946 and 1949 eight New Towns were designated for Greater London, including Stevenage, Crawley, and Hemel Hempstead. The "web" of approaches comprising doctrine in this planning effort for London has been described by Donald L. Foley (1963). At the national level the main social goal was maintenance of full employment, from which were obtained derivative program goals such as the redistribution of employment to areas of high unemployment, made possible by the supporting approach of industrial development certificates. At the regional planning level the main goal related to the physical organization of the London region, while at the town planning level the goal was to create the best possible living conditions for the population. A strong intent of "containment" permeated the effort; crowded inner areas were to be thinned, the growth of the core of London was to be contained, and new growth was to be directed into socially and economically balanced New Towns of limited scale.

Since the initial plans were drawn, the number of New Towns actively under development has been increased to 22, including a later stage of higher-density towns such as Cumbernauld outside Glasgow, and a new phase of massive extensions to such existing towns as Peterborough and Northampton (28 in all are now designated). These will be supplemented in the future by substantial additions to some of the existing towns, and by very large new communities developed as the outcome of the regional strategy plans of the last decade which will be discussed later.

Related development controls

Although they are the most conspicuous parts of British planning, only a small part of total public-sector activity is to be found in the new towns.

Local-authority public housing estates still provide half of all new housing for Britons. Private-sector housing caters in large measure to families of at least middle class. Slum clearance has proceeded rapidly. Control of building licenses (subsequently replaced by the requirement of planning permission to build), in combination with the development controls provided by the act of 1947, provided for an effective public counterpoint to speculative profit-seeking private development through strict land use zoning principles. The 1947 act made it obligatory for local authorities to control land use. Housing subsidies have encouraged higher-density developments, of both apartments and offices. Industrial decentralization has been substantial, although massive office growth has led to continued employment growth in central London. The Clean Air Acts have eliminated much of the smoky atmosphere of British towns. But Gordon Cherry (1972) argues that although the results have been impressive, they also have been surprisingly conservative:

> The success of local authority intervention has been to see that development has occurred more or less at the right time in the right places. Schools have been built in accordance with housing programs; open space has been reserved; building in certain areas has been restricted; elsewhere it has been encouraged; unfit houses have been demolished and new ones built; shopping centers rather than ribbons have been created; the worst excesses of badly sited industry have been avoided. This represents a high point in public intervention in the shaping of an urban environment.... A coherent strategy for the planned distribution of population and employment which was linked to a policy of containing the further growth of large cities was pursued.... But the total influence of planning intervention has to be seen in even wider terms than the security of order and a sense of rationality. Its greatest impact has been to bolster the traditional view of the West European city...that cities, as ancient repositories of culture, should be protected from decay. In this sense, planning has been essentially a conservative movement aiming to retain traditional ideas about urban society and urban functions.

In other words, a nostalgic sense of attachment to urban forms, perceived to be functional in the past, has pervaded Britain's development controls.

The new towns assessed

If the development controls were conservative in their application, what of the radical alternative, the New Towns ? In many respects, London's New Towns have been great successes, housing over 470,000 people. Peter Hall (1966) argues that they have become magnets for over 400 industrial plants, places of employment for over 250,000 workers, attractive shopping centers, and major market areas. But, in a deeper sense, they reflect a failure of Abercrombie's plan. The initial idea was to create self-contained communities that would receive a population of 400,000 and a corresponding amount of employment, developed by public corporations and instrumental to the removal of Londoners from a London prevented from growing and attracting jobs from outside.

The Abercrombie proposals were modified in several ways. Only eight new towns were designated instead of ten, and three of these were already well-established population centers. Many of the modifications were needed because Abercrombie's population forecasts proved to be in error. In the 1940s it was thought that the country faced a period of population stagnation. It was not expected that the population of the London region would increase, and plans for the early postwar period were concerned almost entirely with decentralization, envisaging a movement of about one million people from the Greater London Conurbation to the Outer Metropolitan Area. The New Towns were to play a major role in casting decentralization into new forms.

Between 1951 and 1971 the Greater London Conurbation had, indeed, decreased in population by some 3.0 per cent per decade. But the Outer Metropolitan Area had increased in population far more rapidly than this— between 1951 and 1961 alone, by nearly the million envisaged as the "ultimate" target in Abercrombie's plans. Thus, London's New Towns had much less of a total impact on decentralization than envisaged. By 1971 their population approached 470,000 people, with their ultimate targeted population 650,000, but this has represented only 18 per cent of population growth in the outer ring of southeastern England.

The most important idea in the creation of the New Towns was that they should be self-contained and balanced communities for working and living (Thomas, 1969), thus differing from the local-authority housing estates and from private enterprise's version of the garden city, the commuting suburb attached to an existing settlement. Each New Town was to be made up of neighborhood units centering on schools, shops, and other local facilities, following the concepts of Clarence Perry discussed earlier. The neighborhoods, together with a major shopping center, industrial estate, and a range of community-wide educational, cultural and recreational facilities, were to constitute a self-contained entity, with girdling open space. In this way the ideals of Ebenezer Howard would be met, combining the advantages of town and country with none of the disadvantages. Balance was to be achieved in several ways: between residence and employment, between facilities and needs, and also between social classes. For example, one of the charges to Lord Reith's committee in 1946 was as follows: "If the community is to be truly balanced, so long as social classes exist, all must be represented in it... the community will be poorer if all are not there." Thus, subsidized rental housing was to be provided alongside homes for higher-income groups, and different family sizes and age groups would be catered to. It had been felt that, at Letchworth and Welwyn Garden City, a mixture of social classes had brought about a sense of community that was absent in most towns. It was hoped that a similar sense of community, in which class distinctions were minimized, could be created in the New Towns by means of physical planning. In most of the New Towns, there was a mixture of housing for different groups in each neighborhood.

The New Towns have served as magnets for modern industry and have provided vastly improved physical conditions and housing for their residents, predominantly young Britons in the child-rearing stages of the family cycle. Slightly more than half of their residents came from London, but the Londoners tended to stay; 96 per cent of the net immigration came from London (Thomas, 1969). For the others, the New Towns acted as staging points in migration to parts of the country other than London.

In large measure the self-containment goal has been achieved. A far higher proportion of the residents of the New Towns work in them than elsewhere. Whereas in other parts of the Outer Metropolitan Zone of London, where population growth has been accommodated by the more conventional suburbs and housing estates, growth of car ownership and long-distance commuting to London have gone hand in hand. In the New Towns, on the other hand, increasing entry of women into the labor force has increased rather than decreased their self-containment, and commuting to London has decreased.

The question of social balance in the New Towns is much more debatable. A significant proportion of the residents in the New Town of Harlow had other family members in the town, frequently of the same generation, but also including retired parents because of the housing created for retirees. Critics have spoken of "New Town blues" of working-class families cut off from traditional social networks of the increasing social homogeneity of towns from which professionals and managers move to private housing developed in surrounding villages. Yet studies comparing residents of New Towns with those of old city boroughs have failed to find differences in psychosomatic or psychiatric symptoms attributable to the new environment (Pahl, 1970). To be sure, feelings of social isolation tended to be greater among those just separated by long distance from kin and friends, but they were soon offset by involvement in local circles. The greatest strains in the new towns appear to have been among the blue-collar workers displaced from inner-London neighborhoods in which family and kinship ties were strongest (Young and Willmott, 1957). Where kin have moved to the same New Town—in Harlow this was the case for 47 per cent of the families—these strains have been absent. And whereas there is a general feeling of satisfaction with the new community, there is little evidence that the life style of the New Town residents changed significantly as a result of the move. Nor is there any evidence, in Britain's class-conscious society, that communications across class lines have improved.

It was possible to build socially mixed neighborhoods in Letchworth and Welwyn Garden City exactly because the class system that pervaded interwar Britain preserved social distances and specified the etiquette of social relationships across class boundaries. But in an increasingly mobile postwar society, the planners' ideal of the social mixing of housing has foundered. It had limited success in the early New Towns because of the greater propensity of the middle class to move out to private housing estates adjacent to villages in surrounding areas. This led to stickiness in the market for the higher-priced dwellings. Thus, more recently, there has been a much greater degree of spatial segregation of housing types in the new towns. Social idealism has given way to the influences of the market (Heraud, 1968).

At the level of the towns as a whole, although they do have higher proportions of managers, professionals, and higher-skilled manual workers than the country as a whole—a consequence of their industrial structure—the New Towns are far from the one-class communities that characterize North American suburbia. Instead, they have many neighborhoods catering to the variety of housing needs of the persons employed in their industrial estates and commercial enterprises.

But now in Britain there are many calls for a shift in strategy. Thomas (1969) notes that the New Towns

> have ridden the crest of the wave. They have accommodated manufacturing industry...much originated from London. But manufacturing industry has been decentralizing from London anyway. [They] have accommodated thousands of young mobile migrants from London. But so has the rest of the Outer Metropolitan Area. The fact that the new towns have moved with rather than against the tide has probably been the most vital ingredient in their successful growth.

Thomas, among others, calls for a new move with the newer tides of change in Britain.

New trends and new policies

London has continued to grow, and with rising real incomes and increased mobility, long-distance commuting is on the increase, creating commuting fields much like those in North America. Between 1961 and 1971, there was a consistent pattern of growth differentials favoring the outer rings of metropolitan areas in England and Wales. Of 52 metropolitan areas, 49 experienced relative decentralization and, in 27 of them, central-area loss was coupled with outer-area growth to give absolute decentralization (Kivell, 1972).

The basis of London's continued employment growth has been the office, not the factory. The "office boom" of 1955–1962, in particular, produced rapid increases in the number of jobs in central London, increasing congestion. Great concern about this resulted in growing pressure on the government to control office growth and engage in broader development strategies than those contained in the New Town Policy (Cowan et al., 1969). The government responded: a 1963 White Paper on Central Scotland suggested major expansion of East Kilbride, Cumbernauld, Livingston, and Glen-Rothes into major growth zones, transforming the new towns into economic weapons to accelerate growth, reduce unemployment, and change migration patterns. The South-East Study of 1964 and the later Strategic Plan for the South-East of 1970 recommended rethinking London's green-belt strategy and the diversion of growth away from London to major new urban complexes near Southampton, at Crawley, around Reading, in South Essex, and at Northampton. These complexes, of at least 250,000 people, are regarded as weapons for spurring growth of the national economy and reducing the infrastructure costs of continued expansion of outer London. Thus, a policy originally aimed at controlling London's growth has been turned around to become a policy for national growth and development. The push to greater size and variety represents a new phase of New Town planning, in which traditional decentralization and housing goals have been superseded by the attempt to use major New Town developments as the leading edges of national growth policies. By developing those towns, the public has entered into a new role of developmental leadership, supplemented by the controls of the town and country planning machinery.

The Soviet Union

The Communist Revolution of 1917 marked the beginning in Russia, and later in Eastern Europe, of a substantially different form of planned urban development. What was sought in Soviet urban development was far more radical than in Britain: what Lenin had called "a new pattern of settlement for mankind," the city of socialist man. The classic writings of Marxism-Leninism suggested ways in which the goal might be achieved: planning was to create cities without social or economic divisions; there was to be a commitment to the socially integrative value of housing and a wide range of social services; city planning was to be responsive to economic planning, which would determine industrial location and set limits to the rate of urbanization in developed regions and major urban complexes; and thus city planning per se was to be restricted to a basic physical engineering-architectural profession, providing high-density new developments in approved styles.

The accomplishments of the Soviets in urban development are un-questionable. During the Soviet period the USSR has been transformed from a rural society to a predominantly urban one, through a combined process of industrialization and urbanization achieved as the outcome of a series of five-year plans. The population was 82 per cent rural in 1926 but 56 per cent urban in 1969, at which time the USSR had 209 cities of more than 100,000 people. Chauncy D. Harris (1970) has revealed several things about the urban system that has been created. First, size and economic power within the urban network are closely related; each of 24 major urban regions has a relatively complete urban hierarchy of the "command" economy. Second, growth has been led by economic policies, while the succession of economic plans has brought this growth and its associated urbanization to regions successively more remote from Moscow.

Other conditions have also affected the urban development process, particularly after the Second World War: wartime devastation and the high cost of bringing existing cities up to minimal standards; the political rigidity of the Stalinist era, its single-minded focus on specified planning standards, and its peculiar baroque architectural manifestations; the necessity to develop basic industrialized construction skills, emphasizing quantity rather than quality; the basic emphasis on heavy industry in investment allocations, with only small percentages going to housing, urban development, and service facilities.

The authoritarian role of the central government and the priority of the economic goals of the state have been expressed at all levels of urbanization, down to the precise physical nature of the new urban developments that have been built. The procedure is as follows. The State Planning Commission determines the economic norms for the city and, therefore, the basic employment required. Given this basic employment figure as a base, the city planner's role is simply to implement existing norms, also determined by the central planning authorities and laid out in basic books of standards that specify the physical layout, densities, street patterns, utilities networks, and so on, of the settlements to be built (Fisher, 1962). Building proceeds in the Soviet Union on the basis of micro-rayons (neighborhood units), built for 6,000 to 12,000 people, together with whatever services are specified for inclusion in them.

The automatic nature of the process is a result of the directed nature of the Soviet state, as well as of the need to provide massive quantities of housing and urban services after years of neglect of urban needs. This neglect was caused by a preoccupation with industrialization, and by the long period during which the effects of wartime devastation were felt. A major goal of standardization was to industrialize housing construction through factory production of prefabricated and precast materials and forms, both to reduce costs and to speed construction that could meet the immense needs for housing. Because quantity was so important, quality was sacrificed, and nowhere does one find more drab and monotonous modern cities or buildings with poorer internal design than in the USSR building styles, apartment sizes, and qualities of developments. The industrialized methods produce standard apartment blocks almost exclusively. Movement in the cities is by public transportation. Services and facilities are the minimum necessary. An elaborate, often monumental, political-cultural administrative core is provided for the city, surrounded by a succession of self-contained neighborhood units, undifferentiated socially.

The individual city plans, as in the case of Moscow, nonetheless have a distinctive new town flavor. Official Soviet policy is to restrain Moscow's growth and to channel such outlying growth as occurs into satellite towns. Like Abercrombie's Greater London Plan, that for Moscow is formulated in a series of rings. Within the inner belt highway, growth will be restrained. Focus will be on major rehousing schemes, facilitated by industrialization of a standardized housing industry. This is encircled by a green belt, a ten-kilometer-wide ring being developed into a recreation area, in which residential dacha construction is being restrained. Finally, there are the satellite towns (*goroda-sputnika*), accommodating new growth, and built according to the standardized micro-rayon physical design specifications outlined earlier.

The resulting spatial patterns are held to be consistent with socialist principles of urban development, the antithesis of the European industrial urbanization of the nineteenth century that so angered and repelled Marx and Engels. One Polish planner's version of the principles is as follows (Fisher, 1962). The principles of social justice are realized by using the official norms and standards which determine per capita living spaces, population density, and quantity of services adjusted to projected population limits without class distinctions. The only basis for differentiation of available environment among urban families are the biological characteristics of the families. The functional and the spatial structure of new residential areas and towns correspond to one another, because of the development of functionally similar neighborhood units and the social conception of a socialist urban community.

The impact of Russian socialism on Eastern Europe's cities

These principles have been applied throughout Eastern Europe since the Second World War, although distinctively different styles of urban development are distinguishable: (1) the postwar reconstruction period, when (with the exception of the meticulous reconstruction of Warsaw as it had existed in 1939, as Poland's national symbol) housing was to be built as quickly as possible; (2) a period of "Stalinesque" massive developments; and (3) the "post-Stalin modern" period.

During the Stalinist period, there was a strong ideological overtone that the socialist city should be significantly different from that of the West, and uniform in its characteristics, consistent with the goal of social equality. Eastern European planning at that time showed heavy Soviet influence, and there was consistent application of Soviet procedures, norms, and plans. The buildings that were constructed were massive and the town planning "absolutist baroque," typified by the Stalin Allee in East Berlin, or North Avenue in Bucharest, along which the Open Air Museum of Folklore and the Academy of Sciences greet visitors. In this period, too, some of the socialist new towns were started, designed to represent complete departures from Western experience. Many are showplaces of the new socialist city: Nowa Huta in Poland, Dunaujvaros in Hungary, and Titograd in Yugoslavia, to mention a few. The cities accomplish a specific economic purpose—to house steel-mill workers, or those in a regional administrative center. Their form is simple: a square of administrative-cultural composition at the center, with radiating streets flanked by massive residential developments comprising five- to nine-story apartment buildings containing two- or three-room apartments for families and dormitories for unmarried workers. Adjacent to the complex is the economic unit that provides the employment. In this period, too, the application of socialist principles to existing cities began to have a significant impact on their spatial form and social structure.

For example, the effect of Russian-style socialism on the cities of East Germany, differentiating them from their counterparts of the West, has been to reduce areal specialization and segregation by substituting for market processes direct planning by the state (Elkins, 1973). The expansion of the central office-shopping core has ceased, and the type of retailing has changed. Land vacated by departing industry has been occupied by apartment blocks, and people have been brought back to the central city. The high prestige assigned to the manufacturing industry has resulted in the dispersion throughout the city of large industrial complexes with associated housing areas. Because new residential development is all in apartments, significant density differences among parts of the city have been ironed out, and social segregation has been largely abolished. Housing policy gives priority in new dwellings to young families with children, employees in key jobs, and families living in bad conditions. Thus, there is a strong correlation between age of structures and age of inhabitants, producing socially mixed neighborhoods in which social status as a differentiating factor has virtually disappeared. Only intellectuals and party leaders enjoy a distinctive residential area. And because there is little reliance on the automobile, all residential areas are closely related to public transportation facilities. Similar changes are reported throughout Eastern Europe.

Since Stalin, however, greater variety in building styles has emerged, and the planning process and goals have come to differ from one country to another. The philosophical tenets of Marxist-Leninist dogma and their implications for operational city planning are being re-examined, but several features do stand out and persist: standardization, concern for the proper size of cities, a particular concept of the city center, and development by neighborhood units (Fisher, 1962).

But the nature of these concerns has begun to diverge. Only where the scale of construction has been great enough—in the major cities of Poland, Hungary, and Czechoslovakia—has prefabricated construction proved to be economical. Elsewhere, there has been greater reliance on traditional building techniques. And although apartment houses and tenements were nationalized throughout Eastern Europe, this is still not the case everywhere with urban land, much of which remains in private ownership, albeit strictly controlled by zoning and subject to alienation for public purposes. In Hungary, for example, a considerable amount of public land is divided each year into lots and sold by the government for construction of condominiums, single-family housing and weekend homes; in Budapest this is producing a certain amount of social differentiation between the elite neighborhoods to the west in old Buda and the working-class communities of old Pest to the east.

Most Eastern European planners are seeking to create city centers with other than the monumental and political-administrative functions of the Stalinesque baroque. All planners agree on the necessity for neighborhood unit development, but they differ on the nature of these units. Finally, in all countries there is concern for the "proper" or "balanced" size of the city, based on the primary economic functions of the place; and in all cases there is the underlying conviction that large cities must have their growth contained and that new towns and satellite cities should be built around large urban agglomerations.

The United States

At the opposite end of the ideological continuum, the American planning style has tended to be supportive of privatism and the mosaic culture rather than productive of alternative urban futures. Consider the U.S. experience with the construction of new towns. The first of these, dating from before the twentieth century, were company towns to serve industrial enterprises. Later came planned suburban developments established to capture the residential spin-off from large cities. Around the time of the First World War, the federal government established a number of emergency housing communities in industrial areas. During the 1920s some efforts were made to create American garden cities patterned after Letchworth in England. The most notable examples are a series of communities designed by Clarence Stein, although most of these were only planned areas within a larger city, as in the case of Sunnyside Gardens in New York City (1924–1928) and Chatham Village in Pittsburgh (1932). The first community that actually started as an independent garden city was Radburn, New Jersey (some 16 miles from New York City), which was begun in 1928.

During the New Deal in the 1930s serious attention was given to creating a number of new towns, the Greenbelt Towns, under federal government auspices. Rexford G. Tugwell, the program administrator, wanted to build three thousand. Twenty-five were selected by the Resettlement Administration. President Roosevelt approved eight of these, and Congress reduced the number to five. Finally, only three—Greenbelt, Maryland; Greenhills, Ohio; and Greendale, Wisconsin—were ever built (Conkin, 1959; Arnold,

1971). In addition, the federal government became involved in new town development in the 1930s in connection with a number of large-scale power and reclamation projects, and in the 1940s through its atomic energy program.

One recent study has found that a total of 376 urban developments of 950 acres or more, using nearly 1.5 million acres of land, were started in the United States between 1960 and 1967. Of these, 43 could be classified by the survey as new towns, mainly located in areas of rapid growth and warm weather. The builders of all these new town projects have been called "the new entrepreneurs." They include (1) builder-developers with a real estate and homebuilding background; (2) large national corporations interested in product promotion and financial diversification; (3) large landowners looking for a way to increase the value of property originally acquired for other purposes, such as for farming or mining; and (4) the big mortgage lenders, such as banks, insurance companies, and savings and loan associations, as well as an occasional independent developer who enters the field more or less by accident. The important thing to note about all of these groups is that they are private. American new town development is explicitly entrepreneurial, exploiting the profit-making potentialities latent in urban growth and change. To the extent that there is public involvement, it functions for reducing the risks to the entrepreneur in exchange for some mild regulation of the development style.

On occasions, the result is of high quality. Reston, Virginia, and Columbia, Maryland, are certainly the two most highly publicized examples of privately built new towns in the United States. Reston has attracted a good deal of attention for its efforts at preserving the great natural beauty of its site and for its high quality in architectural design. Columbia has made a definite effort to attract black residents, has explored federal programs for providing low-cost housing for the poor, and has shown considerable interest in the sociology of urban development; yet it has also been challenged by local resident interest groups as it has grown, on issues of corporate paternalism. Nonetheless, recent research indicates that residents of Reston and Columbia rate both their communities and their micro-neighborhoods more highly than residents of less-planned suburbs rate theirs (a finding that has been repeated elsewhere, for example in Britain's New Towns), and indeed, in many cases the concept and planning were the attractions leading to the initial move-in. Among the important features appear to be the adequacy and location of open spaces for family activities that wind as sinews through the residential areas of both towns, the lower noise levels, and the superior maintenance levels. Along with quality of the schools, the greatest sources of satisfaction with planned communities relate to the environment (Zehner, 1972).

Housing policy

Much the same story can be told about U.S. housing policy. Recall that the Federal Housing Administration (FHA) was created as part of the New

Deal's efforts to insure mortgages under certain circumstances, to eliminate the worst housing, to provide public housing for the poor, and, above all, to help in priming the pump of the economy. By 1950, 170,000 dwellings had been provided for low-income families.

Program development following the Second World War was additive. To the FHA were added the mortgage assistance programs of the Veterans Administration, which ultimately were taken advantage of by more than 6.8 million returning veterans seeking new homes. The Housing Act of 1949 strengthened and extended the slum clearance and public housing programs, to be followed by later acts that created the workable program for community improvement as the prerequisite to housing assistance and urban renewal (1954), and the community renewal (1959) and model cities programs, progressively moving towards comprehensive socioeconomic as well as physical goals for the cities. Even more comprehensive enactments came in 1968 and 1970, leaving the United States not simply with a Department of Housing and Urban Development, but with a "New Communities" program too. And many other public investment programs have materially affected the urban scene, most notably the massive Interstate Highway Program, but including a host of others, such as airports, sewage systems, recreation and open space facilities, and hospitals.

Federal housing programs contributed to the suburbanization of America in important ways, creating hundreds of standardized "Levittowns." In these developments, federal policy combined with local planning to maintain and support neighborhood homogeneity, and specifically to exclude the black and the poor.

In city planning practice in the interwar and immediate postwar periods, one of the most influential ideas was Clarence Perry's "Neighborhood Unit" concept. Perry (1939) felt it important that cities be built up of sharply bounded neighborhood units, physically distinctive, and possessing local unity by virtue of organization around a shared focus of community activity. Perry's model had a profound influence on local planning commissions and zoning boards. The idea that well-designed neighborhoods would bring about social cohesiveness, neighborliness, and the virtues of the small community within the large city was consistent with the dominant social philosophy. Associated was the belief in the necessity of maintaining neighborhood homogeneity: an "incompatible mix" referred to different racial or ethnic groups as much as to industrial activity in residential areas. Mixture of racial and cultural groups, in particular, was considered detrimental to the neighborhood. Restrictive covenants explicitly excluded members of minority groups. The National Association of Real Estate Boards' "code of ethics" made it unethical for a realtor to introduce "incompatible" groups to a neighborhood. Early editions of the FHA Appraisers Handbook forbade social integration, and demanded that mortage institutions follow suit. The suburbanization of the white middle class, and the ghettoization of the poor and minorities in the central city implied thereby, were not creations of federal policy—residential segregation and subcommunity formation date at least from the early nineteenth century—but they certainly were promoted by federal activity in the postwar years.

Emergence of broader concepts of urban development

With the 1960s came a search for a new comprehensiveness in the federal approach to urban problems (Scott, 1969). Part of the shift was one of emphasis, from planmaking to planning as a process; part involved recognition of the need to orchestrate physical and social programs in the central cities if a significant impact was to be made on the pockets of poverty in the ghettos. Such efforts as the Community Renewal Program and the Model Cities Program were launched. The most significant development came in the Housing Act of 1968. This act confirmed growing disaffection with the pace of housing starts for the poor, and provided new programs that produced 300,000 new low-income housing units in the years 1968–1970, raising the low-income proportion of total housing starts from 3 per cent in 1961 to 16 per cent in 1971 (Kristof, 1972). A parallel debate on national growth policy culminated in 1970 with the passage of Title VI of the Housing and Urban Development Act, requiring the president

> in order to assist in the development of a national urban policy . . . to transmit to the Congress during the month of February in every even-numbered year beginning with 1972, a Report on Urban Growth.

The first such report was the White House Domestic Council Report on National Growth 1972, which in proposing policy direction concluded that in the United States privatism should prevail:

> Patterns of growth are influenced by countless decisions made by individuals, families and businesses . . . aimed at achieving the personal goals of those who make them. . . . [Such] decisions cannot be dictated. . . . In many nations, the central government has undertaken forceful, comprehensive policies to control the process of growth. Similar policies have not been adopted in the United States for several reasons. Among the most important is the distinctive form of government which we value so highly . . . it is not feasible for the highest level of government to design policies for development that can operate successfully in all parts of the nation.

Likewise, in Canada, N.H. Lithwick noted in 1970 that impetus for urban growth derives not from planning policy but rather from Canadians' preoccupation with economic achievement. He noted that Canada has specific economic goals such as growth, full employment, and rising levels of income which are recognized and accepted by government, labor unions, agriculture and business. Urban policy served these economic goals, providing education, roads, and utilities and supplying public housing, welfare, and protective services. But this limited urban planning to an amerliorative problem-solving role, which is reinforced by an attitude that accepts the inevitability of a continuation of the processes inherent in the present. "Because," Lithwick noted, "these processes are abstract and powerful, and have served the needs of those groups who have benefited most, there is great pressure not to tamper with them." Thus, he came to the "inescapable conclusion that of all [Canadian] urban problems, the one most likely to deter any major improvement is [the] urban policy problem. [The] first priority is thus not what urban policy to follow, but an agreement that any urban policy is needed."

In short, no governmental bodies in North America can be credited with the development and execution of national urban policy at the present time. What substitutes for it is a complex set of uncoordinated, often contradictory, essentially random public policies and programs provided in the wake of strong economic forces which set the agenda for urban growth. Thus, if in the past urbanization has been governed by any conscious public objectives at all, these have been, on the one hand, to encourage growth, apparently for its own sake; and on the other, to provide public works and public-welfare programs to support piecemeal, spontaneous development impelled primarily by private initiative. In contrast, development of a national urban policy suggests a shift in the locus of initiative, imposing on public authorities an obligation to orient, rationalize, and plan the physical, economic, and textural character of urban life. Such an urban policy would include a *complementary* set of policies and programs based on an explicit statement of the purpose of urbanization, its pace, its character, and the values that are to prevail.

Chapter 18
The social consequences of
third-world urbanization:
the culture of poverty revisited

A transformation of the urbanization process as profound as that described for Western Europe and North America has been experienced in the countries of the Third World in recent decades, producing different urban forms and social consequences. Adna Weber (1899) showed urbanization beyond Western Europe and North America to be limited in the nineteenth century to the tentacles of colonial expansion. During the twentieth century this situation has changed dramatically. As part of the quadrupling of the world's urban population during the last fifty years, the developed regions increased their urban population by a factor of 2.75 (that is, from 198 to 546 million), while the Third World countries increased their urban population by a factor of 6.75 (from 69 to 464 million). In both Latin America and Africa the urban population increased eightfold. The big-city population of the Third World increased even faster—ninefold—during the period 1920–1960, as compared to 0.6 times for Europe and 2.4 times in other more developed regions (Table 18.1). Clearly, it is the Third World that is experiencing the major thrust of urban growth today. With 25 per cent of the world's urban population in 1920, the Third World will encompass 51 per cent by 1980 (Davis, 1969). Before the end of this century, Third-World urban populations will increase by more than one billion to a total three times the current urban population of developed countries (World Bank, 1972).

The differing context of urban growth

This rapid urbanization is taking place in those countries with the lowest levels of economic development rather than the highest, as was the case when accelerated urbanization began in Western Europe and North America.

Moreover, it involves countries in which the people have the lowest levels of life expectancy at birth, the poorest nutritional levels, the lowest energy consumption levels, and the lowest levels of education.

In the West, urbanization involved gradual innovation and interdependent economic and social change spanning more than a century. Contemporary Third-World urbanization involves greater numbers of people than it did in the West. Migration is greater in volume, and more rapid. Industrialization lags far behind the rate of urbanization, so that most of the migrants find at best marginal employment in the cities.

Whereas the new industrial cities in the West were death-traps, the cities of the Third World are usually more healthful than their rural hinterlands and are almost as healthful as cities in the most advanced countries. They have participated disproportionately in the miraculous fall in mortality that has occurred in nonindustrial countries since 1940—a fall that has enabled them to make death-control gains in twenty years that industrial countries, starting at a similar level, required seventy to eighty years to achieve. The cities have been the main recipients of this new death control because they are the places to which the medical and scientific techniques, expert personnel, and funds from the advanced nations are first imported and where the most people are reached at least cost.

Nor do conditions in cities of nonindustrial countries seem as hostile to reproduction as those of the nineteenth and early twentieth centuries were. Urban fertility remains lower than rural fertility, but not much lower; and in both cases net reproduction rates are higher than they ever were in most of the industrial countries. To some extent this urban fertility is a function of good health and low mortality, but it is also a function of some of the very changes that make better health possible. Economic improvement, public welfare, international aid, subsidized housing, and free education keep penalties for having children from being as great as they once were. Giving priority in housing to larger families, maintaining maternal- and child-health clinics, and discouraging labor-force participation by married women are additional props to urban reproduction. Equally important are old institutional structures with built-in incentives for prolific breeding—structures that persist in the cities because the paternalism of the times treats them as sacred.

For the new urban residents of the burgeoning Third World cities, such improvements as are taking place create a gulf between where they are and where prior Western experience suggests they might aspire to be—a revolution of rising expectations. As a result, pressures for rapid social change are greater than they were in the West. Lacking an effective capacity to respond, national governments are increasingly confronted by people seeking more revolutionary solutions.

The political circumstances conducive to revolutionary take-overs are present as a result of the recent colonial or neocolonial status of most of the

Third World nations. The imprint of colonialism was, first, that many of the major cities originated as administrative centers for the colonizing nations. The recency of independence from colonialism means that most of the developed countries have inherited an intentionally centralized administration, with the result that government involvement in urban development is more likely in the Third World countries today than in the West. In nineteenth-century Europe it was the craftsman and small entrepreneur who promoted industrial development, not the university-educated national bureaucrat. Lacking private development capital and an entrepreneurial class, more Third-World development is governmental, involving foreign economic and technical assistance and requiring an assertive governmental role in international diplomacy. As a consequence of governmental leadership in the development process, public goals have priority over private goals.

The colonial powers did not permit effective indigenous leadership to develop. Whereas the collection, bulking, and exporting of industrial raw material required a high degree of organizational and business skills, the logic of the colonial relation was such that the opportunities for doing these tasks were jealously guarded and restricted to citizens of the colonizing country. Instead, much postindependence leadership developed within the framework of nationalist or revolutionary movements oriented to independent self-determination, albeit on the foundations of Western educational experiences, and often with the Western model of nineteenth-century urbanization in mind.

The change in the colonial world came at the end of the Second World War. Initially, the industrial powers relinquished controls to parliamentary governments they had created, and to the Westernized bureaucratic elites who had served them. Only in a few cases has parliamentary democracy worked. Most ex-colonial states have moved very quickly to single-party government and to substantially authoritarian control—either by the revolutionary elites, who set modernization as their task, generally with an underlying socialist ideology that leads them to try to create basic structural changes in society and to change the entire social context within which urbanization is proceeding; or else by military juntas whose perspective leads them to seek national efficiency without social change.

The continuing problem resulting from these circumstances is that the cities are becoming the main centers of the social and political changes that the new political elites are attempting to produce. This new centrality is a force that attracts people to the cities. The cities have become symbols, drawing in massive immigrant streams, especially of young men, from overcrowded rural areas; the immigrants only found rural poverty replaced by urban poverty. To be sure, economic development has the highest national priority in the new governments, but it has failed to keep pace with the growth of urban populations.

In all of these complex changes three themes stand out as revealing most about Third-World urbanization and its social consequences: (1) the nature of migration and the role of peripheral settlements in facilitating the transformation of rural societies into urban ones (indeed, some say that the city of the future is now emerging in these peripheral settlements); (2) problems associated with absorption of labor into the urban economies, with attendant effects upon the spatial diffusion of growth, upon class structures and class conflict, and upon the integration of developing subcultural mosaics; (3) the efforts of Third World governments to control the pace, scale, and direction of urbanization, initially with "Western" concepts and ideology as a base, and now by more radical means. These themes will be the focus of what follows in this and the following chapter. The conclusion we reach is that Third World urbanization is a fundamentally different process than that which occurred in the West, with social consequences that do not conform to traditional ecological theory.

Migration and the growth of peripheral urban settlements

Two demographic factors are of fundamental importance in Third-World urbanization. First is an elevated natural growth rate, caused by the fact that birth rates have remained relatively stable for several decades while death rates have been in steady decline. Second is the heavy migration from rural areas and more traditional country towns and cities to the principal urban centers of each country, especially to peripheral settlements located in and around the national capitals and regional industrial centers. The impact of internal migration on urbanization varies according to country and region, but in most cases exceeds 50 per cent of the total population increase, as Table 18.1 reveals.

The consequences of migration

Given the scope of migration to urban areas in the Third World, it should not be surprising that a substantial amount of research has been conducted on its social consequences. This research began with propositions borrowed from theorists who codified the nineteenth and early twentieth century migration experience in the West. Increasingly, researchers of Third World cities have come to question these propositions and to suggest drastic revisions in traditional theories, particularly regarding the role of the peripheral settlements in the urbanization process.

The borrowed propositions center on three themes, according to W. A. Cornelius, Jr. (1971): material deprivation and frustration of mobility expectations; personal and social disorganization; and political radicalization and disruptive behavior. Migrants are expected to experience the first two

Table 18.1

Estimates of migrants as a percentage of recent population increases

City	Period	Total population increase (thousands)	Migrants as a percentage of total population increase
Abidjan	1955–1963	129	76
Bombay	1951–1961	1207	52
Caracas	1950–1960	587	54
	1960–1966	501	50
Djakarta	1961–1968	1528	59
Istanbul	1950–1960	672	68
	1960–1965	428	65
Lagos	1952–1962	393	75
Nairobi	1961–1969	162	50
São Paulo	1950–1960	2163	72
	1960–1967	2543	68
Seoul	1955–1965	1697	63

Source: World Bank, *Urbanization Sector Working Paper* (Washington, D.C., The World Bank, June 1972), p. 80.

conditions and graduate into the third. Continuing driving forces are held to be the high rates and volumes of immigration and the limited absorption capacity of the cities, caused by the discrepancy between urban and industrial growth rates. Such theory is distinctively Wirthian, with roots in Toennies and Simmel, as well as in the works of Karl Marx.

But migration research completed since 1960 provides startlingly contradictory evidence. Migration, it has been found, does not necessarily result in severe frustration of expectations for socioeconomic improvement or in widespread personal and social disorganization; even when these latter conditions are present, they do not necessarily lead to political alienation. Nor, apparently, does alienation lead to political radicalization or disruptive behavior. The urban migrants fail in most respects to conform to the usual conception of a highly politicized, disposable mass. Residents of peripheral settlements rather frequently acquiesce to regimes that sustain the status quo. The migrants' dominant perception is one of improved living conditions and life chances experienced as a result of the migration to the cities, as well as of a fundamental belief in the potential for future betterment and a low tolerance for political risks. Political violence has tended to be restricted to student and military elites, with little mass participation.

Why are traditional theories today so deficient? It is now apparent that Wirth's concept of urbanism as a way of life was both time- and culture-bound to the immigrant city of North America at the turn of the twentieth century. Moreover, as noted in Chapter 3, his conception was strongly influenced by Toennies' idealized conception of the people of "folk societies" as socially cohesive, personally contented, nonconflictual, and well adjusted, leading to the inference that urbanization destroyed this idyllic folk culture.

The limitation of this conventional wisdom is its conceptualization of urban migrants as an undifferentiated mass responding in uniform fashion to a given set of conditions to which all migrants to large cities are presumably exposed. In reality, migrants comprise a large and disparate array of social types both before and after migration. Distinct migrant subcultures have developed with widely varying life styles, value orientations, and levels of subjective political competence. Moreover, a rural-urban dichotomy does not exist; rather, there is a broad range of continuity of rural traditions within these urban subcultures (Redfield, 1941). Institutions, values, and behavior patterns have persisted or have been adapted to the specific requirements of the urban setting. Social organization and mutual-aid networks continue to function in the urban scene.

Africanists, for example, have described the rise of tribal consciousness in the new African metropolises (Little, 1965; Miner, 1967). This is probably caused by conflict with other groups, as social scientists have long observed, such conflict typically strengthens the internal ties of the collectivity. Most of these studies report that rapid migration has not produced the alienation, anomie, or other symptoms of social disorganization that, according to the Wirthian model, are hallmarks of rapid urbanization. This is not to say there is no poverty, unemployment, crime, or other social problems; these exist in abundance. But for the vast majority of African migrants the ties of the extended family and those between city and village have been maintained. Far from being a "detribalizing" process, much of the rich associational life of African cities is based upon common interests, mutual aid, and the need for fellowship of people in the towns who are members of the same tribe or ethnic group, speak the same language, or have come from the same region. Much the same conclusions have been reached in Islamic and Asian urban studies and in investigations of Indian migration to the cities in Latin America. In all cases, ethnic competition remains high and racial confrontation frequent.

Squatter settlements: the "culture of poverty" rejected

Most Third-World urban growth is concentrated in the so-called "squatter" or "uncontrolled" peripheral settlements, which account for a substantial share of city populations throughout the Third World (see Table 18.2). The names vary: in Latin America they are *barrios, barriadas, favelas, ranchos, colonias proletarias,* or *callampas*; in North Africa, *bidonvilles* or *gourbivilles*; in India, *bustees*; in Turkey, *gecekondu* districts; in Malaya, *kampongs*; and in the Philippines, *barung-barongs*. The inherited conventional wisdom leads to an interpretation of such settlements as physically decrepit slums, chaotic, disorganized and lacking in basic amenities—an attitude that persists in much of the urban planning community, which tends to interpret such settlements as obstacles to good civic design.

To cite one example, Morris Juppenlatz, a former UN official, describes them (1970) as a "spreading malady," "fungus," or "plague" of "excessive squalor, filth and poverty...human depravity, deprivation, illiteracy, epidemics and sickness" with

> growing crime rates and juvenile delinquency...land grabbing and disrespect for property rights by a growing number of squatters...in the mounting social disorder and tension in the cities, in the weakening and breaking down of the administrative discipline of the authorities, in the unsightly human depravity in the midst of the affluent established urban society, and in the inadequacy of the essential public services.... As the political control of the cities...passes from the presently established urban society...into the hands of the emergent urban squatter society who have little or no heritage of city-dwelling...it can be expected that essential services will diminish until they finally break down and collapse.

Such views have been given apparent intellectual support by scholars like the American anthropologist Oscar Lewis, who, in his research in the 1950s and 1960s, described a subculture with a life style that he felt transcended national boundaries and regional and rural-urban differences with nations. This subculture he called the *culture of poverty*—of the slum, the ghetto, the squatter settlement. Wherever it occurs, Lewis argued, its practitioners exhibit remarkable similarity in the structure of their families, in interpersonal relations, in spending habits, in their value systems, and in their orientation in time.

His studies identified a large number of traits that characterize the culture of poverty. The principal ones may be described along four dimensions of the system: the relationship between the subculture and the larger society; the nature of the "ghetto" community; the nature of the family; and the attitudes, values, and character structure of the individual.

The disengagement—the nonintegration—of the poor with respect to the major institutions of society Lewis felt to be a crucial element in the culture of poverty. It reflects the combined effect of a variety of factors: including poverty, to begin with, but also segregation and discrimination, fear, suspicion and apathy, and the development of alternative institutions and procedures in the slum community. The people do not belong to labor unions or political parties and make little use of banks, hospitals, department stores, or museums. Such involvement as there is in the institutions of the larger society—in the jails, the army, and the public-welfare system—does little to suppress the traits of the culture of poverty.

People in a culture of poverty, Lewis argued, produce little wealth and receive little in return. Chronic unemployment and underemployment, low wages, lack of property, lack of savings, absence of food reserves in the home, and chronic shortage of cash imprison the family and the individual in a

Table 18.2

Extent of uncontrolled peripheral settlements

Country	City	Year	City population (thousands)	Uncontrolled settlement	
				Total (thousands)	As percentage of city population
Africa					
Senegal	Dakar	1969	500	150	30
Tanzania	Dar es Salaam	1967	273	98	36
Zambia	Lusaka	1967	194	53	27
Asia					
China (Taiwan)	Taipei	1966	1300	325	25
India	Calcutta	1961	6700	2220	33
Indonesia	Djakarta	1961	2906	725	25
Iraq	Baghdad	1965	1745	500	29
Malaysia	Kuala Lumpur	1961	400	100	25
Pakistan	Karachi	1964	2280	752	33
Republic of Korea	Seoul	1970	440 (d.u.)*	137 (d.u.)*	30
Singapore	Singapore	1966	1870	980	15
Europe					
Turkey	Total urban population	1965	10,800	2365	22
	Ankara	1965	979	460	47
		1970	1250	750	60
	Izmir	1970	640	416	65
North and South America					
Brazil	Rio de Janeiro	1947	2050	400	20
		1957	2940	650	22
		1961	3326	900	27
	Brasilia	1962	148	60	41
Chile	Santiago	1964	2184	546	25
Colombia	Cali	1964	813	243	30
	Buenaventura	1964	111	88	80
Mexico	Mexico City	1952	2372	330	14
		1966	3287	1500	46
Peru	Lima	1957	1261	114	9
		1961	1716	360	21
		1969	2800	1000	36
Venezuela	Caracas	1961	1330	280	21
		1964	1590	556	35
	Maracaibo	1966	559	280	50

Source: U.N. General Assembly, *Housing, Building and Planning: Problems and Priorities in Human Settlements*, Report of the Secretary-General, August 1970, Annex III, p. 55. Definitions vary. Additional details are given in the source quoted.

* Dwelling units.

vicious circle. Thus, for lack of cash, the slum householder makes frequent purchases of small quantities of food at higher prices. The slum economy turns inward; it shows a high incidence of pawning of personal goods, borrowing at usurious rates of interest, informal credit arrangements among neighbors, and use of secondhand clothing and furniture.

There is awareness of middle-class values. People talk about them and even claim some of them as their own. On the whole, however, they do not live by them. They will declare that marriage by the law, by the church, or by both is the ideal form of marriage—but few will marry. For men who have no steady jobs, no property, and no prospect of wealth to pass on to their children; who live in the present without expectations of the future; who want to avoid the expense and legal difficulties involved in marriage and divorce; a free union or consensual marriage makes good sense. The women, for their part, will turn down offers of marriage from men who are likely to be immature or generally unreliable. They feel that a consensual union gives them some of the freedom and flexibility men have. By not giving the fathers of their children legal status as husbands, the women have a stronger claim on the children. They also maintain exclusive rights to their own property.

Along with the disengagement from the larger society, Lewis felt there to be a hostility to the basic institutions of what are regarded as the dominant classes: hatred of police, mistrust of government and of those in high positions, and a cynicism that extends to the church. The culture of poverty thus, he felt, holds a certain potential for protest and for entrainment in political movements aimed against the existing order.

With its poor housing and overcrowding, the community of the culture of poverty is high in gregariousness, but Lewis argued that it has a minimum of organization beyond the nuclear and the extended family. Occasionally slumdwellers come together in temporary informal groupings; neighborhood gangs that cut across slum settlements represent a considerable advance beyond the zero point of the continuum. It is the low level of organization that gives the culture of poverty its marginal and anomalous quality. Lewis felt that this culture, so often noted in literature on the American black, was equally strong elsewhere in the world, for slumdwellers who are not segregated or discriminated against as a distinct ethnic or racial group.

Lewis has been criticized by his peers (Valentine, 1968; Leacock, 1971), and what is emerging throughout the world is an alternative interpretation: *that under conditions of rapid urbanization the uncontrolled new settlements play an important functional role.* William Mangin (1967) has written that squatter settlements represent a *solution* to the complex problem of urbanization and migration, combined with a housing shortage. Although there are indeed some "slums of despair," displaying the traits listed by Lewis, in *both* advanced and Third World countries, many studies have shown that characteristics of so-called squatter settlements vary widely. Clinard (1966) writes that although some squatter settlements show disorganization, it is important that each be examined in the light of its own distinctive subculture, which is the dominant influence on the life pattern of its inhabitants. Of critical significance is how the settlement was formed—by organized squatter invasion, gradual accretion, or government initiation, for example—and whether or not there is a sense of ownership or control of property.

John Turner (1968) has distinguished three different economic levels of the transitional urban settlements:

1. *The low-income bridgeheads.* Populated by recent arrivals to the city, with few marketable skills, the need to obtain and hold work dominates. Modern standards have low priority, and access is all important. Thus, such bridgeheads tend to be in decaying old homes in the central city or in the centrally located "slums of despair." They are poor, and lack essential services.
2. *The low-income consolidators.* As stability of a permanent income is achieved, access is no longer so critical (often because public transport is cheap), although conventional housing is still out of reach. Money is available for other than necessities, and family orientation begins to dominate in more peripheral squatter "slums of hope." Many such settlements are highly organized and planned, and upgrade in quality through self-help efforts over the years, providing shelter and security without imposing on the residents the middle class's priorities for housing.
3. *The middle-income status seekers.* Those with economic security then seek social status through choice of location—in Rio de Janeiro, this was adjacent to the high-status "conventional" residential areas. Upgrading of housing quality assumes priority, along with education and quality of services.

Thus, no universal form of decaying slum settlement is to be found, but instead a range of markedly different settlement types with fundamentally different subcultures. According to Robert J. Crooks, Director of the United Nations Center for Housing, Building, and Planning, they are better called *transitional urban settlements*, which, he feels, demonstrate remarkable vigor and ingenuity in improving their living conditions.

One of the most important variables confounding traditional theory about the experience of the new migrant is the transitional networks of social relations, which have facilitated successful assimilation of many migrants into urban life. In Africa, as we noted earlier, these networks involve a continuity of rural ways in the city, and E. M. Bruner (1961) has shown in a study of northern Sumatra that

> the cultural premises and roots of urban Batak life are to be found in village society.... Most urban Batak have more meaningful associations with their rural residents in the highlands than with their non-Batak neighbors.

Similarly, Janet Abu-Lughod (1971) concluded in studies of Cairo that the

Wirthian model of anonymity, secondary contacts, and anomie is not appropriate. L. Alan Eyre (1972) has described the "shantytown" resident of Montego Bay as "the poor suburbanite of the developing world... upwardly mobile...industrious...a saver...more often a conservative than a radical."

Even Calcutta's teeming bustees have been described as performing six functions that are of major importance to the urbanization process as a whole. They provide housing at rents that are within the means of the lowest income groups. They act as reception centers for migrants, providing a mechanism to assist in adaptation to urban life. They provide within the bustee a wide variety of employment in marginal and small-scale enterprises. They provide a means of finding accommodation in close proximity to work. Their social and communal organization provides essential social support in unemployment and other occasions of difficulty. Finally, they encourage and reward small-scale private entrepreneurship in the field of housing.

But the conventional wisdom still dominates attitudes of many planners and policy makers, stereotyping all transitional settlements as social aberrations, "cancers" overwhelming an otherwise healthy municipal body. In many cases, governments have responded by the expulsion of squatters and the costly and disruptive clearance of slum areas, resulting in a net reduction of housing available to low-income groups—as when, in 1963, inner-city Manila squatters were relocated to an unprepared site deep in the countryside at Sapaney Palay; or as, more recently, the Brazilian government has eliminated the *favelas* from Rio de Janeiro. Where clearance together with rehousing has been attempted, it has generally resulted in the unproductive use of scarce public resources, meager improvements if any, and unequal treatment of the families inhabiting transitional areas.

This is because public-housing programs frequently attempt to follow the pattern set in the more developed countries. Complete dwellings are constructed prior to occupancy, at minimum space and material standards. But even these standards tend to be too costly in relation to the total needs and total resources that are available. Rarely are they a realistic solution for most squatters and slumdwellers, as they involve heavy interest and maintenance costs, apart from the high capital costs per family. If rents are high enough to amortize investments, they are likely to be far above the ability of the people to pay, creating a high rate of default. However, the most serious criticism of prebuilt housing projects is that they remain beyond the financial grasp of precisely the group most in need of housing assistance, those with the lowest incomes occupying transitional settlement areas. One of the few successful examples of resettlement by means of low-cost building programs may be in Singapore, where the Housing and Development Board created in 1960 has tackled the problem with all the energy of the welfare socialism of that city-state's government.

Because of the self-improving nature of the transitional settlements, the United Nations Center for Housing, Building, and Planning now stresses the importance of achieving a major shift in attitude and emphasis from the current norms of national and international policies and programs that attempt to deal with transitional urban settlements. The most basic policy and program directions are the acceptance and support of the long-term existence of transitional areas and adequate preplanning for future transitional settlement growth. The Center notes that, in many cases, transitional urban settlements constitute valuable actual or potential additions to the urban housing stock and fixed capital investment at city and national scales. In conditions of rapid urbanization and even more rapid growth of transitional areas through migration and natural population growth, vast clearance schemes, with or without high-cost public housing, can only aggravate the problems of people's living in these areas. It therefore recommends that, consistent with a positive supportive attitude, governments should take action to make normal urban utilities and community services available to these areas, according to priorities established through the involvement of the residents themselves in the development process. Because of the importance of the degree to which the residents feel a secure right to the land they occupy, the Center recommends that supportive programs should treat this issue as a matter of high priority. And because the forces leading to the rapid growth of transitional urban settlements will continue, they recommend preplanning for transitional settlement growth, in a manner that will emphasize the positive aspects of these areas. They note that clearance of slum and squatter areas is a waste of popular resource investment and often results in a net destruction of the living environment. Governments and international organizations, they say, must develop and use legal and administrative mechanisms that will make possible planned land acquisition and development in urban areas in advance of need, taking into account not only the possibility of extending utilities and community facilities to the areas, but also other key aspects such as transportation and location in relation to jobs.

Absorption of labor and the structure of the urban economy

The most pressing problems associated with Third-World urbanization arise because, despite accelerated industrialization, the rapidly increasing labor force of the cities is not being absorbed into full and productive employment (Friedmann and Sullivan, 1972). With urban growth rates typically running at least twice the rate of natural increase, frequently in excess of 5.0 per cent per annum (Davis, 1969), but with industrial employment increasing at 4.4 per cent per annum (Turnham and Jaeger, 1971), the bulk of new manpower is absorbed by small-scale enterprise, personal services and open unemployment. Moreover, spurts in urban investment tend only to bring more migrants to the city.

The problems resulting may be understood in terms of the structure of the urban economies, which are made up of three separate sectors. The *individual enterprise* sector comprises the unemployed workers of the "street economy" of the city, including the offspring of urban residents, recent migrants to the city, those laid off from other jobs, street hawkers, casual construction workers, prostitutes and panderers, professional beggers and petty thieves. It accounts for between 25 and 40 per cent of the urban labor force. Few earn more than the subsistence minimum, and those who do most frequently share the surplus with their kin. There is intense competition for work and this keeps earnings at the subsistence minimum. Any growth in the urban economy simply brings in more migrants and keeps rewards to individual enterprise at the lowest possible level. Most people engaged in this sector live marginal lives in the bridgehead settlements, or simply live on the streets, and under the worst conditions experience large-scale misery in its full harshness. At night the sidewalks of Calcutta, for example, become public dormitories, heaped with emaciated men, women, old people, and children. The poor of Calcutta lack the most elementary belongings, owning neither pillow, mattress, nor blanket; their bodies stink and are covered by soiled rags. At dawn, before the city awakens, carts collect the corpses of those who have died in the night.

The second sector of the Third World's urban economies is that devoted to *family enterprise* in the traditional bazaar-type economies. The land use patterns of the parts of the city in which this sector dominates are chaotic. Such family enterprise accounts for 35–45 per cent of the labor force in small trade and service establishments and industrial workshops. By and large, traditional commodities are produced for the low-income mass market using local raw materials and lacking quality control and standardization. Production is dependent on the utilization of the entrepreneur's family (the extended kin group), for whom the end-product is a condition of shared poverty. Because pricing is competitive and the activities are labor-intensive, returns seldom provide for more than the subsistence requirements of the family.

The third sector is the *corporate*, including capital-intensive businesses, the government, and the professions. Depending upon the particular city and country, this sector provides between 15 and 50 per cent of the urban employment. Economic units are larger, people work regular hours, capital investment is on the large scale, levels of technology and productivity are high. There is continuing pressure to provide all the perquisites of similar occupations in the developed countries. Education is required for entry to the sector, and employment in it automatically conveys middle-class status as a minimum, and produces the professional-managerial urban elite at its upper echelons, together with the juxtaposition of luxury and poverty that is one of the striking physical characteristics of the Third World cities.

A particular set of labor market dynamics between the three sectors maintains workers in the urban economy at base survival levels. Any growth in the corporate or family enterprise sectors, for example, filters immediately to the individual enterprise sector. The prospect of gaining employment brings more new migrants into the city and intense competition for jobs keeps wage rates at their minimum. To add to the downward pressure on wages, expansion of mass production by the corporate sector frequently forces family sector enterprises producing inferior goods out of business. The higher productivity of the corporate sector means that job gains are less than production gains, while job losses in the family sector are substantial.

This structure of the urban economy is argued by T. G. McGee (1967) to be a direct product of the colonial or neocolonial experience of most Third World countries. Early European contact saw the establishment of embryonic colonial urban networks designed largely to aid European control of indigenous trade. Western controls expanded in the nineteenth century, involving creation of far more extensive urban and communications networks than had existed in the area before. But the colonial city remained essentially a conservative force, economically subordinate to the developed nations and world trade. The culture, way of life, and population of the city were alien to the indigenous inhabitants, for the cities were populated by heterogeneous populations, many of whom were migrants or foreigners. Occupational and residential segregation existed according to ethnic groups, as did social stratification; Europeans at the top, commercial groups (frequently Asiatic) in the middle with the local Western-educated elites, and finally the indigenous immigrant population at the bottom. Much of the indigenous population was transitory, young, and male. Even today, McGee argues, with inherited colonial economic structures displaying excessive specialization in the production of raw materials for the industries of the international powers, major cities of the Third World still tend to function as "head links" between the industrialized powers and their sources of raw materials. In a sense, then, they remain transplants more closely related to the external industrial world than to the countryside of the Third World.

Three basic consequences have resulted from the colonial heritage of many Third World nations: (1) the maintenance of a minimal "survival" economy; (2) the reinforcement of traditional subcultures in the city; and (3) the concentration of economic growth and development in a few large cities that often leads to a "parasitic" draining of the society at large. The third consequence, which has come to be known as "primacy," is a critical factor for the development of many Third World societies. Let us now turn to this important issue and address it in some depth.

Chapter 19
Urban hierarchies and economic development: decolonization and the primacy question

Are many of the societal inadequacies in the developing countries attributable to a dysfunctional relationship between the urban system and spatial development that is the continuing legacy of colonialism? Many of the Third World's policy makers apparently believe so, characterizing the malaise as "primacy." It is the purpose of this chapter to assess the evidence bearing on this belief. The central ingredient in the discussion is the role of the urban hierarchy in the spatial organization of developing countries.

Development role of the urban hierarchy in the west

The point of departure for most of the debates is the development role of urban hierarchies in the Western world. This role is usually assumed to be the desirable norm against which the functional or dysfunctional character of the urban system is judged in other contexts. The argument proceeds in several steps.

Common growth characteristics of the developed countries

That there is currently a correlation between the level of economic development of countries (however indexed), the degree to which they are urbanized, and, correspondingly, the extent to which their populations are concentrated in large cities is unquestioned. Table 19.1 provides clear evidence that these systematic differences have emerged because a series of common growth characteristics have served to differentiate the "developed" countries from more "traditional" polities and from their own prior traditional states. All the developed countries have experienced high rates of increase in per capita product (15–30 per cent per decade) accompanied by

Table 19.1

Indicators of living conditions. Countries grouped by national income per capita, post-World War II years

	Groups of countries by per capita income (U.S. $)					
	1,000 and over	575–1,000	350–575	200–350	100–200	Under 100
Urbanization						
Percentage of total population in urban areas (recent census)	68.2	65.8	49.9	36.0	32.0	22.9
Percentage of population in communities of more than 100,000, about 1955	43	39	35	26	14	9
Mortality						
Expectation of life at birth, 1955–1958 (years)	70.6	67.7	65.4	57.4	50.0	41.7
Infant mortality per 1,000, 1955–1958	24.9	41.9	56.8	97.2	131.1	180.0
Food consumption						
Percentage of private consumption expenditures spent on food, 1960 or late 1950s (36 countries)	26.2	30.5	36.1	37.6	45.8	55.0
Per capita calorie consumption latest year (40 countries)	3,153	2,944	2,920	2,510	2,240	2,070
Percentage of starchy staples in total calories, latest year (40 countries)	45	53	60	70	74	77
Energy consumption						
Per capita kilos of coal equivalent, 1956–1958	3,900	2,710	1,861	536	265	114
Education						
Percentage of population illiterate, 15 years and over, 1950	2	6	19	30	49	71
Percentage of school enrollment to four-fifths of the 5–19 age group, latest year	91	84	75	60	48	37

substantial rates of population growth (over 10 per cent per decade). Much of the economic growth has been sustained as a result of improved production techniques, largely improvements in the quality of inputs and greater efficiency traceable to increases in useful knowledge and improved institutional arrangements. All sectors of the developed countries' economies, manufacturing in particular, have participated in the increasing efficiency, while at the same time sustained growth has involved changes in the relative sector importance. A declining share of total product is attributable to agriculture, rising shares to manufacturing and public utilities, and rapidly increasing shares to personal, professional, and governmental services. Significant changes have occurred in the structure of final demand. These changes have been both the effect and the cause of changes in the productive process, and have included important shifts in the regional allocation of resources and increasingly large production units, as both product and labor have shifted from smaller firms and organizations to larger ones. These shifts, of course, have led to increased concentration of people in cities. Shifts in capital allocation, in product, and in labor, have in turn depended upon rapid institutional adjustments and mobility in factor inputs, and it is here that urbanization has played a critical role in facilitating shifts in population and labor force, both between and within regions, and by type.

In short, as the developed countries of the world have modernized, the size of their production units has increased along with the number and complexity of production decisions. Market mechanisms have expanded in scope and extent to bring together production and consumption through use of transportation, communication, financing, policing, and related services, providing order for increasingly impersonal processes. Progressive division of labor, increasing scale, regional specialization, and the complexities of articulating intensifying marketing networks have all bred greater population concentration in larger cities and higher degrees of urbanization than is characteristic of more traditional societies.

Regional implications of the growth process

Rapid increases in total product have also implied greater pressure on natural resources, whereas increasing scale and concentration imply wide differentials in types and rates of growth among different social and economic groups in different regions, the interregional mobility of labor and capital, and the emergence of a spatial pattern of core (heartland) and periphery (hinterlands).

Large-scale industry has tended to concentrate in a limited number of cities in a limited region that serves as a polity's industrial heartland and, because of the large numbers of industrial workers employed, as the center of national demand. Such a concentration develops a self-generating momentum as complementary services and activities are established, each helping the other to pyramid the productive process: increasing numbers of workers further concentrate the scale of the local market and even more strongly pull to themselves activities seeking optimal national market access.

This cumulative causation extends outwards to the hinterlands, for once the core-periphery pattern is set the core region becomes the lever for development of peripheral regions, reaching out to them for their resources as its input requirements increase and stimulating their growth differentially in accordance with its resource demands and the resource endowment of the regions.

The result of this core-centered pattern of growth and expansion is regional differentiation—the specialization of regional roles in the national economy. Specialization, in turn, determines the entire content and direction of regional growth. Because regional economic growth is externally determined by national demands for regional specialties, and organized geographically in a pattern of industrial heartland—together with hinterlands specialized in resource subsectors—the nature of these specialties, alternative sources of them, and changes in the structure of demand therefore determine in large measure the nature and extent of regional growth. This extends to the secondary support needed by export industries—housing, public facilities, retail establishments, service facilities, and the like. The size of the multiplier-effect exports on "local" or "residentiary" growth, however, depends upon local expenditure patterns and income distributions, patterns of ownership, and political organization. Among the relevant issues raised are whether earnings are retained locally or transferred outside, and whether the basic industry generates a middle class. Local decisions can of course also help shape the future of a regional economy—when, for example, the local economy is "closed" from external influence or when local decision making becomes an element in national decisions. But basically, it is external factors that create growth opportunities or lead to decline. Opportunities have to be perceived and seized by imaginative leaders; otherwise they are lost, for ultimately growth can be traced back to individual location decisions about particular business establishments, and the size and nature of the entrepreneurial class thus assume a critical role.

Cities articulate relations among regions

Cities are the instruments whereby specialized subregions are articulated in a national space economy. They are the centers of activity and of innovation, focal points of the transport network, and locations of superior accessibility at which firms can most easily reap scale economies and at which industrial complexes can obtain the economies of localization and urbanization. Agricultural enterprise is more efficient in the vicinity of cities. The more prosperous commercialized agricultures encircle the major cities, whereas the inaccessible peripheries of the great urban regions are characterized by backward, subsistence economic systems.

In Chapter 14 we observed two major elements characterizing this spatial organization of the developed countries:

1. A system of cities, arranged in a hierarchy according to the functions performed by each.
2. Corresponding areas of urban influence or urban fields surrounding each of the cities in the system.

Generally, the size and functions of a city and the extent of its urban field are proportional. Each region within the national economy focuses upon a center of metropolitan rank, and it is the network of intermetropolitan connections that articulates the whole. The spatial incidence of economic growth is a function of distance from the metropolis. Troughs of economic backwardness lie in the most inaccessible areas along the intermetropolitan peripheries. Further subregional articulation is provided by successively smaller centers at progressively lower levels of the hierarchy—smaller cities, towns, villages, etc.

Economic change is patterned by the urban hierarchy

Similarly, we noted in Chapter 14 that impulses of economic change are transmitted in such a system simultaneously along three planes:

1. Outward from heartland metropolises to those of the regional hinterlands.
2. From centers of higher to centers of lower level in the hierarchy, in a pattern of "hierarchical diffusion."
3. Outward from urban centers into their surrounding urban fields.

We observed that part of the diffusion mechanism is to be found in the operation of urban labor markets. When growth is sustained over long periods, regional income inequality, for example, should be reduced because the higher the capital-labor ratio in a region, and the higher the employment level of the unskilled at any wage rate and at any given social minimum, the smaller the number of involuntary unemployed. Any general expansion in a high-income area, such as a heartland metropolis, will reach a rising floor to the wage rate first. Some industries will be priced out of the high-income labor market, and there will be a shift of that industry to low-income regions, that is, to smaller urban areas or more peripheral ones. This "filtering" or "trickle-down" process, as we described, will have substantial direct as well as indirect effects. The significance of these diffusion mechanisms is that if economic growth is sustained over long periods it results in progressive integration of the space economy. Regional differences in levels of welfare are progressively eliminated, because demands for and supplies of labor are adjusted by outward flows of growth impulses through the urban hierarchy and by the outward migration of labor to central cities. Troughs of economic backwardness are reduced in intensity, and each area finds itself within the fields of influence of a variety of urban centers of a variety of sizes. Sufficient growth impulses will move through the system so that each region and each city can expect to grow at about the same rate as the nation, although local factors, when viewed from the national perspective, may cause seemingly random variations above and below this national expectation.

Hierarchical organization in the developing countries

In every country, of course, urban centers are organized functionally into a hierarchy; the hierarchy is the instrument whereby society, polity, and economy are integrated over space. In traditional societies, however, hierarchical organization was based as much on purposive sacerdotal, juridical, military, or administrative principles as on economic grounds. But, whatever the principles, status of towns within the hierarchy determines their sphere of influence and depends upon the power residing in their level of the hierarchy. Status and sphere, in turn, vary from level to level depending upon the organizing principle. City size remains the simplest and best index of this power. Differences in the numbers of cities in different population-size classes should therefore reveal differences and variations in the degree of urbanization and the proportion of population concentrated in the largest cities.

In the developed nations, the distribution of cities by size follows the rank-size distribution; that is, the distribution is approximately lognormal. The rank-size pattern has thus been interpreted as the "functional" outcome of economic growth, with the result that it is deviations from the rank-size distribution that have excited the greatest interest in the developing nations. The argument is that a rank-size distribution should be characteristic of any well-defined system of cities in which growth has, as a minimum, obeyed the law of proportionate effect for some period of time.

One particular deviation has been of overwhelming concern. Where the actual population of the largest city exceeds that which is expected on the basis of the rank-size rule, a condition of "primacy" is said to exist. Such deviation is thought by many to be dysfunctional. Colin Clark (1967) uses the additional term *oligarchy* to describe situations such as Japan, India, Australia, or Brazil, where the towns over 100,000 population have a bigger share of the total urban population than would be expected from the straightline relationship, but where at the same time the primacy of the leading city is kept in check. A particularly striking example is provided by the Portuguese colonial system, in which major urban "headlinks" form one oligarchic rank-size regime and centers functioning at the local levels of the hierarchy form another (see Figure 19.1). Clark also defines as

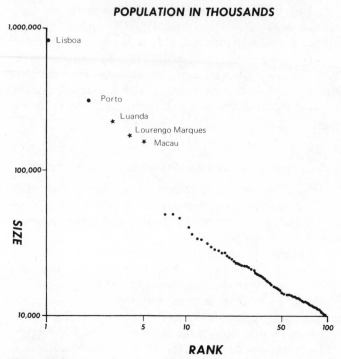

Figure 19.1
Rank-size distribution for the cities of the Portuguese colonial
system (prepared by Gerald F. Pyle).

counter-primacy either declining primacy of the largest city through time, or
increasing negative deviation of the population of the largest city from
predictions based on the rank-size rule under conditions of national planning
directed at achieving such outcomes as goals. Good examples of counter-
primacy in which there has been increasing convergence of the city-size
distribution on the rank-size pattern are provided by the Philippines (Figures
19.2, 19.3, 19.6 and 19.7) and, for comparison, Denmark (Figure 19.4). On
the other hand, Buenos Aires has maintained its primacy in Argentina for
the last century (Figure 19.5).

The interpretations of primacy

The idea of primacy, as initially formulated by Mark Jefferson (1939),
was very simple. He argued that everywhere "nationalism crystallizes in
primate cities...supereminent...not merely in size, but in national influence." He assessed the degree of eminence of cities within countries by
computing the ratios of size of the second- and third-ranking cities to that

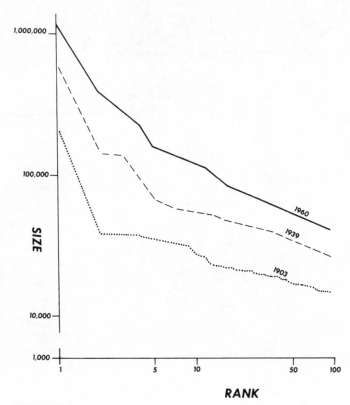

Figure 19.2
Rank-size distributions for the Philippines, 1903–1960 (prepared by Gerald F. Pyle).

of the largest place. But immediately after Jefferson's papers had appeared, Zipf (1941) directed attention to the entire system of cities, focusing in particular on the case of the rank-size rule. Such, he argued, was the situation to be expected in any "homogeneous socioeconomic system" that had reached a harmonious equilibrium state.

It remained for discussants at a series of postwar UNESCO conferences on urbanization in Asia and the Far East, and in Latin America, to put the two together. Cases deviating from the role were said to arise from "overurbanization" of the economies of lesser-developed countries because of "excessive" in-migration and superimposition of limited economic development of a colonial type, creating "dual economies" characterized by "primate cities" that tend to have "paralytic" effects upon the development of smaller urban places and to be "parasitic" in relation to the remainder of the national economy.

Obviously, each of the words in quotation marks involves a value

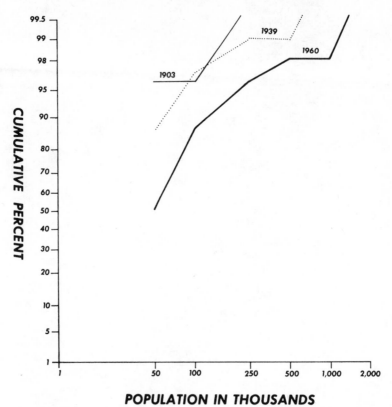

Figure 19.3
Lognormal plots for the Philippines, 1903–1960 (prepared by Gerald
F. Pyle).

judgement, but nonetheless the idea of the primate city from this point on
was firmly established as a malignant deviation from expectations about
hierarchical organization derived from the rank-size rule, with strong
pejorative connotations. The value-laden attitude is illustrated in the
following quotation from Lampard's (1955) work:

> Its growth and maintenance have been somewhat parasitical in the sense that
> profits of trade, capital accumulated in agriculture and other primary pursuits,
> have been dissipated in grandiose urban construction, servicing, and con-
> sumption.... The labor and enterprise which might otherwise have been
> invested in some form of light manufacturing or material processing...are
> drawn off to the great city by the attractive dazzle of a million lights.

Measurement of primacy has subsequently become a textbook exercise in
urban studies (see for example the volume by Jack P. Gibbs, 1961). To
planners such as John Friedmann and Tomas Lackington (1967), primacy
cannot be separated from the idea of "hyperurbanization":

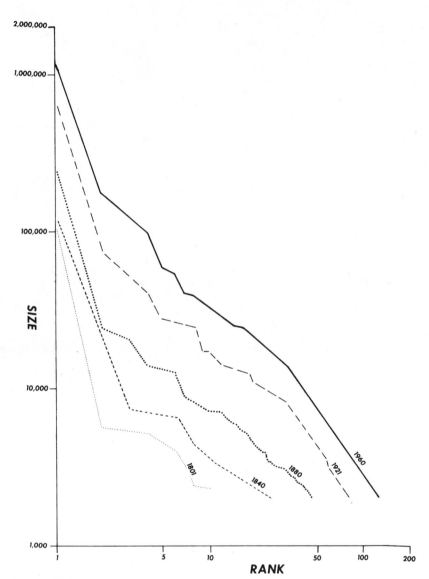

Figure 19.4
Rank-size distributions for Denmark, 1801–1960 (prepared by Roger LeCompte).

a society exhibiting a high measure of disequilibrium between its levels of urbanization and per capita income is likely to experience serious internal tensions...[this] hyperurbanization...takes the form of increasing concentration of urban activities. This may be established in several ways...the degree of urban primacy [may be measured] according to...the relation of the largest to the second urban complex.... Another possible measure of primacy

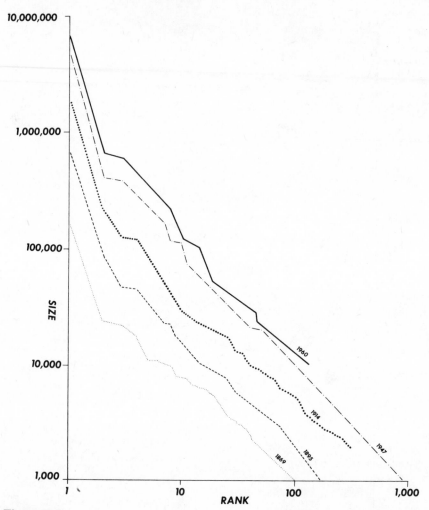

Figure 19.5
Rank-size distributions for Argentina, 1869–1960. Source: Cesar A. Vapnarsky, "Rank Size Distribution of Cities in Argentina." Unpublished M.A. thesis, Cornell University, 1966.

is in relation to the rank-size rule which describes a standardized distribution of a country's population among its cities.

To many planners and policy makers in the developing countries, "gigantism" of the largest cities is a characteristic to be feared, even when the principal city is small relative to cities in other areas; the elimination of primacy is seen as a key ingredient in economic as well as political decolonization.

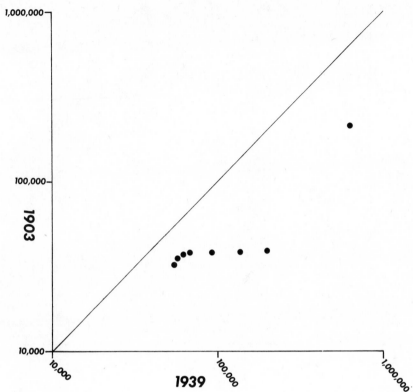

Figure 19.6
Sizes of the eight largest cities in the Philippines, 1903 and 1939. The growth rate of the top four far exceeds that of the next four in rank (prepared by Gerald F. Pyle).

Reasons for primacy

The reason for primacy is to be found in the filtering mechanism that produces hierarchical diffusion. This mechanism works poorly, if at all, in many parts of the world. Instead of development "trickling down" the urban size-ratchet and spreading its effects outward within urban fields, growth is concentrated in a few metropolitan centers, and a wide gulf between metropolis and smaller city is apparent. Rather than articulation there is polarization. Why?

There are two classes of reasons, one institutional, the other functional. Under colonial rule, most empires were controlled by holding key cities and strategic points: "head links" that connected the colonial net. For colonial powers to extend and consolidate their authority in alien social and geographical territories, cities had to be the base of action for those powers. British rule in India, for example, centered on capital and provincial cities

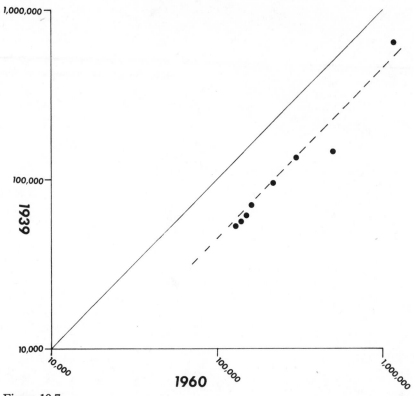

Figure 19.7
Sizes of the eight largest cities in the Philippines, 1939 and 1960. There is much greater conformity to Gibrat's law, and the accelerated growth rate of the smaller places has led the city-size distribution much closer to rank-size (prepared by Gerald F. Pyle).

both for maintaining an integrated and authoritarian administrative structure and for securing the economic base of its power—the collection of taxes and control over the export of raw materials and the import of British manufactured goods. McGee (1967) writes that the structure of the colonial economy elsewhere in Southeast Asia did not permit the cities to be generative of economic growth. The colonial cities were subordinate to international world trade, acting as foci for the alien middle man and effectively inhibiting economic growth.

However, the colonial cities did help create modern indigenous elites and give birth to nationalist movements, and they restructured traditional economic patterns to focus on the cities, initiating massive inward flows of rural migrants. The resulting high rates of population growth outstripped rates of economic growth in the cities after independence, and confounded the trickle-down process. The result was to worsen the way of life outside the metropolis and simultaneously increase the urban labor supply at all levels, but particularly among the unskilled at a pace that has never permitted economic expansion to meet rising floors to metropolitan wage rates; thus growth cannot decentralize naturally. This systematic confounding of trickle-down mechanisms is particularly acute where, either because of the very small size of a political unit or because of a critical dependency of the polity upon external trade relationships, there is a single national locus without sizeable alternative growth centers.

Policies to combat primacy

What can be done to combat primacy? One argument is that it is a temporary phenomenon characteristic of early stages of the growth process when the advantage lies with the developed centers, the large cities of the national heartland or the centers of colonial penetration which enjoy the existence of overhead facilities, external economies, political power, spatial preferences of the decision makers, immigration of the more vigorous and educated elements from the underdeveloped regions, flows of funds from the land-wealthy in the hinterlands to the financial markets in the cities, and a variety of other factors. These factors lead to polarization, to concentration in the large cities and increases in the differences of regional incomes. Trickle-down mechanisms will reassert themselves, this argument proceeds, when the spread of literacy and bureaucratic practices improve knowledge in and about the backward areas; when the opening of transportation routes to reach these areas as markets for the developed centers also opens them as possible locations for productive activities; and when universal education and standardization of all aspects of life permit an integration of the space economy, and, by making externalities more nearly comparable everywhere, make the more distant opportunities more accessible and more interesting for development. Thus, in this view, in the early states of development there will be increasing disparity between developed and underdeveloped regions, but there will be a tendency toward equalization as the economy reaches maturity.

An alternative view is that continued change within the framework of present economic systems is insufficient, and that radical changes in the nature of urban systems must be introduced by fiat to overcome continuing constraints of size, low rates of growth, export orientation, and colonial inheritance that conspire to confine modernizing influences to primate cities with little foreseeable hope for filtering to take place.

The argument is that historical experiences of developed nations are irrelevant in developing countries today. The countries that are now developed were economically ahead of the "rest of the world" when they commenced modern economic growth, and the underdeveloped countries are the worst off economically at the present time. The world of the eighteenth and nineteenth centuries had freer trade policies, freer opportunities for international movements of population, and fewer political and economic barriers than the world of today. In its initial phases, modern economic growth was associated with skewed income distributions, with respect to both wage earners, on the one hand, and agriculturists (as opposed to manufacturers), on the other. Economic growth thus involved a transfer of resources from agriculture to industry, and took place at the expense of agriculture. It involved a transfer of resources to the upper middle classes, which have a greater propensity to save and hence to invest. The transfer was also from urban wage earners. The political feasibility today of resource transfers to a richer class (with a greater propensity to save) is debatable.

Further, in the underdeveloped countries today institutional settings that are the products of modern economic growth of the developed countries have preceded the process of growth. The political institutions of democracy, for instance, often preclude exploitation of the proletariat for the process of growth, a feature which was not uncommon in the initial stages of growth in the West. Welfare measures like minimum wages, regulation of hours of work, and prohibition of child labor are all institutions which in certain ways do not make for efficient economic development of the sort that took place under private entrepreneurship. This is compounded by the co-existence of highly developed and underdeveloped countries with rapid means of communication and transportations, enabling peoples of the underdeveloped countries to copy consumption patterns of the developed countries. This "demonstration effect" is asymmetrical: it applies to consumption but not to investment and savings, thus diverting resources from investment to conspicuous and other types of consumption. Parallel "derived development" projects divert investments to conspicuous and monumental structures rather than to productive ones.

For each of these reasons, an increasing number of urban and regional planners have begun to argue that the only solution will come from programs of decentralization into new growth centers. Decentralized patterns of urbanization, they say, emphasize medium-sized cities and probably will cost less in infrastructural investment while avoiding the social perils of gigantism. Further, strategies of "downward decentralization" are argued to contribute less to equity goals in a restructured society than to rapid growth, serving efficiency goals by bringing new or underutilized resources into the development process. Allocation of public investments to simulate the filtering process bypasses current constraints by creating alternative magnets for migrants and reducing pressures on the large-city labor market, hence enabling the "natural" filtering process to reassert itself as long as overall national growth is maintained at high levels and attempts to reduce the natural rate of population increase are successful. It all sounds very plausible. Whether or not such proposals can be successful we do not know, however; evaluation must now await the test of time and practice.

Chapter 20
Comparative urbanization strategies: an overview

Let us now try to draw together our foregoing chapters on comparative urban structure and planned change. What is implied is that there are in the world today identifiably different policies regarding urban growth and development, that these policies are being translated into workable programs, that the policies and programs lend themselves to worthwhile systematic comparison because they are altering urban-growth directions, and that new avenues of research in urban ecology are thereby being opened up. For this to be so, something significant must have happened in the past quarter century. As Lloyd Rodwin remarked in *Nations and Cities* (1970), "before World War II almost no one wanted the central government to determine how cities should grow." But he did go on to say that "radical changes in technology and in analytical and planning methods now make significant changes in the urban system not only feasible, but to some extent manipulable," and, because of this, "today, national governments throughout the world are adopting or are being implored to adopt urban growth strategies."

Rodwin put his finger on two important bases of comparison of urbanization strategies—the planning method and the political process. They are *ideologically* linked in terms of world view, which involves both the perception of urban problems and the specification of goals to be achieved, and *practically* linked in terms of the power to implement the means thought likely to achieve the ends. Societies differ in their planning capabilities, and this fact introduces one important element of differentiation of urbanization strategies. But more importantly, cultural differences have produced divergent goals, divergent planning methods, divergent plans, and now divergent paths being followed by urban development in the world today.

We thus begin our overview of comparative urbanization strategies by reemphasizing their sociopolitical bases and returning thereafter to the question of planning styles and details of strategies, for they are firmly derivative of the cultural context. It will be evident that we find no reason to change what was set down in 1973 by one of the present authors in *The Human Consequences of Urbanisation*, which concluded as follows:

> The diverse forms that public intervention is taking, the variety of goals being sought, and the differences in manipulation and manipulability from one society to another are combining to produce increasingly divergent paths of deliberate urbanisation. This makes it all the more important to understand the relations between socio-political forms and urbanisation, because the socio-polity determines the public planning style. Urbanisation, in this sense, can only be understood within the broad spectrum of closely interrelated cultural processes; among these processes, planning increases rather than decreases the range of *social* choice as modernisation takes place, while simultaneously restricting the range of *individual* choice to conform to the social path selected. Images of the desirable future are becoming the major determinants of that future in societies that are able to achieve closure between means and ends. Political power is thus becoming a major element of the urbanisation process. Combined with the will to plan and an image of what might be, it can be directed to produce new social forms and outcomes, making it possible for a society to create what it believes *should be* rather than extending what *is* or what *has been* into the future.

Sociopolitical form and urbanization process

Free-enterprise dynamics

At one end of the urbanization spectrum are free-enterprise, decentralized, market-directed societies. Traditionally in such societies, decisions are made by individuals and small groups, and interact in the marketplace through free interplay of the forces of supply and demand. Economic and political power, vested in claims of ownership and property, is widely dispersed and competitively exercised. The instruments of collective or government action are used only to protect and support the central institutions of the market and to maintain the required dispersion of power. Thus, the public role is limited to combating crises that threaten the societal mainstream, as privately initiated innovation produced social change. Legal systems are mainly regulatory, too, functioning to preserve established values; thus, to cite one example, the reliance of American law upon the regulatory approach to city building has brought about the atrophy of city planning as a constructive element in social change.

One consequence is to be seen in North American political science, in which there has developed an explicit belief system concerning the process by which policies change, which in turn influences the way in which a problem is perceived. The dominant mode of thought on this subject in American political science is that of *incrementalism*. As Charles E. Lindbloom said in

1963, "Democracies change their policies almost entirely through incremental adjustments. Policy does not move in leaps and bounds." The political processes of bargaining, log-rolling, and coalition-building are the major factors producing a situation in which past decisions are the best predictors of future ones. Under such conditions, no explicit policy guides urban growth. Rather impetus for growth derives from preoccupation with economic achievement, and this limits urban planning to an ameliorative problem-solving role, which is reinforced by an attitude that accepts the *inevitability* of a continuation of the processes inherent in the present.

What are some of these processes? Increasing organizational scale and concentration of power are dominant characteristics of today's postindustrial democracies. As the scale of economic and bureaucratic organization increases, changes are taking place in the dynamics of social and economic change. Increasing numbers of the more salient developmental decisions are made by negotiation among large-scale autonomous organizations and by voluntary associations, profit-oriented but not necessarily maximizers, countervailing and countervailed against, negotiating together and existing in a context of negotiated relationships, rather than these decisions' being delivered by the guiding hand of an unseen and more neutral market. Power is determined as a matter of policy or agreed upon by counterbalancing powers. Under such conditions there is organization of production by large corporations run for the benefit of stockholders, and wage negotiation by labor through large-scale unions. The consumption of end products is determined partly by individual choice, and partly by governmental policy. The collective power of organizations, the collective power of the government, and the free choice of individuals are all part of the system. Hence, the "market" is no longer the single master. Rather, elaborate negotiation for "satisfactory" solutions tends to prevail; instead of maximization, there is "satisficing" and, increasingly, each of the large-scale organizations engages in planning in terms of corporate goals, often with systems-analysis staffs at their disposal to help them select desirable courses of action. The resulting combination of scale, power, and corporate planning means that developmental leadership, the innovative "cutting edge" of urbanization, is now more frequently in the hands of corporate oligarchs, responding to their particular economic agendas, for in spite of the opportunity for the collective power of the government to be exercised as a significant countervailing force —Canadian and Australian experiments notwithstanding—no governmental bodies in the world's free-enterprise societies have taken the initiative to describe and execute a national urban policy to date. Instead, as we noted earlier, we find a complex set of uncoordinated and often contradictory public policies and programs provided, on the one hand, to encourage economic growth, and on the other, to provide public-works and public-welfare programs to support piecemeal, spontaneous development impelled primarily by private initiative.

Public counterpoints in the redistributive welfare state

It has been the greater radicalism of the welfare states of Western Europe that has produced modification of the free enterprise system and its large-scale twentieth-century heirs by governmental action, aimed at reducing social and economic inequities and at providing every citizen with minimum guarantees for material welfare—medical care, education, employment, housing, and pensions. This has most usually been achieved through differential taxation and welfare payments, but it also increasingly involves extension of more centralized decision making, designed to make the market system satisfy social goals in addition to its traditional economic functions. In this way, there have emerged mixed economies, the hallmarks of which are pluralistic societies with multiparty governments, relatively high levels of development and per capita output, and built-in capicity for continued growth—but with substantial public sectors alongside elaborate private markets and "modern" economic institutions.

Public involvement in urbanization in the mixed economies is to be seen as more than merely a counterpoint to private interests, however. Directing society toward goals of redistribution and equity reorients the competitive drive. By constructing a large share of all housing in existing towns, and by constructing new towns, the public exercises developmental leadership. Urbanization is deliberately led in new directions.

The particular form of policy varies from country to country. Ebenezer Howard's New Towns philosophy and the apparent successes of British planning have induced other countries to attempt to control physical development, direct new settlement away from congested metropolitan centers, and stimulate economic and urban growth in peripheral regions. For example, the French have been concerned for a decade and a half about the steady concentration of people and economic activity in Paris. The Finns have seen the population of the northern half of their country pour into Helsinki, or else leave Finland entirely. The Swedes are worried because most of their people are concentrated in Stockholm and two other metropolitan areas of southern Sweden. They are concerned, not because their metropolitan centers are too big, as the French are about Paris, but because the depopulation of northern Sweden threatens to erode the social structure in that part of their nation. The Hungarians have a similar problem: their national economy is dominated by Budapest. Only the Poles are unconcerned with the outflow of rural people into their metropolitan areas. Lagging behind the rest of Europe in their rate of modernization, the Poles see metropolitanization as a process that is essential to absorb a "surplus" rural population.

Despite these differences in forms, however, the central concern of urban policy in all of Europe is the regional distribution of growth. Economic growth is viewed as the basic means to achieve social objectives such as improved income, housing, education, health, welfare, and recreational opportunity. European growth policies are intended to ameliorate disparities in income and welfare among regions of the country and, to a lesser extent, to minimize deleterious effects of economic growth on the natural environment.

As noted, the goals and objectives of urban growth policies vary from country to country, but to some degree all have common aims:

1. *Balanced welfare.* Achieving a more "balanced" distribution of income and social well-being among the various regions of the country, as well as among social classes.
2. *Centralization/decentralization.* Establishing a linked set of local and national public institutions that make it possible to develop, at the national level, overall growth strategies, integrated with regional or metropolitan planning and implementation, which are partly a product of a reformed local governmental system and are directly accountable to local officials and the affected constituency.
3. *Environmental protection.* Channeling future growth away from areas suffering from environmental overload or possessing qualities worthy of special protection, toward areas where disruption of the environment can be minimized.
4. *Metropolitan development.* Promoting more satisfactory patterns of metropolitan development through new areawide governmental bodies, special land use controls, new towns, housing construction, new transportation systems, and tax incentives and disincentives.
5. *Nonmetropolitan development.* Diverting growth into hitherto bypassed regions by developing "growth centers" in presently nonmetropolitan regions, constructing new transportation links between such regions and centers of economic activity, using various incentives and disincentives to encourage or compel location of economic activity in such areas, and forcibly relocating certain government activities into them.

Given such goals, the urban future that is unfolding in each case represents a delicately orchestrated balance between individual and corporate interests, on the one hand, and the collective power of the state, on the other. Much of this power is exercised in a negative fashion, to constrain individual or corporate drives, but the best appears when the public sets in motion new growth directions by exercising developmental leaderships, thereby initiating new and exciting trends.

Directed change in the socialist state

From a public counterpoint and developmental leadership perspective, the next step is to the command structure of the socialist states, where monolithic governmental systems are dominated by a single party, non-agricultural (and sometimes agricultural) industries are operated by the state, and the economy is centrally directed. Each of the socialist nations shows strong commitment to economic growth, but on the social side there is also a desire for elimination of most of the status differences, based upon economic rewards, that are the hallmark of free-enterprise competition. A greater uniformity and lack of specialization is to be seen in the urban fabric, alongside more highly regimented life styles and building patterns. It is easier to command when there is an explicit set of rules and procedures to be followed; in this way urban development has been both bureaucratized and standardized under socialism.

As was noted in Chapter 17, the ultimate goal sought in Soviet urban development is what Lenin had called "a new pattern of settlement for man," the city of socialist man. More clearly than anywhere else, in socialist societies what men believe (that is, their ideology) determines the urban future they try to create. Nonetheless, one sees in the Soviet literature today major debates about the desired urban future.

One such widely publicized debate is between two scholars influential in the Soviet planning process, geographer B. S. Khorev (1972) and economist V. V. Perevedentsev (1972). Khorev argues that continued attempts must be made in the Soviet Union to develop new forms of urbanization consistent with the tenets of Marxism-Leninism. He writes that the division of labor leads in the first place to a separation of industrial and commercial labor from the labor of agriculture and, thus, to a separation of town from countryside. In a class society this separation produces antagonism between town and countryside. But as classes disappear and society is rebuilt on a communist foundation, he feels that significant differences between town and countryside also gradually disappear.

> The essence of long-range changes in settlement and in growth of cities lies in the gradual erasing of differences between town and countryside, to yield a unified system of settlement, whose planned regulation may help prevent haphazard and uncontrolled city growth.

The classic writings of Marxism-Leninism, according to Khorev, suggest ways of achieving that goal: harmonious development of productive forces according to a single plan; a more uniform distribution of large-scale industry and population throughout the country; achievement of close internal links between industrial and agricultural production; expansion of the means of transportation; reduction of the concentration of population in large cities. To date, he argues,

> we have experienced a concentration of industrial production and urban population. Revolutionary changes are now leading to a single system of settlement, a functionally delimited and structurally interrelated network of places that can be regulated in a planned manner for the benefit of society, and encompassed by a unified system of regional planning, a type of spatial organization that can be conceived only as an integral combination of artificial and natural environments.

Perevedentsev contests this argument saying that the literature on city

planning is strewn with expressions like excessive growth, excessive development, excessive concentration, excessive saturation with industry. "I once
made persistent efforts to find out what the originators of such expressions
had in mind. Alas, I did not succeed in determining this. Here, everything is
excessive." He notes that B. S. Khorev writes: "The excessive, hypertrophic
growth of our large and superlarge cities is inexpedient and undesirable."
Perevedentsev asks, "Where do the hypertrophy and excessiveness start? At
what size?" "One searches in vain," he says, "for the criteria of optimal
size or of excessiveness," noting that the productivity of social labor is
generally considered to be the chief criterion for effectiveness in distributing
production forces. Calculations show that the productivity of labor in large
cities is many times higher than that in the small ones, and in the superlarge
cities it is many times higher than in the large cities. Thus, for example, the
productivity of labor in the industry of cities with populations of more than
1,000,000 persons is 38 per cent higher than in cities with populations of
from 100,000 to 200,000 and the return on assets is more than twice as high.

Perevedentsev also notes the ineffectiveness of attempts to limit growth
by registration requirements placed on migrants. "Many large cities have
long had rigidly limited registration; they grow and grow all the same." By
no means is the problem here attributable to natural growth (the prepon-
derance of births over deaths), which in the largest cities is insignificant. In
Moscow, for example, in 1967 it was 1.7 persons for every 1,000 residents—
that is, 11,000 for all of Moscow—but Moscow's annual growth still amounted
to more than 60,000. The administrative regulation of migration has proved
ineffective in the extreme, and where there has been some success it has led
only to a shortage of manpower—that is, industry and other branches of the
national economy were unable to exploit their potentialities. The manpower
shortage in many cities has only been alleviated by increasing the number of
people who commute to work every day from the suburbs. The price is high.
Many commuters spend an hour and a half or two hours en route, in addition
to the usual city norm, and some spend even more. They commute to work
in Moscow from as far as 100 kilometers or more away.

What Perevedentsev sees in the Soviet Union is continued population
concentration in major cities in the foreseeable future, because major
structural shifts are taking place in the national economy. The proportion of
people employed in the "primary" sphere—in agriculture and extractive
industry—is growing smaller. The proportion of people employed in the
manufacturing and the service industries is growing. But extractive industry
is the industry of small cities, whereas manufacturing industry is the industry of large cities. Economic reforms promoting greater efficiency in
industrial production will contribute to the trend. The increased effectiveness of enterprises situated in the large cities will inevitably promote a
tendency toward priority development of them. And, of course, in the
majority of cases, the reconstruction of enterprises is considerably more

advantageous than the construction of new ones. Further, since science holds first place in the rate of increase of employees, its share in the economy will increase rapidly. And in Soviet conditions science is a big city industry. Finally, the entire service sphere, which lags behind the other branches of the national economy, is inevitably bound to those places where people are concentrated. Its importance for the growth of cities will increase more and more. Social factors will play an increasing role, too. The large city will become increasingly more attractive as people's free time increases. The importance of contacts made outside work will grow, and the conditions for this in a small city simply cannot be compared to those in a well-organized large city. Thus, the state policy of regulating the growth of cities must be based on a precise knowledge of the relative shortcomings and advantages of cities of various sizes and types. To do this it is necessary to study the economic, social, demographic, public health, and other aspects of the growth of cities. Right now, he says that knowledge of these factors is patently inadequate, and what is known speaks in favor of large cities rather than against them. In the Soviet Union, apparently, continuing centralization is most likely, at the time that planned decentralization is changing urban areas in Western Europe and unplanned decentralization is transforming urban regions in the United States.

The third world: fragmented centralization

All of the countries of the Third World aspire to increasingly affirmative and effective planning and action to eliminate problems that are perceived to be the products of colonialism. But the Third World countries constitute a diverse mosaic in which traditional self-perpetuating, self-regulating, semi-autonomous, preindustrial "little" societies, welded by colonial powers into ill-fitting states, coexist with and are being changed by postwar modernization. Traditional forms of authority and the centralized controls of colonialism have been replaced by one-party governments or military dictatorships. There is frequent instability, and limited capacity for public administration. The public sectors are small. Economies are fragmented along geographic, ethnic, and modern-versus-traditional lines; the Third World experiences imperfection of markets, limited development of modern economic institutions, limited industrial development and continued predominance of agriculture, low per capita product, and market dependence on foreign economic relations.

The urge for more and better control of urban development results from accelerated urban growth, a compounding of the scale of the primate cities and their associated peripheral settlements, perceived increases in social pathologies, growing attachment to national urban planning as a means of securing control of social and economic change, and an increasing willingness to experiment with new and radical plans and policies. The countries of the Third World are reaching for powers, controls, and planning best exemplified on the side of innovative planning by the welfare states of Western Europe, and by the command economies of Eastern Europe and the USSR in the sense of more complete and effective controls. At the same time many are seeking to preserve significant elements of their traditional cultures, so that modernization and Westernization are not synonymous.

A variety of radical solutions to urban development are being proposed, but what characterizes most of the planning efforts in the Third World is the absence of will to plan effectively, and more often than not, political smoke-screening. Most urbanization policy is unconscious, partial, uncoordinated, and negative. It is unconscious in the sense that those who effect it are largely unaware of its proportions and features. It is partial in that few of the points at which governments might act to manage urbanization and affect its course and direction are in fact utilized. It is uncoordinated in that national planning tends to be economic, and urban planning tends to be physical; the disjunction often produces competing policies. It is negative in that the ideological perspective of the planners leads them to try to divert, retard, or stop urban growth, and in particular to inhibit the expansion of metropolises and primate cities.

Elsewhere, in Maoist China for example, this antiurban bias is also clear. In China it has obvious historical roots in the history of the Chinese Communist Party and its struggle for power before 1949, and in the modern history of a China dominated by treaty-port colonialists who controlled and shaped nearly all of its large cities. These cities were also the homes of the Chinese bourgeoisie; they were felt to have been reactionary in the past, potentially revisionist now and in the future, and alienating at all times. Thus, their growth in China continues to be controlled by an unprecedented policy which limits their size and which channels new industrial investment into new or smaller cities in previously remote or backward areas, or into rural communes, which are to be made industrially as self-sufficient as possible without acquiring the morally corrupting and alienating qualities of big cities, or their damaging effects on the environment. Citydwellers, especially white-collar workers, must spend a month or more every year, whatever their status, in productive physical labor in the countryside, where they may regain "correct" values. The distinctions between mental and manual labor, city and countryside, "experts" or bureaucrats, and peasants or workers are to be eliminated. The benefits and the experience of in-dustrialization and modernization are to be diffused uniformly over the landscape and to all of the people, while the destructive, dehumanizing, corrupting aspects of overconcentration in cities are to be avoided.

Planning style and urbanization strategy

The keys to the Maoist reconstruction of China have been a will to plan, clear objectives, and totalitarian powers; indeed, wherever there has been affirmative pursuit and a modicum of success with an urbanization strategy, the necessary ingredients have been orientation toward the future, agreement upon goals, and the power to act.

An urbanization strategy is, axiomatically, concerned with the future. It involves goals; it involves motivated and informed decision makers; it involves the will to act and the power to achieve. A society with an urbanization strategy is, necessarily, a planning society.

But the nature of planning varies with sociopolitical structure, as do the nature and degree of future-orientation and the capability to achieve consensus on goals. In consequence, it is possible to identify a sequence of planning styles—and by extension, of urbanization strategies and determinants of the future—roughly paralleling the sequence from free-enterprise conditions to the directed state.

The simplest, as outlined in Table 20.1, is simply *ameliorative problem-solving*—the natural tendency to do nothing until problems arise or undesirable dysfunctions are perceived to exist in sufficient amounts to demand corrective or ameliorative action. Such "reactive" or "curative" planning proceeds by studying "problems," setting standards for acceptable levels of tolerance of the dysfunctions, and devising means for scaling the problems back down to acceptable proportions. The focus is upon present problems, which implies continually reacting to processes that have already worked themselves out in the past; in a processual sense, then, such planning is past-oriented. And the implied goal is the preservation of the "mainstream" values of the past by smoothing out the problems that arise along the way.

A second style of planning is *allocative* or *trend-modifying*. This is the future-oriented version of reactive problem solving. Present trends are projected into the future, and likely problems are forecast. The planning procedure involves devising regulatory mechanisms to modify the trends in ways that preserve existing values into the future, while avoiding the predicted future problems. Such is Keynesian economic planning, highway building designed to accommodate predicted future travel demands, Master Planning using the public counterpoint of zoning ordinance and building regulations, or, most effectively, immunological programs of public-health authorities.

The third planning style is *exploitive* or *opportunity-seeking*. Analysis is performed not to identify future problems, but to seek out new growth opportunities. The actions that follow pursue those opportunities most favorably ranked in terms of returns arrayed against feasibility and risk. Such is the entrepreneurial world of corporate planning, the real-estate developer, the industrialist, the private risk-taker—and also of the public entrepreneur acting at the behest of private interests, or the national leader concerned with exercising *developmental leadership*, as when Ataturk built Ankara, or as the Brazilians are developing Amazonia today. It is in this latter context in already developed situations that the concept of strategy planning was developed.

Table 20.1
Four policy-making styles

	Planning for present concerns	Planning for the future		
	Reacting to past problems	Responding to predicted futures		Creating desired future
	Ameliorative, problem-solving	Allocative, trend-modifying	Exploitive, opportunity-seeking	Normative, goal-oriented
	Planning for the present	Planning toward the future	Planning with the future	Planning from the future
Planning mode	Analyze problems, design interventions, allocate resources accordingly.	Determine and make the *best* of trends and allocate resources in accordance with desires to promote or alter them.	Determine and make the *most* of trends and allocate resources so as to take advantage of what is to come.	Decide on the *future desired* and allocate resources so that trends are changed or created accordingly. Desired future may be based on present, predicted, or new values.
"Present" or short range results	*Ameliorate present problems*	*A sense of hope* New allocations shift activities.	*A sense of triumphing over fate* New allocations shift activities.	*A sense of creating destiny* New allocations shift activities.
"Future" or long-range results of actions	*Haphazardly modify the future* by reducing the future burden and consequences of present problems.	*Gently balance and modify the future* by avoiding predicted problems and achieving a "balanced" progress to avoid creating major bottlenecks and new problems.	*Unbalance and modify the future* by taking advantage of predicted happenings, avoiding some problems and cashing in on others, without major concern for emergence of new problems.	*Extensively modify the future* by aiming for what could be. "Change the predictions" by changing values or goals, match outcomes to desires, avoid or change problems to ones easier to handle or tolerate.

Finally, the fourth mode of planning involves explicitly *normative goal-orientation*. Goals are set, based upon images of the desired future, and policies are designed and plans implemented to guide the system toward the goals, or to change the existing system if it cannot achieve the goals. This style of planning involves the cybernetic world of the systems analyst, and is only possible when a society can achieve closure of means and ends—that is, acquire sufficient control and coercive power to ensure that inputs will produce desired outputs.

The four different planning styles have significantly different long-range results, ranging from haphazard modifications of the future produced by reactive problem solving, through gentle modification of trends by regulatory procedures to enhance existing values, to significant unbalancing changes introduced by entrepreneurial profit seeking, to creation of a desired future specified *ex ante*. Clearly, in any country there is bound to be some mixture of all styles present, but equally, predominant value systems so determine the preferred policy-making and planning style that significantly different processes assume key roles in determining the future in different societies.

The publicly supported private developmental style that characterizes the American scene, incorporating bargaining among major interest groups, serves mainly to protect developmental interests by reactive or regulatory planning, ensuring that the American urban future will be a continuation of present trends, changing only as a result of the impact of change produced by the exploitive opportunity-seeking planning of American corporations.

On the other hand, hierarchical social and political systems, in which the governing class is accustomed to govern, other classes are accustomed to acquiesce, and private interests have relatively less power, can more readily evolve urban and regional growth policies at the national level than can systems under the sway of the market, local political jurisdictions, or egalitarian political processes. This is one reason that urban growth policies burgeoned earlier in Britain than in the United States. Controls are of several kinds. Most basically, use of the land is effectively regulated in conformity to a plan that codifies some public concept of the desirable future and welcomes private profit-seeking development only to the extent that it conforms to the public plan. Such is the underpinning of urban development in Britain, in Sweden, in France, in the Netherlands, in Israel's limited privately owned segments, or within the designated white areas of South Africa. Such a situation also obtains, it might be added, in the planning of Australia's new capital, Canberra. To understand the developmental outcome in these circumstances, one must understand the aspirations of private developers or of public agencies involved in the development process, on the one hand, and the images of the planners built into the Master Plan, on the other. It is the resolution of the two forces that ultimately shapes the urban scene. In Britain, the planners' images of the desirable future have been essentially conservative, aiming to project into the future a belief that centrality is an immutable necessity for urban order, leading to the preservation of urban forms that are fast vanishing in North America. Thus, the key elements are the utopian image that becomes embedded in the specific plan and the efficacy with which the public counterpoint functions to constrain private interests.

Nowhere has the imagery of the social reformers of the nineteenth century been more apparent than in Soviet planning for the "city of socialist man." Reflecting the reactions against the human consequences of nineteenth century industrial urbanization, the public counterpoint of the "mixed" economies has been replaced by the goal-oriented planning of the socialist states and other directed societies. If, to understand urbanization and its consequences in the mixed economies, one has to understand the nature and resolution of private and public forces, in directed societies one has to understand national goals and the ideologies of the planners, for one of our most important realizations of the past quarter-century is that such sought-after goals can be achieved.

Conclusion

Adna Ferrin Weber concluded in his masterpiece, *The Growth of Cities in the Nineteenth Century* (1899), that the history of attempts to change the nature of urbanization was, to his day, a history of failure. Today we must conclude differently. Urban planning successes in those societies that have been able to achieve closure between means and ends clearly demonstrate that the capacity to affect desired change exists. In these societies, public intervention has effectively operated to ensure that sought-after human consequences of urbanization become part of the urbanization process itself. Such successful public intervention, in turn, has contributed to urban social and spatial structures that differ markedly from those evolving in the context of spontaneous ("natural") ecological forces. And it is in this ability to influence future urban social and spatial structure that we find the roots of modern urban planning which seeks to direct and regulate contemporary urbanization so as to improve the overall prosperity and well-being of the individuals and societies it affects.

PART VII
TOWARD AN ECOLOGY
OF THE FUTURE

Chapter 21
Deliberate change in ecological systems: an avenue for future inquiry and research

It should now be apparent that one of the most important social changes of our time is the emergence of a range of processes of policy formulation and planning for direct and deliberate contrivance of change itself. Increasingly, we try to anticipate change, measure the course of its direction, and shape it for predetermined ends. As one indication of how far the change has gone, the White House established a National Goals Research Staff in 1968, and assigned it the following functions:

1. Forecasting future developments, and assessing longer-range consequences of present social trends.
2. Measuring the probable future impact of alternative courses of action.
3. Estimating the actual range of social choice, indicating what alternative sets of goals might be attainable in terms of available resources and possible rates of progress.

One of the things being sought by White House policy makers, in a sociocultural system which values most highly its democratic pluralism, was a continuing base of knowledge to which decisive appeal can be made in situations where conflict arises. Such research, undertaken to illuminate or influence a particular policy issue, has an honorable and long tradition in the social sciences, going back at least as far as the economists' research on the Corn Laws, providing needed information to enable policy makers to estimate various parameters of a problem, or to help them choose among competing hypotheses. It is a mischaracterization to think of research as merely providing data or information, however. Perhaps the most important influence is perceptual, through its effect on the way that policy makers look at the world. It influences what they regard as fact or fiction; the problems they see and do not see; the interpretations they regard as plausible or nonsensical; the judgments they make as to whether a policy is potentially effective, irrelevant, or worse. Much of this influence may occur before a specific policy issue arises, in how people are educated before they become policy makers. The conceptions that a policy maker brings to a problem may loom large in importance relative to his efforts to learn about it by consciously surveying the state of relevant knowledge when confronted by a particular problem. Cognitive differences centered in the planning process and derived from differences in both national ideologies and personal values are becoming, we feel, important sources of variance in ecological processes in the world today, and are thus a critical arena for future ecological research.

The influence of research on policy also takes place as knowledge is marshalled in the course of the policy-formation process, of course, and increasingly we find such marshalling of knowledge taking the form of explicit public policy analysis, the future-oriented inquiry into the optimum means of achieving a given set of social objectives, encompassing, also, studies designed to extend the range of social alternatives perceived by decision makers, and suggesting the long-range and implicit consequences of various sets of value priorities. This may involve inquiry into alternative descriptions of the problem context, ways of conceptualizing the problem, sets of goals and objectives, courses of action for achieving selected objectives, predictions of probable outcomes of the courses of action considered (including alternative models), strategies for assuring preferred outcomes, and methods of appraising the implementation of a selected course of action. Ever more explicitly, the evaluative function is expressed in such terms as *technology assessment, systems analysis, social accounting, continuous appraisal,* or some such designative terminology.

What is implied, as a consequence, is progressive abandonment of what Carson (1972) calls the "aimless pursuit of policies" stemming from cultural lag, the "Garden of Eden complex," and various cure-all approaches. By *cultural lag,* Carson means our habit of identifying "problems" and reacting to them long after they have ceased to be important. The Garden of Eden complex is the yearning for some dimly-perceived ideal. It can be a longing for the past, or some vague search for stability that stems from antipathy to city life and results in plans for "optimum" population distribution patterns, or in prescriptions for "balanced" regional growth. The cure-all approach involves heaping purpose upon purpose in our comprehensive policy and set of programs in a confusion of comprehensive planning with universal remedy. Like the traveling medicine man's snake-oil nostrum, much is claimed and little achieved by such devices.

The change has two sources: first, the development of cybernetics and systems analysis, and second, the beginnings of serious study of the future. In what follows we will describe each of these approaches as a preface for reiterating the central importance of planning behavior in setting policy. We have already argued in the previous chapter that there are several significantly different modes of planning, and that fundamentally different uses of systems analysis and choices among alternative futures flow from the preference for one planning style as opposed to another. This enables us to sketch along the way some examples of the likely divergencies in the future form and dynamics of ecological systems from one society to another, as a result of ideological differences in the concept of policy analysis and rational planning. The principal conclusion is the suggestion that productive insights into ecological dynamics in the future are less likely to flow from analysis of past processes that have generated current ecological patterns than they are from evaluating the nature and impacts of salient decisions made in the public sector, as determined by the value (sociocultural) system within which the decisions are made.

Cybernetic rationality

What is the cybernetic approach ? The term *cybernetics* derives from the Greek word *kybernetes* ("steersman"). Plato uses it to describe the prudential aspect of the art of government. Ampère in his *Essay on the Philosophy of Science* (1838) used the term *cybernetique* for the science of civil government. The Latin term *gubernator* is derived from the Greek, and hence also our word *governor*. In English we use the term *governor* in at least two ways: first in the traditional sense of a public steersman or political decision maker; second to refer to the self-adjusting valve mechanism on a steam engine that keeps the engine at a constant speed under varying conditions of load. In the steam-engine governor, a valve linked to the engine's output shaft increases steam flow into the engine as the output speed decreases, raising the speed to the level desired, or reduces steam flow if the speed exceeds the pre-established level. Maxwell analyzed this control phenomenon mathematically in his paper on governors published in 1868. What is essentially involved in steering behavior or controlling behavior of the type illustrated by the steam-engine governor is a feedback loop through which the *output* of the *system* is linked to its *input* in such a way that variations in output from some pre-established or "programmed" norm or goal result in compensatory behavior that tends to restore the system output to that goal. In language, then, policy and technique are linked.

The observation that analogous processes also characterize living systems led Norbert Wiener and his associates to write a paper in 1943 that became the watershed in cybernetic thinking. There they made an important distinction between the "functional analysis" of an entity and a "behavioristic approach." In the former, "the main goal is the intrinsic organization of the entity studied, its structure and its properties.... The behavioristic approach consists of the examination of the output of the object and of the relations of this output to the input." Adequate knowledge of any system requires, of course, both structural-functional analysis and behavioral analysis. The two are obviously linked, as we shall see later, but in talking about planning we will emphasize the behavioral component. Wiener et al. went on to describe purposeful behavior in self-regulating systems as being directed at "a final condition in which the behaving object reaches a definite correlation in time or space with respect to a relatively specific goal." Self-regulation involves, in turn, a feedback loop whereby, following comparison of performance of the system with the goals that have been set, information is returned to the input channels so that inputs can be modified if necessary. Thus self-regulation requires a functional distinction among perception of deviation from goals, decision making, and action to correct the deviation. This is normally achieved by a structural distinction between perceptor elements, control elements, and effector elements in the system.

The significance of Soviet cybernetics

Cybernetics has become the basis of modern systems analysis and scientific planning—in effect, of those disciplines whose purpose is to determine how policies (inputs) can be devised and effected to achieve goals (outputs). What Wiener et al. set in motion was an attempt to provide rational understanding and the basis for controlling the behavior of complex systems. And although the American literature on systems analysis has grown by leaps and bounds, it is in the Soviet Union that a massive experiment has been set in motion to test the applicability of cybernetics to total social-system transition (Ford, 1966).

First, effort has been addressed to the automation of many dimensions of social reality, based upon the belief that such automation is required to achieve the increased organizational complexity necessary for social progress. Second, this effort is to serve the perceived need for highly perfected systems of automatic control. Cybernetics has been tied to the goal-seeking activity of a centrally directed state; it has been linked as a science and technology to the concept of controlled social progress; and it has been linked ideologically by means of a theory of development to the fundamental tenets of communism.

Figure 21.1 depicts in a very abbreviated form the essential elements in purposeful systems that are objects of research for the Soviet cybernetics program. The "real world" is made up of people, nations, factories, transportation systems, mines, and so on. Obviously, sensors are needed that are appropriate to detect changes in each system in the real world; eyes, radar, nerve endings, and pattern-recognition devices are examples. The information processors also differ depending on the type of information processed. Comparators receive processed information about some aspect of the behaving world and compare it with the kind of behavior called for by the reference model. The results of the comparison are transmitted to

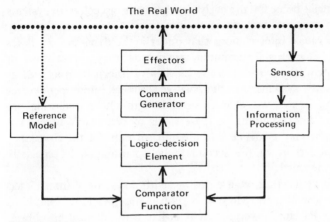

Figure 21.1
Simplified block diagram of a cybernetic system.

the command element, which then decides whether to leave the "real world" as it is or whether commands should be transmitted to the effectors to change the behavior of the real world. If the latter course is elected, information about the ensuing change is sensed, processed, compared with the reference model, and so on around the feedback loop. Obviously an indefinitely large number of interconnected loops would be necessary to describe fully the organization of a system for the purposeful control of a total social system in the real world. But Figure 21.1 does convey the essential notion that information about the real world is a necessary input to the effectors if the resulting control of the rate and direction of change is to be optimal in relation to the purpose dictated by the reference model of what "ought" to be.

The system will be made operational in the Unified Information Network of the USSR, scheduled for completion before 1980. This is to be a "nervous system" tying together the systems' "sensors" of internal and external environments at all organizational levels with the highest decision centers. These can then determine optimal courses of action and transmit information to the effector organs of the social system—ministries, production complexes, schools, defense installations, and so on. The new behavior of the system is transmitted to the decision makers, and new actions are undertaken in a continuous process analogous to that by which a helmsman steers a ship toward its destination.

Modern systems concepts

What the Soviet experiments suggest is that thinking in ecological systems analysis has progressed far beyond the initial homeostatic formulations of Wiener and his associates. Whereas the mere fact of interdependence of parts was a crucial point for the early development of ecological systems analysis, modern systems analysis goes further to distinguish the very different kinds of causal interrelations that underlie the dynamics or statics of different kinds of systems. The mutual, reversible energy interrelations of an equilibrial system, the negative feedback flow of information as well as energy in homeostatic systems, and the positive feedback interdependencies additionally found in systems characterized by growth or evolution are fundamental to an understanding of these types of systems. Whereas the sociocultural system may evidence interrelations of parts of the first kind, and certainly shows those of the second to be operating, it is especially characterized by the third.

The question of whether we are dealing with a system of interrelated parts, then, is not a vital issue. More important is the question: What kind of a system is it? Pareto answered that it is like a mechanical equilibrial system. Spencer thought it was like an organism. Traditional social-system theorists have tended to view it as an equilibrial or homeostatic organismic system. The modern systems perspective suggests that it is a morphogenic system. Equilibrial systems are generally closed to environmental variety or easily disordered by it; homeostatic systems continually fight it in acting to maintain structures and processes within certain limits; morphogenic systems use it as a pattern for growth, adaptation, or structural development. It is suggested that a sociocultural system—even in its organized or institutionalized aspects, and no matter how "stable" or homeostatic—is also morphogenic, continually generating variety by virtue of its normal dynamic interrelations of parts and selectively mapping that variety against the variety of its external environmental and internal milieu.

It follows from the above two points that the sociocultural system is in the same class, not with an equilibrial or an organismic homeostatic system, but with a species-environment system or an ecological learning system. The latter are uniquely capable of changing or elaborating their structures as a condition of maintaining their viability and continuity as systems. The study of complexly organized systems requires a study of the various types of causal interrelations that underlie their complexity.

Decision making in ecological systems

The latter case can be illustrated in the case of ecological systems, to provide linkage to our disciplinary concerns. No modern social scientist will disagree with the notion that the structure of a social system operates on behavior through the mediation of cognition, and that, feeding back, structure in its turn is composed of those aggregations of behavior that we call "processes." As Peter Drucker says in *Landmarks of Tomorrow*, such a world view assumes process...embodying the idea of growth, development, rhythm or becoming. These are all irreversible processess...because the process changes its own character; it is in other words self-generated change.

This basic feedback relationship is the starting point for Figure 21.2. In the model, it is assumed that the "system" of concern to ecologists is the ecosystem, a functioning system in which living organisms, including man, interact with their effective physical, biological, and cultural environments— the creation of interlocking natural and cultural processes in the biosphere. Tracing back the long feedback loop, the processes are in turn made up of spatial behavior, based upon environmental and locational decision-making sequences and planning, made in prior environmental contexts, and driven by combination of biological needs (survival, maintenance, reproduction) and cultural drives (such as the need for achievement built into the value system of societies displaying economic and technological progress).

In the system, decisions relate to the ecosystem through intervening perceptual filters that bias the feelings of need for action, goals to be achieved, and capacity to effect change, as they shape and color the actor's mental maps of the ecosystem and his action space within it. Perceptions, in turn, are a product of biological needs and constraints, natural endowments, and the actor's world view and cognitive structure, based on the values of his culture and the roles, expectations, and aspirations imposed on its members, together with the fruits of learning based on experience with the results of prior decision making and action. Indeed, beliefs and perceptions may be among the most critical elements because what men believe determines what men do. This is why Figure 21.2 explicitly includes as a variable the decision maker's world view, whether past- or future-oriented, and his concept of the role of self in change, ranging from fundamentalist, accepting whatever happens, to voluntarist, convinced that he can create the future.

Decisions translate into action only when conflicts with other actors have been resolved; thus a conflict-resolution and re-evaluative feedback loop is provided. Once conflict is resolved, however, the resulting actions taken give rise to locational and environmental behavior: movements, locations, relocations, adjustments, resource use, and so on. But these change the spatial system, so that if decision making is a biologically and culturally driven and perceptually biased *creature* of environment, it is also a *creator* of environment through impacts on natural and cultural processes resulting from spatial behavior, as is shown in the long feedback loop in Figure 21.2.

Individual actions are, of course, myriad, and it is useful to think of them as events which, in repetitive, rhythmic, or cumulative sequences, contribute to ecological processes of one of three kinds:

1. *System-maintaining.* This involves repetitive events that either keep the system functioning or, in a cybernetic sense, seek to eliminate perceived

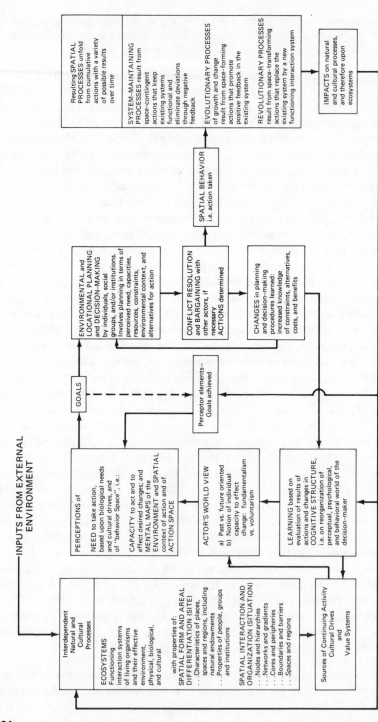

INPUTS FROM EXTERNAL ENVIRONMENT

Interdependent Natural and Cultural Processes

ECOSYSTEMS
Functioning interaction systems of living organisms and their effective environment, physical, biological, and cultural

with properties of:
SPATIAL FORM AND AREAL DIFFERENTIATION (SITE)
...Characteristics of places, spaces and regions, including natural endowments
...Properties of people, groups and institutions

SPATIAL INTERACTION AND ORGANIZATION (SITUATION)
...Nodes and hierarchies
...Networks and gradients
...Cores and peripheries
...Boundaries and barriers
...Spaces and regions

Sources of Continuing Activity Cultural Drives and Value Systems

PERCEPTIONS of
NEED to take action, based upon biological needs and cultural drives, and of "behavior Space", i.e.:
CAPACITY to act and to effect desired changes; and MENTAL MAPS of the ENVIRONMENT and SPATIAL context of action and of ACTION SPACE

ACTOR'S WORLD VIEW
a) Past vs. future oriented
b) Notion of individual capacity to effect change: fundamentalism vs. voluntarism

LEARNING based on evaluation of results of actions and changes in COGNITIVE STRUCTURE, i.e. on reorganization of perceptual, psychological, and behavioral world of the decision-maker

ENVIRONMENTAL and LOCATIONAL PLANNING and DECISION-MAKING by individuals, social groups, and/or institutions. Involves planning in terms of perceived need, capacities, resources, constraints, environmental context, and alternatives for action

CONFLICT RESOLUTION and BARGAINING with other actors, if necessary ACTIONS determined

CHANGES in planning and decision-making procedures learned: increased knowledge of constraints, alternatives, costs, and benefits

GOALS

Perceptor elements— Goals achieved

SPATIAL BEHAVIOR i.e. action taken

Resulting SPATIAL PROCESSES unfold from cumulative actions with a variety of possible results over time

SYSTEM-MAINTAINING PROCESSES result from space-contingent actions that keep existing systems functional and eliminate deviations through negative feedback

EVOLUTIONARY PROCESSES of growth and change result from space-forming actions that promote positive feedback in the existing system

REVOLUTIONARY PROCESSES result from space-transforming actions that replace the existing system by a new functioning interaction system

IMPACTS on natural and cultural processes, and therefore upon ecosystems

Figure 21.2
An ecosystem model of sociospatial processes.

424

dysfunctions and deviationist tendencies through negative feedback. It is such processes that maintain steady-state patterns.

2. *Evolutionary*. These events are those which, in cumulative morphogenic sequences, produce growth and progressive change by amplifying positive feedback in the system.

3. *Revolutionary*. Such events set in motion sequences that transform the system by redefining its members, limits, and styles and types of interactions.

These three kinds of ecological processes have been called by modern geographers, in their focus on the spatial aspects of ecological systems, space-contingent, space-forming and space-transforming. An important distinction lies between the first, which involves those rhythmic repetitive events that characterize a given system and are essentially processes *in* history (such as the daily ebb and flow of commuters or the seasonal variations in demands for home heating), and the second and third, which refer explicitly to change and to processes that *are* history (as when industry and people relocate). If the first provides complex systems with powerful self-organization that tends to suppress change, it is the latter two which embed in such systems the capacity for self-transformation into new and different states through introducing the variety that permits constructive change.

The problem of means and ends

No one will disagree if we say that the current state of any system describable in the foregoing terms is determined by its antecedent states. In the same way, what we have said earlier implies that future states of sociospatial systems depend upon the current state, external inputs from the environment, and, because self-regulation is present, upon internal impulses to achieve desired future states (goals). To this extent, the future can be an input into decision making. And if actions are then taken to achieve that future state which is perceived as being desirable, the concept of the future is to some degree a self-fulfilling one. It follows that in the reality of human affairs—and in planning for sociospatial systems—means and ends can therefore never be separated, but are part of interwoven processes.

The study of the future and the significance of planning style

All of this brings systematic planning squarely within the realm of "futurology." Based on the ideas of such scholars as Ossip Flechteheim (1966, citing uses as early as 1949) Dennis Gabor (1964), and Bertrand de Jouvenel (1963), the "futurist" argument is that the future is contained within a domain of alternative possibilities that may be specified by logical means and laid out in balance sheets that facilitate selection among the alternatives, and the formulation of policies to realize the preferences effectively.

A variety of forecasting methods for laying out alternative futures can be seen in the literature:

1. *Genius forecasting:* a process of individual intuition and judgment.
2. *Trend extrapolation:* based on the concept that the historical changes and forces of change will continue into at least the near future.
3. *Modeling and simulation techniques:* refers to a conceptual image depicting the interrelationship of several parameters involved in a particular process or subsystem.
4. *Historical analogy:* involves the study of historical events and circumstances surrounding certain periods of change in an effort to generalize about what might happen in the future under similar circumstances.
5. *Morphological analyses:* all of the variables of a particular problem are defined and combined in all possible ways. The less promising combinations are weeded out, and goals or potential courses of action are selected from among the remaining alternatives.
6. *Cross-impact matrix technique:* attempts to assess the likely interrelationships between future policies and likely socioeconomic developments. Using questions of relationship and their impact on one another, an estimate of probabilities is derived of the items being considered with one another.
7. *Delphi:* an interactive survey process whereby group intuitive judgment is the basis for plausibility ratings for future developments.
8. *Scenario writing:* a process that demonstrates the possibility of future developments by exhibiting a chain of events that might lead them, and the interaction of complex factors in this chain.

The consequences of using these techniques differ, and the selection of one or the other depends upon the particular planning style that is preferred by the policy maker. Four such styles of policy making and planning were suggested in Chapter 20: the ameliorative, the allocative, the exploitive, and the normative. These four different planning styles, we observed, have markedly variant long-range results, ranging from haphazard modifications produced by reactive problem solving, through gentle modification of trends by regulatory procedures to enhance existing mainstream values, to significant changes introduced by entrepreneurial profit-seeking, to creation of a future specified *ex ante*. We also pointed out that in any societal context there will likely be some combination of the four styles, but equally clearly, predominant value systems so determine the preferred policy-making and planning style that substantially different processes assume fundamental roles in determining the future in different societies. We have already seen how a directed society, the USSR, has moved rapidly toward the world of goal-oriented cybernetic controls. In stark contrast is traditional Islamic fatalism. What of the United States?

The White House National Goals Research Staff turned out to be a dismal failure, in spite of substantial thinking by academics and investment by the federal government in the development of systems of social accounts. No one quarreled with the idea of social accounts—the development of measurement devices that "enable us to assess where we stand and are going with respect to our values and goals, and to evaluate specific programs and determine their impact" (Bauer, 1966, p. 1). Surely, it was thought, better statistics of direct normative interest would help us make balanced, comprehensive, and concise judgments about the condition of society. Surely we would benefit if, as many have said, we "apply real science to social affairs," thus eliminating the corruptions of the principle of rationality that arise when decisions about social affairs are made on the basis of beliefs about facts, rather than "true knowledge."

What was ignored were the multiple and competing interest groups in American society. First, in contrast to Soviet academia, in which public purpose has priority, there is a basic conflict in the United States between most disciplinary research and policy research. The object of the former is to advance knowledge in the particular scholarly concerns of a discipline by arriving at empirically valid and theoretically significant conclusions about the state of affairs. Such research begins with an intellectual problem posed by previous research or theory, and proceeds at a pace dictated only by the demands of scholarship. The intended audience is the discipline's teachers and researchers; and the self-corrective method employed is the well-known "adversary proceedings" of replicative studies and scholarly reviews.

Policy research is significantly different from disciplinary research in the following respects: its object is to provide information immediately useful to policy makers in grappling with the problems they face. It begins outside a discipline with a social problem defined by a decision maker. The pace of the research is forced by the policy maker's need to make a decision dictated by nondisciplinary imperatives. The intended audience is the decision maker, to whom it must be made intelligible and convincing if it is to be useful. Generally, little time is available to collect new data or to engage in prolonged analysis. To make matters worse, there is in the United States no single academic stance, but as Peter Drucker has pointed out in *Landmarks of Tomorrow*, a "maddening confusion of tongues among the various disciplines," as well as reward systems that promote narrow specialization rather than broader speculation. Further, the policy maker must contend every day with the differing and changing goals of competing interest groups in a society that values democratic pluralism as an end in itself. There are not, we submit, the necessary preconditions in the United States today for the development of effective goal-oriented systems of planning on either the academic or policy-making side. All national debates notwithstanding, there can be no agreement either on goals or on societal relationships that link program inputs to sought-after outputs, simply lowest-common-denominator compromises between competing interest groups.

In brief, American society, like any other democratic pluralistic society, is inherently incapable of being goal-oriented for basic ideological reasons. To repeat what we said earlier, the future, instead, is likely to be an outgrowth of present processes, as regulated by legal devices to preserve "mainstream" values, even though it is subject to the possibility of major transformations produced not by design, but by major entrepreneurial decisions made in the private sector or carried out by the government at the behest of powerful private lobbying interests. The modus operandi of our governmental agencies is that accountable officials should have the decisive role in determining what programs are developed and how money is spent, and they are the officials who must respond to interest-group and political pressures.

Under such pressures, applied rationality presents fundamental challenges to the traditional American decision-making style. To cite one example, the machine-boss politician whose control is based on a hierarchic structure of authority and communication and who remains in power by manipulating interest-group politics and dispensing patronage would be severely challenged because the very utility of future-oriented planning is to provide a basis for decision making more rational than interest-group politics. Indeed, as one political commentator remarked recently, "no computer-based simulation model jazzed-up analysis scheme is really going to get very far in terms of adoption if the policy-maker has the feeling that he can't control it." Deliberate fuzziness and clouding of the perception of reality seems an end in itself, an alternative technique for survival under conditions of increasing complexity in the absence of an agreed-upon direction for society.

Michael (1972) has suggested that application of cybernetic techniques serves, in American society, to intensify the individual's feelings of anxiety and uncertainty, even though such techniques can help him cope with the barriers to understanding in data-glutted circumstances. Michael is correct in suggesting that the vast majority are not receptive to cybernetic techniques, actively opposing them or having a fixed and limited view of exactly what information is legitimate and useful. At the same time, he notes the "deep personal frustrations resulting from the contrast between (1) the policy-maker's sense of potential power to generate and implement better policies based on more information, and (2) his sense of vulnerability to exposure of personal and organizational insufficiencies." As one of our colleagues said recently, "I often feel that politicians don't want issues crystallized. There's a certain kind of defense mechanism that operates in avoiding the crystallization of issues that would conflict and sharpen when they get crystallized." Or to return to Michael, more information would only "be used by other groups to challenge the policy-maker's policies and programs." Fear of vulnerability is such as to restrict policy makers' willingness to expose their policies to accountability in a social system in which planning can never be part of a centrally directed societal control mechanism.

Values and ecologies of the future

Where does all this leave us ? First, it is clear that students of ecology in the future will have to be increasingly concerned with the role of ideologies and values in determining the nature of ecological processes. The idea of a "scientific," value-free ecology is a contradiction in terms. And whether or not futuristic studies are undertaken in order that man can control the future, these value implications have direct consequences in at least three ways, revealing the frequently perverse behavior of complex systems in the face of planning. One is the self-fulfilling prophecy, in which prediction of events increases their probability of occurring; such is the case in directed socio-economies in which the map is shaped by central design. A second is the opposite, when prediction of an event stimulates a marked change in social behavior to avoid it; such is very likely where interest-group politics prevails in pluralistic democracies. And thirdly, of course, the planner, by producing images of the future, may be quite influential in producing that future, as he indicates a range of alternatives that people can strive to achieve or to avoid. Thus, to return full circle to the point at which we began this final chapter, the most important social change of our time is the spread of awareness that we have the ability to strive and deliberately to contrive change itself. This fact alone calls for a future discipline of ecology that is more actively part of the creation of the ecologies of the future.

References

Abrams, Charles

1955 *Forbidden Neighbors*. New York: Harper and Row.

1964 *Man's Struggle for Shelter in an Urbanizing World*. Cambridge, Mass.: The M.I.T. Press.

Abu-Lughod, Janet L.

1961 "Migration Adjustments to City Life: The Egyptian Case." *American Journal of Sociology* 67:22–32.

1968a "The City Is Dead, Long Live the City." C.P.D.R. Monograph 12. Berkeley: University of California.

1968b "The Factorial Ecology of Cairo." Unpublished paper, Northwestern University.

1971 *Cairo*. Princeton, N.J.: Princeton University Press.

Ackoff, Russell L.

1961 *Scientific Method: Optimizing Applied Research Decisions*. New York: John Wiley and Sons, Inc.

Ahmad, Qazi

1965 "Indian Cities: Characteristics and Correlates." Department of Geography Research Paper No. 102. Chicago: The University of Chicago.

Akers, Ronald, and Frederick L. Campbell

1970 "Size and the Administrative Component in Occupational Associations." *Pacific Sociological Review* 13 (Fall):241–251.

Alihan, Milla

1938 *Social Ecology: A Critical Analysis*. New York: Columbia University Press.

Allport, Gordon W.

1954   *The Nature of Prejudice.* Cambridge, Mass.: Addison-Wesley.

Alonso, William

1960   "A Theory of the Urban Land Market." *Papers and Proceedings of the Regional Science Association* 6:149–158.

Ampère, A. M.

1838   *Essay on the Philosophy of Science.* Paris: Bachelier.

Anderson, Charles H.

1970   *White Protestant Americans.* Englewood Cliffs, N.J.: Prentice-Hall.

Anderson, T. R., and L. L. Bean

1961   "The Shevky-Bell Social Areas: Confirmation of Results and Reinterpretation." *Social Forces* 40:119–124.

Anderson, T. R., and J. A. Egeland

1961   "Spatial Aspects of Social Area Analysis." *American Sociological Review* 26:392–398.

Appalachian Regional Commission

1970   *The Urban-Rural Growth Strategy in Appalachia: A Commission Staff Summary Report.*

Armor, David J.

1972   "The Evidence of Busing." *The Public Interest* 28:90–126.

Armstrong, Regina B.

1972   *The Office Industry: Patterns of Growth and Location.* Cambridge, Mass.: The M.I.T. Press.

Arnold, J. L.

1971   *The New Deal in the Suburbs: A History of the Greenbelt Town Program.* Columbus: Ohio State University Press.

Axelrod, Morris

1956   "Urban Structure and Social Participation." *American Sociological Review* 21:13–18.

Babchuk, Nicholas, and John Edwards

1965   "Voluntary Associations and the Integration Hypothesis." *Sociological Inquiry* 2:149–162.

Bailey, Kenneth D., and Patrick Mulcahy

1972   "Sociocultural Versus Neoclassical Ecology: A Contribution to the Problem of Scope in Sociology." *The Sociological Quarterly* 13:37–48.

Banfield, Edward C.

1968   *The Unheavenly City.* Boston: Little, Brown.

Barlow Report

1940   *Report of the Royal Commission on the Distribution of Industrial Population.* Cmd. 6153. HMSO.

Bauer, Catherine

1951   "Social Questions in Housing and Community Planning." *Journal of Social Issues* 7 (1):1–34.

Bauer, Raymond A. (ed.)

1966   *Social Indicators.* Cambridge, Mass.: The M.I.T. Press.

Beale, Calvin L.

1969   "Statement before the Subcommittee on Urban Growth." House Committee on Banking and Currency, June 24, 1969.

1971   "Statement on Population and Migration Trends in Rural and Nonmetropolitan Areas." Submitted to the U.S. Senate Committee on Government Operations, April 27, 1971. Economic Research Service, U.S. Department of Agriculture. Mimeographed.

1975   "The Revival of Population Growth in Nonmetropolitan America." Economic Development Division, Economic Research Service, U.S. Department of Agriculture. ERS-605.

Beale, Calvin L., and Glenn V. Fuguitt

1975   "The New Pattern of Nonmetropolitan Population Change." Center for Demography and Ecology. University of Wisconsin—Madison. Working paper 75-22.

Beckman, Martin J.

1957   "On the Distribution of Rent and Residential Density in Cities." Paper presented to the Interdepartmental Seminar on Mathematical Applications in the Social Sciences, Yale University.

Beer, S.

1961   "Below the Twilight Arch: A Mythology of Systems." In D. P. Eckman (ed.), *Systems: Research and Design,* pp. 1–25. New York: John Wiley and Sons, Inc.

Bell, G.

1962   "Change in City Size Distribution in Israel." *Ekistics* 13 (76):103.

Bell, Wendell

1953   "The Social Areas of the San Francisco Bay Region." *American Sociological Review* 18:29–47.

1955   "Economic, Family and Ethnic Status: An Empirical Test." *American Sociological Review* 22:45–52.

1959   "Social Areas: Typology of Urban Neighborhoods." In M. B. Sussman (ed.), *Community Structure and Analysis,* pp. 61–92. New York: Thomas Y. Crowell Company.

Bell, Wendell, and Marion D. Boat

1957   "Urban Neighborhoods and Informal Social Relations." *American Journal of Sociology* 62:391–398.

Berry, Brian J. L.

1961a   "City Size Distribution and Economic Development." *Economic Development and Cultural Change* 9 (4):573–588.

1961b   "A Method for Deriving Multi-Factor Uniform Regions." *Przeglad Geograficzny* 33:263–279.

1962   "Comparative Studies of Central Place Systems." Final Report NONR 2121–18 NR 389–126. Geography Branch, U.S. Office of Naval Research. Department of Geography, The University of Chicago.

1963   "Commercial Structure and Commercial Blight." Department of Geography Research Paper No. 85. Chicago: The University of Chicago.

1965a   "Internal Structure of the City." *Law and Contemporary Problems* 30 (1):111–119.

1965b   "Research Frontiers in Urban Geography." In P. Hauser and L. Schnore (eds.), *The Study of Urbanization*, pp. 403–430. New York: John Wiley and Sons, Inc.

1966a   "Essays on Commodity Flows and the Spatial Structure of the Indian Economy." Department of Geography Research Paper No. 111. Chicago: The University of Chicago.

1966b   "Strategies, Models and Economic Theories of Development in Rural Regions." Washington, D.C.: Agriculture Economic Report No. 127.

1967a   *Geography of Market Centers and Retail Distribution.* Englewood Cliffs, N.J.: Prentice-Hall, Inc.

1967b   "Spatial Organization and Levels of Welfare." Paper presented to first Economic Development Administration Research Program Conference.

1968   "Metropolitan Area Definition: A Re-Evaluation of Concepts and Statistical Practice." U.S. Bureau of the Census, Working Paper No. 28.

1969a   *Growth Centers and Their Potentials in the Upper Great Lakes Region.* Washington, D.C.: Upper Great Lakes Regional Commission.

1969b   "Relationships Between Regional Economic Development and the Urban System: The Case of Chile." *Tijdschrift voor Economische en Sociale Geografie* 60:283–307.

1970   "The Geography of the United States in the Year 2000." *Transactions of the Institute of British Geographers* 51:21–53.

1971   "DIDO Data Analysis: GIGO or Pattern Recognition?" In H. D. McConnell and D. Yaseen (eds.), *Models of Spatial Variation*, pp. 105–131. DeKalb, Ill.: Northern Illinois University Press.

1972   "Latent Structure of the American Urban System, with International Comparisons," In B. J. L. Berry (ed.), *City Classification Handbook*, pp. 11–60. New York: John Wiley and Sons, Inc.

1973   "A Paradigm for Modern Geography." In R. J. Chorley (ed.), *Directions in Geography*, pp. 3–21. London: Methuen and Co., Ltd.

1976   "Geographic Theories of Social Change," In Daniel Bell (ed.), *Theories of Social Change*. New York: The Twentieth Century Fund.

Berry, Brian J. L., and H. G. Barnum

1963   "Aggregate Relations and Elemental Component of Central Place Systems." *Journal of Regional Science* 4 (1):35–68.

Berry, Brian J. L., H. G. Barnum, and R. J. Tennant

1962   "Retail Location and Consumer Behavior." *Papers and Proceedings of the Regional Science Association* 9:65–106.

Berry, Brian J. L., and W. L. Garrison

1958   "Alternate Explanations of Urban Rank-Size Relationships." *Annals of the Association of American Geographers* 48 (1):83–91.

Berry, Brian J. L., Peter G. Goheen, and Harold Goldstein

1968   *Metropolitan Area Definition: A Re-Evaluation of Concept and Statistical Practice*. Washington, D.C.: U.S. Government Printing Office.

Berry, Brian J. L., and Frank Horton

1970   *Geographic Perspectives on Urban Systems*. Englewood Cliffs, N.J.: Prentice-Hall, Inc.

Berry, Brian J. L., and Elaine Neils

1969   "Location, Size and Shape of Cities as Influenced by Environmental Factors." In Harvey S. Perloff (ed.), *The Quality of the Urban Environment*, pp. 257–302. Baltimore: Johns Hopkins University Press.

Berry, Brian J. L., and Allen Pred

1961   *Central Place Studies: A Bibliography of Theory and Applications*. Philadelphia: Regional Science Research Institute.

Berry, Brian J. L., James W. Simmons, and Robert J. Tennant

1963   "Urban Population Densities: Structure and Change." *Geographical Review* 53:389–405.

Bertalanffy, Ludwig von

1950   "An Outline of General System Theory." *British Journal for the Philosophy of Science* 1 (2):134–165.

1956   "General System Theory." *General Systems* 1:1–10.

1962   "General System Theory: A Critical Review." *General Systems* 7:1–20.

Beshers, James A.

1962   *Urban Social Structure*. Glencoe, Ill.: The Free Press.

Bidwell, Charles E., and John D. Kasarda

1975   "School District Organization and Student Achievement." *American Sociological Review* 40 (February):55–70.

Bird, Alan R.

1968 *Growth Areas and Development Districts of the Upper Great Lakes Economic Development Region.* Washington, D.C.: Upper Great Lakes Regional Commission. Mimeographed.

Blake, Judith

1965 "Demographic Science and the Redirection of Population Policy." *Journal of Chronic Diseases* 18:1181–1200.

Blalock, Hubert M.

1959 "Status Consciousness: A Dimensional Analysis." *Social Forces* 37 (3):243–248.

1964 *Causal Inferences in Nonexperimental Research.* Chapel Hill: The University of North Carolina Press.

1967 *Toward a Theory of Minority-Group Relations.* New York: John Wiley and Sons, Inc.

1971 "Path Analysis Revisited: The Decomposition of Unstandardized Coefficients." Unpublished research note. Department of Sociology, University of Washington, Seattle.

Blau, Peter M.

1970 "A Formal Theory of Differentiation in Organizations." *American Sociological Review* 35 (April):201–218.

1972 "Interdependence and Hierarchy in Organizations." *Social Science Research* 1 (March):1–24.

Blau, Peter M., and Otis D. Duncan

1967 *The American Occupational Structure.* New York: John Wiley and Sons, Inc.

Blau, Peter M., and Richard Schoenherr

1971 *The Structure of Organizations.* New York: Basic Books.

Bloom, Leonard

1971 *The Social Psychology of Race Relations.* London: George Allen and Unwin, Ltd.

Bluestone, Herman

1970 *Focus for Urban Development Analysis: Urban Orientation of Counties.* Washington, D.C.: Economic Research Service, U.S. Department of Agriculture.

Blumberg, Leonard

1964 "Segregated Housing, Marginal Location, and the Crisis of Confidence." *Phylon* 25 (4):321–330.

Blumenfeld, Hans

1959 "Are Land Use Patterns Predictable?" *Journal of the American Institute of Planners* 25:61–66.

Bogue, Donald J.

1949   *The Structure of the Metropolitan Community: A Study of Dominance and Subdominance.* Ann Arbor: University of Michigan Press.

1953   *Population Growth in Standard Metropolitan Areas, 1900–1950.* Washington, D.C.: U.S. Government Printing Office.

1957   *Components of Population Change in Standard Metropolitan Statistical Areas.* Oxford, Ohio: Scripps Foundation.

Bohm, Robert A., and David A. Patterson

1972   "Interstate Highway Location and County Population Growth." Urban Research Section, Oak Ridge National Laboratory.

Bopegamage, A.

1966   "A Methodological Problem in Indian Urban Sociological Research." *Sociology and Social Research* 50:236–240.

Borchert, John R.

1967   "American Metropolitan Evolution." *Geographical Review* 57 (3):301–332.

Borchert, John R., and Russell B. Adams

1963   "Trade Centers and Tributary Areas of the Upper Midwest." Upper Midwest Economic Study Urban Report No. 3.

Borts, George H.

1967   "Patterns of Regional Economic Development in the United States, and the Relations to Rural Poverty." Report to the National Advisory Commission on Rural Poverty, U.S. Department of Agriculture.

Bose, Nirmal Kumar

1965   "Calcutta: A Premature Metropolis." *Scientific American* 213 (3):90–102.

1968   *Calcutta, 1964: A Social Survey.* Calcutta: Lalvani Publishing House.

Boulding, Kenneth E.

1953   "Toward a General Theory of Growth." *The Canadian Journal of Economics and Political Science* 19 (August):326–340.

1956a   "General Systems Theory: The Skeleton of Science." *Management Science* 2:197–208.

1956b   "Toward a General Theory of Growth." In Joseph J. Spengler and Otis Dudley Duncan (eds.), *Population Theory and Policy*, pp. 109–204. Glencoe, Ill.: The Free Press.

Boyce, Ronald R.

1963   "Changing Patterns of Urban Land Consumption." *Professional Geographer* 15 (2):19–24.

Bradburn, Norman, Seymoure Sudman, and Galen L. Gockel
1970   *Racial Integration in American Neighborhoods: A Comparative Study.* Chicago: National Opinion Research Corporation.

Breckenfeld, Gurney
1972   "Downtown Has Fled to the Suburbs." *Fortune* 86 (4):80ff.

Bressler, Marvin
1960   "The Meyers Case: An Instance of Successful Racial Invasion." *Social Problems* 8:126–142.

Bruner, E. M.
1961   "Urbanization and Ethnic Identity in Northern Sumatra." *American Anthropologist* 63:508–521.

Bruner, J. S.
1951   "Personality Dynamics and the Process of Perceiving." In R. R. Blake and G. V. Ramsey (eds.), *Perception: An Approach to Personality*, pp. 121–147. New York: Ronald Press.

Brush, John E.
1962   "The Morphology of Indian Cities." In Roy Turner (ed.), *India's Urban Future*, pp. 57–70. Berkeley: University of California Press.

Burgess, Ernest W.
1925   "The Growth of the City: An Introduction to a Research Project." In Robert Park, Ernest Burgess, and R. D. McKenzie (eds.), *The City*, pp. 47–62. Chicago: The University of Chicago Press.

Butler, Edgar W., George Sabagh, and Maurice D. Van Arsdol
1964   "Demographic and Social Psychological Factors in Residential Mobility." *Sociology and Social Research* 48 (2):139–154.

Campbell, Angus, and Howard Schuman
1969   *Racial Attitudes in Fifteen American Cities.* Ann Arbor: Institute for Social Research, University of Michigan.

Caplan, E. K., and E. P. Wolf
1960   "Factors Affecting Racial Change in Two Middle Class Income Housing Areas." *Phylon* 21:225–233.

Carneiro, Robert L.
1967   "On the Relationship Between Size of Population and Complexity of Social Organization." *Southwestern Journal of Anthropology* 23 (Autumn): 234–242.

Carroll, Robert L.
1963   "The Metropolitan Influence of the 168 Standard Metropolitan Area Central Cities." *Social Forces* 42:166–173.

Carruth, Eleanor

1969   "Manhattan's Office Building Binge." *Fortune* 80 (5):114ff.

Carson, John

1972   "A National Urban Growth Policy." *Urban Land* 31 (2):3–10.

Cassidy, Robert

1972   "Moving to the Suburbs." *New Republic* 166 (4):20–23.

Caswell, Bruce

1968   "The Urban Ecology of Miami, Florida." Unpublished paper, Center for Urban Studies, The University of Chicago.

Chapin, F. S., and S. F. Weiss

1962   *Factors Influencing Land Development*. Chapel Hill, N.C.: Institute for Research in Social Science, University of North Carolina.

Chatterjee, A. B.

1960   "Howrah: An Urban Study." Unpublished Ph.D. dissertation, University of London.

Cherry, G. E.

1972   *Urban Change and Planning*. Henley-on-Thames: G. T. Foulis.

Chicago Area Transportation Study

1959   *Final Report, I: Survey Findings*. Chicago: Western Engraving and Embossing Company.

Chicago Daily News

1972   "Chicago's Top Ten Neighborhoods: How Neighborhoods Rate on Status Scale." Tuesday, November 28, 1972.

Chinitz, Benjamin (ed.)

1964   *City and Suburb*. Englewood Cliffs, N.J.: Prentice-Hall.

Chisholm, Michael

1962   *Rural Settlement and Land Use*. London: Hutchinson University Library.

Choay, Françoise

1965   *L'Urbanisme, Utopie et Réalités*. Paris: Éditions du Seuil.

Christaller, Walter

1933   *Die zentralen Orte in Süddeutschland*. Jena: Gustav Fischer. Translated into English by C. W. Baskin, *Central Places in Southern Germany*. Englewood Cliffs, N.J.: Prentice-Hall, 1966.

1938   "Rapports Fonctionels entre les Agglomérations Urbaines et les Campagnes." *Congrès International de Géographie*, Comptes Rendus, Section 3a, 2:123–138.

Cicourel, Aaron V.

1964   *Method and Measurement in Sociology.* New York: The Free Press.

City Rating Guide

1964   *City Rating Guide.* Chicago: Rand McNally and Co.

Clark, Colin

1945   "The Economic Functions of a City in Relation to Its Size." *Econometrica* 13 (2):97–113.

1951   "Urban Population Densities." *Journal of the Royal Statistical Society,* Series A, 114:490–496.

1958   "Urban Population Densities." *Bulletin de l'Institut International de Statistique* 36, Part 4:60–68.

1967   *Population Growth and Land Use.* New York: St. Martin's Press.

Clark, Kenneth

1965   *Dark Ghetto: Dilemmas of Social Power.* New York: Harper and Row.

Clark, Terry N.

1968   "Community Structure, Decision-Making, Budget Expenditures, and Urban Renewal in 51 American Communities." *American Sociological Review* 33 (4):576–593.

Claval, P.

1962   *Geographie Generale des Marches.* Paris: Les Belles Lettres.

Clemente, Frank, and Richard B. Sturgis

1972   "The Division of Labor in America: An Ecological Analysis." *Social Forces* 51 (December):176–182.

Cliffe-Phillips, Geoffrey, John Mercer, and Yue Man Yeung

1968   "The Spatial Structure of Urban Areas: A Case Study of the Montreal Metropolitan Area." Unpublished paper, Center for Urban Studies, The University of Chicago.

Clinard, M. B.

1966   *Slums and Community Development.* New York: The Free Press.

Cohen, Yehoshua S.

1968   "The Urban Ecology of Birmingham, Alabama." Unpublished paper, Center for Urban Studies, The University of Chicago.

1972   "Diffusion of an Innovation in an Urban System." Department of Geography Research Paper No. 140. Chicago: The University of Chicago.

Colorado Department of Education

1971   *Annual Reports of Superintendents of Colorado School Districts, 1969–70.* Denver, Colorado.

Commission on Population Growth and the American Future

1972   *Population and the American Future.* Washington, D.C.: U.S. Government Printing Office.

Conkin, P. K.

1959   *Tomorrow A New World: The New Deal Community Program.* Ithaca, N.Y.: Cornell University Press.

Cornelius, W. A., Jr.

1971   "The Political Sociology of Cityward Migration in Latin America." In F. F. Rabinovitz and F. M. Trueblood (eds.), *Latin American Urban Research,* Vol. 1, pp. 95–147. Beverly Hills, Calif.: Sage Publications.

Cowan, Peter, et al.

1969   *The Office: A Facet of Urban Growth.* New York: American Elsevier Publishing Company.

Cressey, Paul

1938   "Population Succession in Chicago: 1898–1930." *The American Journal of Sociology* 44:59–69.

Cumberland, John H.

1971   *Regional Development: Experiences and Prospects in the United States of America.* The Hague: Mouton.

Curry, Leslie

1964   "The Random Spatial Economy: An Exploration in Settlement Theory." *Annals of the Association of American Geographers* 54 (1):138–146.

Dacey, M. F.

1962   "Another Explanation for Rank-Size Regularity." Unpublished paper, Northwestern University.

Dacey, M. F., and T. H. Lung

1963   "The Identification of Randomness in Point Patterns." *Journal of Regional Science* 4 (1):83–96.

Davie, Maurice R.

1937   "The Pattern of Urban Growth." In George P. Murdock (ed.), *Studies in the Science of Society,* pp. 133–161. New Haven: Yale University Press.

Davis, Allison, Burleigh B. Gardner, and Mary R. Gardner

1941   *Deep South.* Chicago: The University of Chicago Press.

Davis, Kingsley

1967   "Population Policy: Will Current Programs Succeed?" *Science* 158:730–739.

1969   *World Urbanization, 1950–70.* Berkeley: University of California.

De, Sambhunath N.

1961 "Cholera in Calcutta." In S. N. De (ed.), *Cholera, Its Pathology and Pathogenesis*, pp. 41–61. London: Oliver and Boyd.

Deutsch, Martin, and Kay Steele

1959 "Attitude Dissonance Among Southville's Influentials." *Journal of Social Issues* 15:44–52.

Downs, Anthony

1973 *Opening Up the Suburbs: An Urban Strategy for America.* New Haven: Yale University Press.

Drucker, Peter F.

1959 *Landmarks of Tomorrow.* New York: Harper and Row.

1968 *The Age of Discontinuity.* New York: Harper and Row.

1970 *Technology, Management and Society.* New York: Harper and Row.

Duncan, Beverly

1964 "Variables in Urban Morphology." In E. W. Burgess and D. J. Bogue (eds.), *Contributions to Urban Sociology*, pp. 17–30. Chicago: The University of Chicago Press.

Duncan, Beverly, and O. D. Duncan

1960 "The Measurement of Intra-City Locational and Residential Patterns." *Journal of Regional Science* 2 (2):37–54.

Duncan, Beverly, and Stanley Lieberson

1970 *Metropolis and Region in Transition.* New York: Sage Publications.

Duncan, Beverly, George Sabagh, and Maurice D. Van Arsdol

1962 "Patterns of City Growth." *American Journal of Sociology* 67 (4):418–429.

Duncan, Otis Dudley

1955 Untitled review of *Social Area Analysis*, by Eshref Shevky and Wendell Bell. *American Journal of Sociology* 61:84–85.

1959 "Human Ecology and Population Studies." In Philip M. Hauser and Otis Dudley Duncan (eds.), *The Study of Population*, pp. 678–716. Chicago: The University of Chicago Press.

1961 "From Social System to Ecosystem." *Sociological Inquiry* 31 (2):140–149.

1964 "Social Organization and the Ecosystem." In Robert E. L. Faris (ed.), *Handbook of Modern Sociology*, pp. 37–82. Chicago: Rand-McNally and Co.

1965 "Farm Background and Differential Fertility." *Demography* 2:240–249.

Duncan, Otis Dudley, Ray P. Cuzzort, and Beverly Duncan

1961 *Statistical Geography.* Glencoe, Ill.: The Free Press.

Duncan, Otis Dudley, and Beverly Duncan

1957   *The Negro Population of Chicago: A Study of Residential Succession.* Chicago: The University of Chicago Press.

Duncan, O. D., and A. J. Reiss

1956   *Social Characteristics of Urban and Rural Communities, 1950.* New York: John Wiley and Sons, Inc.

Duncan, Otis Dudley, and Leo F. Schnore

1959   "Cultural, Behavioral, and Ecological Perspectives in the Study of Social Organization." *The American Journal of Sociology* 65:132–146.

Duncan, Otis Dudley, W. R. Scott, Stanley Lieberson, Beverly Duncan, and Hal Winsborough

1960   *Metropolis and Region.* Baltimore: Johns Hopkins University Press.

Durkheim, Emile

1893   *De la Division du Travail Social.* Paris: Alcan. Translated by George Simpson as *The Division of Labor in Society.* New York: Macmillan, 1933.

Dutt, Ashok K.

1966   "Daily Shopping in Calcutta." *Town Planning Review* 37:207–216.

Eberhard, J.

1966   "Technology for the City." *International Science and Technology* 57:18–31.

Economic Development Administration

1967   *Directions of the District Program.* Office of District and Area Planning. Mimeographed.

1972   *The Economic Development Administration Growth Center Strategy.* Washington, D.C.: U.S. Department of Commerce.

Edwards, John, and Alan Booth (eds.)

1973   *Social Participation in Urban Society.* Cambridge: Schenkman Publishing Company.

Eisenstadt, S. N. (ed.)

1968   *Max Weber on Charisma and Institution Building.* Chicago: The University of Chicago Press.

Elkins, T. H.

1973   *The Urban Explosion.* New York: Macmillan.

Engels, F.

1844   *The Condition of the English Working Classes in 1844.* London: Allen and Unwin.

Eyre, L. A.

1972   "The Shantytowns of Montego Bay, Jamaica." *The Geographical Review* 62:394–413.

Faris, Robert E. L., and H. Warren Dunham

1939   *Mental Disorders in Urban Areas.* Chicago: The University of Chicago Press.

Feldt, Allen G.

1965   "The Metropolitan Area Concept: An Evaluation of the 1950 SMA's." *Journal of the American Statistical Association* 60:617–636.

Festinger, Leon, Stanley Schachter, and Kurt Back

1963   *Social Pressures in Informal Groups.* Stanford: Stanford University Press.

Firey, Walter

1945   "Sentiment and Symbolism as Ecological Variables." *American Sociological Review* 10 (2):140–148.

1947   *Land Use in Central Boston.* Cambridge, Mass.: Harvard University Press.

Fischer, Claude S.

1972a   "The Experience of Living in Cities." Paper prepared for the National Research Council, National Academy of Sciences.

1972b   "Urbanism as a Way of Life: A Review and an Agenda." *Sociological Methods and Research* 1:187–242.

1973   "On Urban Alienations and Anomie: Powerlessness and Social Isolation." *American Sociological Review* 38:311–326.

1975   "Toward a Subcultural Theory of Urbanism." *American Journal of Sociology* 80 (May):1319–1341.

Fishbein, M., and I. Ajzen

1972   "Attitudes and Opinions." *Annual Review of Psychology* 23:487–544.

Fisher, J. C.

1962   "Planning the City of Socialist Man." *Journal of the American Institute of Planners* 28 (4):251–265.

1966   *Yugoslavia: A Multinational State.* San Francisco: Chandler Publishing Company.

Fishman, Joshua A.

1961   "Some Social and Psychological Determinants of Intergroup Relations in Changing Neighborhoods: An Introduction to the Bridgeview Study." *Social Forces* 40 (1):42–52.

Flechtheim, Ossip K.

1966   *History and Futurology.* Meisenheim am Glan: Hain.

Foley, D. L.

1963   *Controlling London's Growth.* Berkeley: University of California Press.

Ford, John J.

1966   "Soviet Cybernetics and International Development." In John Diebold (ed.), *The Social Impact of Cybernetics.* Lafayette, Ind.: Purdue University Press.

Forstall, Richard L.

1967   "Economic Classification of Places Over 10,000, 1960–1963." In *The Municipal Year Book 1967*, pp. 30–65. Chicago: International City Managers' Association.

1970   "A New Social and Economic Grouping of Cities." In *The Municipal Year Book 1970*, pp. 102–159. Washington, D.C.: International City Managers' Association.

1971   "Applications of the New Social and Economic Grouping of Cities." *Urban Data Service* 3.

Frazier, E. Franklin

1935   "The Status of the Negro in the American Social Order." *Journal of Negro Education* 4 (3):293–307.

Freedman, Ronald, et al.

1963   "Current Fertility Expectations of Married Couples in the U.S." *Population Index* 29:366–391.

Friedlander, Stanley

1972   *Unemployment in the Urban Core.* New York: Frederick A. Praeger.

Friedmann, John J., and F. Sullivan

1972   "The Absorption of Labor in the Urban Economy: The Case of Developing Economies." Los Angeles: University of California, School of Architecture and Planning.

Friedmann, John R. P.

1955   *The Spatial Structure of Economic Development in the Tennessee Valley.* Department of Geography Research Paper No. 39. Chicago: The University of Chicago.

1963   "Economic Growth and Urban Structure in Venezuela." Cuadernos de la Sociedad Venezolana de Planificacion.

1966   *Regional Development Policy.* Cambridge, Mass.: The M.I.T. Press.

Friedmann, John R. P., and W. Alonso

1964   *Regional Development and Planning.* Cambridge, Mass.: The M.I.T. Press.

Friedmann, John R. P., and Thomas Lackington

1967   "Hyperurbanization and National Development in Chile: Some Hypotheses." *Urban Affairs Quarterly* 2 (June):3–29.

Friedmann, John R. P., and John Miller

1965 "The Urban Field." *Journal of the American Institute of Planners* 31:312–319.

Frisbie, W. Parker, and Dudley L. Poston, Jr.

1975 "Components of Sustenance Organization and Nonmetropolitan Population Change: A Human Ecological Investigation." *American Sociological Review* 40 (December):773–784.

1977 *Sustenance Organization and Population Redistribution in Nonmetropolitan America.* Beverly Hills, Calif.: Sage Publications.

Fuller, Stephen S.

1970 *Impact of Appalachian Regional Commission Assisted Public Investment in Selected Growth Centers.* Report to the Appalachian Regional Commission.

Gabor, Dennis

1964 *Inventing the Future.* New York: Alfred A. Knopf.

Gans, Herbert J.

1962a *The Urban Villagers.* New York: The Free Press.

1962b "Urbanism and Suburbanism as Ways of Life: A Re-evaluation of Definitions." In Arnold Rose (ed.), *Human Behavior and Social Processes,* pp. 625–648. Boston: Houghton-Mifflin Company.

1967 *The Levittowners.* New York: Random House.

1968 *People and Plans.* New York: Basic Books.

Garrison, William L.

1962 "Toward a Simulation Model of Urban Growth and Development." Proceedings of the IGU Symposium in Urban Geography, Lund, 1960. Lund: Gleerup.

Garrison, William L., et al.

1959 *Studies of Highway Development and Geographic Change.* Seattle: University of Washington Press.

Gettys, Warner E.

1940 "Human Ecology and Social Theory." *Social Forces* 18:469–476.

Ghosh, A.

1961 *Calcutta, The Primate City.* Monograph Series No. 2, Census of India.

Gibbard, Harold A.

1941 "The Status Factor in Residential Succession." *American Journal of Sociology* 46 (6):835–842.

Gibbs, Jack P.

1961 *Urban Research Methods.* Princeton: Van Nostrand.

Gibbs, Jack P., and Harley Browning

1961   "Systems of Cities." In Jack P. Gibbs (ed.), *Urban Research Methods*, pp. 436–461. Princeton: Van Nostrand.

Ginsburg, Norton

1961   *An Atlas of Economic Development.* Chicago: The University of Chicago Press.

Gist, Noel P.

1968a   "The Ecology of Bangalore, India." In S. F. Fava (ed.), *Urbanism in World Perspective*, pp. 177–188. New York: Thomas Y. Crowell Company.

1968b   "Urbanism in India." In S. F. Fava (ed.), *Urbanism in World Prespective*, pp. 22–32. New York: Thomas Y. Crowell Company.

Goldstein, Sidney, and Calvin Goldscheider

1968   *Jewish Americans.* Englewood Cliffs, N.J.: Prentice-Hall.

Gooding, J.

1972   "Roadblocks Ahead for the Great Corporate Move-Out." *Fortune* 85 (6):78ff.

Goodman, Leo A.

1970   "The Multivariate Analysis of Qualitative Data: Interactions Among Multiple Classifications." *Journal of the American Statistical Association* 65:226–256.

1971   "The Analysis of Multidimensional Contingency Tables: Stepwise Procedures and Direct Estimation Methods for Building Models for Multiple Classifications." *Technometrics* 13:33–61.

1972a   "A General Model for the Analysis of Surveys." *American Journal of Sociology* 77:1035–1086.

1972b   "A Modified Multiple Regression Approach to the Analysis of Dichotomous Variables." *American Sociological Review* 33:28–46.

1973   "Causal Analysis of Data from Panel Studies and Other Kinds of Surveys." *American Journal of Sociology* 78:1135–1191.

Gordon, M. M.

1964   *Assimilation in American Life.* New York: Oxford University Press.

Gottman, Jean

1961   *Megalopolis: The Urbanized Northeastern Seaboard of the United States.* New York: The Twentieth Century Fund.

Gower, J. C.

1966   "Some Distance Properties of Latent Root and Vector Methods Used in Multivariate Analysis." *Biometrika* 53:325–338.

Gradmann, Robert

1913a "Das ländliche Siedlungswesen des Königreichs Württemberg." *Forschungen zur Deutschen Landes- und Volkskunde* 21 (1):1–136.

1913b "Die städtischen Siedlungen des Königsreichs Württemberg." *Forschungen zur Deutschen Landes- und Volkskunde* 21 (2):137–215.

Gras, N. S. B.

1922 *Introduction to Economic History.* New York: Harper.

Greenwood, M. J., and Douglas Sweetland

1972 "The Determinants of Migration Between Standard Metropolitan Statistical Areas." *Demography* 9:665–681.

Greer, Scott

1962 *The Emerging City.* Glencoe, Ill.: The Free Press.

1968 *The New Urbanization.* New York: St. Martin's Press.

1972 *The Urbane View.* New York: Oxford University Press.

Grier, Eunice, and George Grier

1960 *Privately Developed Interracial Housing.* Berkeley: University of California Press.

Guest, Avery M.

1971 "Retesting the Burgess Zonal Hypothesis: The Location of White-Collar Workers." *American Journal of Sociology* 76:1094–1108.

1973 "Urban Growth and Population Densities." *Demography* 10:53–69.

Gustavson, Neil C.

1973 *Recent Trends/Future Prospects: A Look at Upper Midwest Population Changes.* Minneapolis: Upper Midwest Council.

Guterman, Stanley S.

1969 "In Defense of Wirth's 'Urbanism as a Way of Life'." *American Journal of Sociology* 74:492–499.

Hadden, J. K., and E. F. Borgatta

1965 *American Cities: Their Social Characteristics.* Chicago: Rand McNally and Company.

Haggerty, Lee J.

1971 "Another Look at the Burgess Hypothesis: Time as an Important Variable." *American Journal of Sociology* 76:1084–1093.

Hall, P.

1966 *The World Cities.* London: Weidenfeld and Nicolson.

Handlin, Oscar, and John Burchard (eds.)

1963 *The Historian and the City.* Cambridge, Mass.: M.I.T. Press.

Hansen, Willard B.

1961 "An Approach to the Analysis of Metropolitan Residential Expansion." *Journal of Regional Science* 3:37–56.

Haren, Claude

1972 "Current Spatial Organization of Industrial Productivity and Distribution Activity." Economic Research Service, U.S. Department of Agriculture.

Harman, Harry

1960 *Modern Factor Analysis*. Chicago: The University of Chicago Press.

Harris, Chauncey

1943 "A Functional Classification of Cities in the United States." *Geographical Review* 33 (1):86–99.

1954 "The Market as a Factor in the Localization of Industry in the United States." *Annals of the Association of American Geographers* 44:315–348.

1970 *Cities of the Soviet Union*. Chicago: Rand McNally and Co.

Harris, Chauncey, and Edward L. Ullman

1945 "The Nature of Cities." *Annals of the American Academy of Political and Social Science* 242:7–17.

Harrison, Bennett

1972 *Education, Training, and the Urban Ghetto*. Baltimore: Johns Hopkins University Press.

Hart, John Fraser

1955 "Functions and Occupational Structures of Cities of the American South." *Annals of the Association of American Geographers* 45 (3):269–286.

Hartnett, H. D.

1971 "A Locational Analysis of Those Manufacturing Firms That Have Located and Relocated Within the City of Chicago, 1955–1968." Ph.D. dissertation, University of Illinois, Champaign–Urbana, Illinois.

Hatt, Paul

1945 "The Relation of Ecological Location to Status Position and Housing of Ethnic Minorities." *American Sociological Review* 10 (4):481–485.

1946 "The Concept of Natural Area." *The American Sociological Review* 11 (4):423–427.

Hattori, K., K. Kagaya, and S. Inanaga

1960 "The Regional Structure of Surrounding Areas of Tokyo." Chirigaku Hyoron. *Geographical Review of Japan*.

Hauser, Philip M.

1965a "Observations on the Urban-Folk and Urban-Rural Dichotomies as Forms of Western Ethnocentrism." In P. M. Hauser and Leo F. Schnore (eds.), *The Study of Urbanization*, pp. 503–517. New York: John Wiley and Sons, Inc.

1965b "Urbanization: An Overview." In P. M. Hauser and Leo F. Schnore (eds.), *The Study of Urbanization*, pp. 1–48. New York: John Wiley and Sons, Inc.

1969 "The Chaotic Society: Product of the Social Morphological Revolution." *American Sociological Review* 34:1–18.

Hauser, Philip M., and Leo F. Schnore (eds.)

1965 *The Study of Urbanization.* New York: John Wiley and Sons, Inc.

Hawley, Amos H.

1941 "An Ecological Study of Urban Service Institutions." *American Sociological Review* 6:629–639.

1950 *Human Ecology: A Theory of Community Structure.* New York: Ronald Press.

1956 *The Changing Shape of Metropolitan America.* Glencoe, Ill.: The Free Press.

1957 "Metropolitan Population and Municipal Government Expenditures in Central Cities." In Paul K. Hatt and Albert J. Reiss (eds.), *Cities and Society*, pp. 773–782. New York: The Free Press.

1967 "Population and Society: An Essay on Growth." In S. J. Behrman (ed.), *Fertility and Family Planning: A World View*, pp. 189–209. Ann Arbor: University of Michigan Press.

1968 "Human Ecology." In *International Encyclopedia of Social Sciences*, Vol. 4, pp. 328–337. New York: Macmillan.

1970 "Environment, Population, and Social System." Unpublished manuscript, Department of Sociology, University of North Carolina.

1971 *Urban Society: An Ecological Approach.* New York: Ronald Press.

Hawley, Amos H., Walter Boland, and Margaret Boland

1965 "Population Size and Administration in Institutions of Higher Education." *American Sociological Review* 36 (April):252–255.

Hawley, Amos H., and Otis D. Duncan

1957 "Social Area Analysis: A Critical Appraisal." *Land Economics* 33:337–345.

Hawley, Amos H., and Basil Zimmer

1956a "Home Owners and Attitude Toward Tax Increase." *Journal of the American Institute of Planners* 21:65–74.

1956b "Property Taxes and Solutions to Fringe Problems." *Land Economics* 32:369–376.

1970 *The Metropolitan Community.* Beverly Hills, Calif.: Sage Publications.

Hendershot, G. E., and T. F. James

1972 "Size and Growth as Determinants of Administrative-Production Ratios in Organizations." *American Sociological Review* 37 (April):149–153.

Henderson, George

1964 "Twelfth Street: An Analysis of a Changed Neighborhood." *Phylon* 25 (1):91–96.

Henderson, James M., and Anne O. Krueger

1965 *National Growth and Economic Change in the Upper Midwest.* Minneapolis: University of Minnesota Press.

Heraud, B. J.

1968 "Social Class and the New Town." *Urban Studies* 5:33–58.

Hettner, Alfred

1895 "Die Lage der menschlichen Ansiedlungen." *Geographische Zeitschrift* 1:361–375.

Hillery, George A.

1968 *Communal Organizations: A Study of Local Societies.* Chicago: The University of Chicago Press.

Hodge, Gerald

1966 *The Identification of "Growth Poles" in Eastern Ontario.* Ontario Department of Economics and Development.

1968 "Urban Structure and Regional Development." *Papers and Proceedings of the Regional Science Association* 21:101–123.

Hofstaetter, Peter R.

1952 "Your City Revisited—A Factorial Ecology of Cultural Patterns." *American Catholic Sociological Review* 13:159–168.

Holdaway, Edward, and Thomas Blowers

1971 "Administrative Ratios and Organization Size: A Longitudinal Examination." *American Sociological Review* 36 (April):278–286.

Homans, George

1961 *Social Behavior: Its Elementary Forms.* New York: Harcourt, Brace and World.

Hoskin, Fran P.

1973 *The Functions of Cities.* Cambridge: Schenkman Publishing Co.

Household Goods Carriers Bureau

1967 *Mileage Guide No. 9.* Chicago: Rand McNally and Company.

Hovland, C. I., I. L. Janis, and H. H. Kelley

1953 *Communication and Persuasion.* New Haven: Yale University Press.

Howard, Alan, and R. A. Scott

1965 "A Proposed Framework for the Analysis of Stress in the Human Organism." *Behavioral Science* 10:141–160.

Howard, Ebenezer

1898   *Tomorrow: A Peaceful Path to Real Reform*. London: Faber and Faber.

Hoyt, Homer

1933   *One Hundred Years of Land Values in Chicago*. Chicago: The University of Chicago Press.

1939   *The Structure and Growth of Residential Neighborhoods in American Cities*. Washington, D.C.: U.S. Government Printing Office.

Hughes, Everett C.

1945   "Dilemmas and Contradictions of Status." *American Journal of Sociology* 50 (5):353–359.

Hunt, C. L.

1959   "Negro-White Perceptions of Interracial Housing." *Journal of Social Issues* 15:24–29.

Hunter, Albert

1971   "Symbolic Communities: A Study of Chicago's Local Communities." Paper presented at the Metropolitan Forum, Center for Urban Studies' Conference on Social and Structural Change for the Chicago Metropolitan Area.

Hurd, Richard M.

1903   *Principles of City Land Values*. New York: The Record and Guide.

Hyderabad Metropolitan Research Project

1966   *Social Area Analysis of Metropolitan Hyderabad*. Hyderabad: Osmania University.

International Labour Office

1969   *Year Book of Labour Statistics, 1969*. Geneva: International Labour Office.

1970   *Year Book of Labour Statistics, 1970*. Geneva: International Labour Office.

Isaacs, Reginald R.

1948a   "The Neighborhood Unit as an Instrument for Segregation." *Journal of Housing* 5 (7):177–180, 5 (8):215–219.

1948b   "The Neighborhood Theory: An Analysis of its Adequacy." *Journal of the American Institute of Planners* 14 (2):15–23.

Isard, Walter

1956   *Location and Space Economy*. New York: John Wiley and Sons, Inc.

Jackson, Kenneth T.

1972   "Metropolitan Government Versus Suburban Autonomy: Politics on the

Crabgrass Frontier." In Kenneth T. Jackson and Stanley K. Schultz (eds.), *Cities in American History*, pp. 442–462. New York: Knopf.

1973   "The Crabgrass Frontier: 150 Years of Suburban Growth in America." In Raymond A. Mohl and James F. Richardson (eds.), *The Urban Experience: Themes in American History*, pp. 196–221. Belmont, Calif.: Wadsworth.

1975   "Urban Deconcentration in the Nineteenth Century: A Statistical Inquiry." In Leo F. Schnore (ed.), *The New Urban History: Quantitative Explorations by American Historians*, pp. 110–142. Princeton, N.J.: Princeton University Press.

Janis, Irving L.

1962   "Psychological Effects of Warnings." In G. W. Baker and D. W. Chapman (eds.), *Man and Society in Disaster*, pp. 55–92. New York: Basic Books.

Janowitz, Morris

1951   *The Community Press in an Urban Setting*. Glencoe, Ill.: The Free
(1967)   Press.

Jefferson, Mark

1939   "The Law of the Primate City." *Geographical Review* 29:226–232.

Jefferson, Richard

1909   "The Anthropography of Some Great Cities." *Bulletin of the American Geographical Society* 41:537–566.

Jencks, Christopher, and David Riesman

1968   "On Class in America." *Public Interest* 10:65–85.

Jones, Victor, and Richard L. Forstall

1963   "Economic and Social Classification of Metropolitan Areas." In *The Municipal Year Book 1963*, pp. 31–44. Chicago: International City Managers' Association.

Jouvenel, Bertrand de (ed.)

1963   *Futuribles I and II*. Geneva: Droz.

1967   *The Art of Conjecture*. New York: Basic Books.

Juppenlatz, M.

1970   *Cities in Transformation: The Urban Squatter Problem of the Developing World*. St. Lucia: University of Queensland Press.

Kain, John F.

1968   "Housing Segregation, Negro Employment, and Metropolitan Decentralization." *Quarterly Journal of Economics* 82:175–197.

Kaplan, Howard B.

1958   "An Empirical Typology for Urban Description." Ph.D. dissertation, New York University.

Kar, N. R.

1962   "Growth, Distribution and Dynamics of the Population Load in Calcutta." Mimeographed.

Karp, Herbert H., and K. Dennis Kelley

1971   *Toward an Ecological Analysis of Intermetropolitan Migration*. Chicago: Markham Publishing Company.

Kasarda, John D.

1971   "Economic Structure and Fertility: A Comparative Analysis." *Demography* 8:307–317.

1972a   "The Impact of Suburban Population Growth on Central City Service Functions." *American Journal of Sociology* 77 (6):1111–1124.

1972b   "The Theory of Ecological Expansion: An Empirical Test." *Social Forces* 51:165–175.

1973   "Effects of Personnel Turnover, Employee Qualifications and Professional Staff Ratios on Administrative Intensity and Overhead." *Sociological Quarterly* 14 (Summer):350–358.

1974   "The Structural Implications of Social System Size: A Three Level-Analysis." *American Sociological Review* 39 (February): 19–28.

1976a   "The Territorial Sources of Metropolitan Growth in the United States." Paper presented at the annual meeting of the Population Association of America, Montreal, Canada, April 1976.

1976b   "The Changing Occupational Structure of the American Metropolis: Apropos the Urban Problem," in Barry Schwartz (ed.), *The Changing Face of the Suburbs*, pp. 113–135. Chicago: The University of Chicago Press.

Kasarda, John D., and Morris Janowitz

1974   "Community Attachments in Mass Society." *American Sociological Review* 39 (3):328–339.

Kasarda, John D., and George Redfearn

1975   "Differential Patterns of Urban and Suburban Growth in the United States." *Journal of Urban History* 2 (November): 43–66.

Kerckhoff, Richard

1957   "A Study of Racially Changing Neighborhoods." *Merrill-Palmer Quarterly* 4 (1):15–49.

Khorev, B. S., and D. G. Khodzhayev

1972   "The Conception of a Unified System of Settlement and the Planned Regulation of City Growth in the U.S.S.R." *Soviet Geography* 8:90–98.

King, Leslie J.

1966   "Cross-Sectional Analysis of Canadian Urban Dimensions, 1951 and 1961." *Canadian Geographer* 10:205–224.

King County Planning Department

1961   "Residential Density Model, Seattle Metropolitan Area, 1961." King County Planning Department. Mimeographed.

Kitano, H. H. L.

1969   *Japanese Americans*. Englewood Cliffs, N.J.: Prentice-Hall.

Kivell, P. T.

1972   "A Note on Metropolitan Areas, 1961–1971." *Area* 4 (3):179–184.

Klatzky, S. R.

1970   "Relationship of Organization Size to Complexity and Coordination." *Administrative Science Quarterly* 15 (December):428–438.

Kneeland, D. E.

1972   "Quiet Decay Erodes Downtown Areas of Small Cities." *New York Times*, February 8.

Kohl, Johann Georg

1841   *Der Verkehr und die Ansiedlung der Menschen in ihrer Abhängigkeit von der Gestaltung der Erdoberfläche*. Leipzig: Arnoldische Buchhandlung.

Kramer, Carol

1958   "Population Density Patterns." *Chicago Area Transportation Study Research News* 2:3–10.

Kristof, F. S.

1972   "Federal Housing Policies: Subsidized Production, Filtration and Objectives." *Land Economics* 48:309–320.

Kunkel, John H.

1967   "Some Behavioral Aspects of the Ecological Approach to Social Organization." *American Journal of Sociology* 73 (1):12–29.

Lampard, Eric E.

1955   "The History of Cities in the Economically Advanced Areas." *Economic Development and Cultural Change* 3:81–136.

1968   "The Evolving System of Cities in the United States." In Harvey Perloff and Lowden Wingo (eds.), *Issues in Urban Economics*, pp. 81–139. Baltimore: Johns Hopkins University Press.

Land, Kenneth

1969   "Principles of Path Analysis." In Edgar Borgatta (ed.), *Sociological Methodology 1969*, pp. 3–37. San Francisco: Jossey-Bass.

Laumann, Edward O.

1965   "Subjective Social Distance and Urban Occupational Stratification." *American Journal of Sociology* 71 (1):26–36.

1973   *Bonds of Pluralism*. New York: John Wiley and Sons, Inc.

Laurenti, L. M.

1961   *Property Values and Race: Studies in Seven Cities.* Berkeley: University of California Press.

Lauwe, Paul H. Chombart de

1952   *Paris et l'Agglomération Parisienne.* Paris: Presses Universitaires de France.

Lawton, Richard

1972   "An Age of Great Cities." *Town Planning Review* 43:199–224.

Lazarus, R. S.

1966   *Psychological Stress and the Coping Process.* Toronto: McGraw-Hill Book Company.

Lazerwitz, Bernard

1962   "Membership in Voluntary Associations and Frequency of Church Attendance." *Journal for the Scientific Study of Religion* 2:74–84.

Lazin, Frederick Aaron

1973   "The Failure of Federal Enforcement of Civil Rights Regulations in Public Housing, 1963–1971: The Cooptation of a Federal Agency by Its Local Constituency." *Policy Sciences* 4:263–273.

Leacock, Eleanor (ed.)

1971   *Culture and Poverty.* New York: Simon and Schuster.

Leone, R. A.

1972   "The Role of Data Availability in Intrametropolitan Workplace Location Studies." *Annals of Economic and Social Measurement* 1:171–182.

Lewin, Kurt

1936   *Principles of Topographical Psychology.* New York: McGraw-Hill Book Company.

Lewis, Oscar

1959   *Five Families: Mexican Case Studies in the Culture of Poverty.* New York: Basic Books.

1968   *La Vida.* New York: Random House.

Lichtenberger, E.

1970   "The Nature of European Urbanism." *Geoforum* 4:45–62.

Lindbloom, C. E.

1963   "The Science of Muddling Through." In N. W. Polsby (ed.), *Politics and Social Life*, pp. 339–348. Boston: Houghton Mifflin.

Lithwick, N. H.

1970   *Urban Canada.* Ottawa: Central Mortgage and Housing Corporation.

Little, Kenneth L.

1965   *West African Urbanization*. Cambridge: Cambridge University Press.

Litwak, Eugene

1961   "Voluntary Associations and Neighborhood Cohesion." *American Sociological Review* 26:258–271.

Lowenthal, David

1961   "Geography, Experience and Imagination: Towards a Geographical Epistemology." *Annals of the Association of American Geographers* 51 (3):241–260.

Mabogunge, Akin

1965a   "Economic Implications of the Pattern of Urbanization in Nigeria." *Nigerian Journal of Economic and Social Studies* 7:9–30.

1965b   "Urbanization in Nigeria: A Constraint on Economic Development." *Economic Development and Cultural Change* 13:413–438.

Maine, Sir Henry J. S.

1861   *Ancient Law*. London: Murray.

Mangin, W.

1967   "Latin American Squatter Settlements." *Latin American Research Review* 2:65–98.

Manners, Gerald

1974   "The Office in the Metropolis: An Opportunity for Shaping Metropolitan America." *Economic Geography* 50:93–110.

Marris, Peter

1967   *African City Life*. Nkanga (editions) 1. Kampala, Uganda: Transition Books.

Maruyama, M.

1963   "The Second Cybernetics: Deviation Amplifying Mutual Causal Processes." *American Scientist* 51:164–169.

Masotti, Louis, and Jeffrey Hadden (eds.)

1974   *Suburbia in Transition*. New York: New Viewpoints.

Mayer, Albert J.

1957   "Race and Private Housing: A Social Problem and a Challenge to Understanding Human Behavior." *Journal of Social Issues* 13 (4):3–6.

1960   "Russell Woods: Change Without Conflict: A Case Study of Neighborhood Racial Transition in Detroit." In Nathan Glazer and Davis McEntire (eds.), *Studies in Housing and Minority Groups*, pp. 198–220. Berkeley: University of California Press.

Mayer, Harold M., and Clyde F. Kohn (eds.)

1959   *Readings in Urban Geography.* Chicago: The University of Chicago Press.

Mayhew, Bruce H.

1973   "System Size and Ruling Elites." *American Sociological Review* 38 (August):468–475.

Mayhew, B. H., J. M. McPherson, R. L. Levinger, and T. F. James

1972   "System Size and Structural Differentiation in Formal Organizations: A Baseline Generator for Two Major Theoretical Propositions." *American Sociological Review* 37 (October):629–633.

McClelland, David C.

1961   *The Achieving Society.* New York: Van Nostrand.

McGee, T. G.

1967   *The Southeast Asian City: A Social Geography of the Primate Cities of Southeast Asia.* London: G. Bell.

McGuire, W. J.

1966   "Attitudes and Opinions." *Annual Review of Psychology* 17:475–514.

McIntosh, Robert

1963   "Ecosystems, Evolution and Relational Patterns of Living Organisms." *American Scientist* 51:246–267.

McKay, Henry, and James F. Short, Jr.

1969   *Juvenile Delinquency in Urban Areas.* Chicago: The University of Chicago Press.

McKenzie, Roderick

1929   "Ecological Succession in the Puget Sound Region." *Publications of the American Sociological Society* 23:60–80.

1933a "Industrial Expansion and the Interrelations of Peoples." In E. B. Reuter (ed.), *Race and Cultural Contacts*, pp. 19–33. New York: McGraw-Hill Book Company.

1933b *The Metropolitan Community.* New York: McGraw-Hill Book Company.

McNulty, Michael L.

1969a "Dimensions of Urban Structural Change in Ghana." Unpublished manuscript.

1969b "Urban Structure and Development: The Urban System of Ghana." *Journal of Developing Areas* 3 (2):159–176.

McQuade, Walter

1973   "A Daring New Generation of Skyscrapers." *Fortune* 87 (2):78ff.

Meier, Richard L.

1962   *A Communications Theory of Urban Growth.* Cambridge, Mass.: The M.I.T. Press.

Meltzer, Jack

1972   "Loop Betrays Deeper Ills of City." *Chicago Sun-Times Viewpoint,* October 22, 1972.

Merton, Robert K.

1968   *Social Theory and Social Structure.* New York: The Free Press.

Meyer, David R.

1972   "Classification of U.S. Metropolitan Areas by the Characteristics of Their Nonwhite Populations." In Brian J. L. Berry (ed.), *City Classification Handbook,* pp. 61–93. New York: John Wiley and Sons, Inc.

Meyer, Marshall W.

1972   *Bureaucratic Structure and Authority.* New York: Harper and Row.

Meyerson, Martin, and Edward Banfield

1955   *Politics, Planning and the Public Interest.* Glencoe, Ill.: The Free Press.

Michael, Donald

1972   "The Individual: Enriched or Impoverished? Master or Servant?" In *Information Technology,* pp. 37–60. New York: The Conference Board.

Micklin, Michael

1973   *Population, Environment and Social Organization: Current Issues in Human Ecology.* Hinsdale, Ill.: The Dryden Press.

Miller, J. C.

1965   "Living Systems: Basic Concepts." *Behavioral Science* 10:193–237.

Mills, C. Wright

1951   *White Collar: The American Middle Classes.* New York: Oxford University Press.

Mills, Edwin S.

1970   "Urban Density Functions." *Urban Studies* 7:5–20.

1972   *Urban Economics.* Chicago: Scott, Foresman.

Miner, H. (ed.)

1967   *The City in Modern Africa.* New York: Frederick A. Praeger.

Mitra, Asok

1963   *Calcutta, India's City.* Calcutta: New Age Publishers Ltd.

Molotch, Harvey

1968   "Community Action to Control Racial Change: An Evaluation of Chicago's

South Shore Effort." Unpublished Ph.D. dissertation, Department of Sociology, The University of Chicago.

1969 "Racial Change in a Stable Community." *American Journal of Sociology* 75 (2):226–238.

1972 *Managed Integration: Dilemmas of Doing Good in the City.* Berkeley: University of California Press.

Moore, Eric G.

1972 *Residential Mobility in the City.* Commission on College Geography Resource Paper No. 13. Washington, D.C.: Association of American Geographers.

Moore, Joan W.

1970 *Mexican Americans.* Englewood Cliffs, N.J.: Prentice-Hall.

Morrill, R. L.

1962 "Simulation of Central Place Patterns Over Time." Proceedings of the IGU Symposium in Urban Geography, Lund, 1960. Lund: Gleerup.

1963 "The Development of Spatial Distribution of Towns in Sweden: An Historical Predictive Approach." *Annals of the Association of American Geographers* 53 (1):1–14.

1972 "Geographic Perspective of the Ghetto." Unpublished paper. Department of Geography, University of Washington.

Morris, R. N.

1968 *Urban Sociology.* New York: Frederick A. Praeger.

Mosca, Gaetano

1939 *The Ruling Class.* New York: McGraw-Hill Book Company.

Moser, Charles A., and Wolf Scott

1961 *British Towns: A Statistical Study of Their Social and Economic Differences.* Edinburgh: Oliver and Boyd.

Müller-Wille, Christopher F.

1976 "The Forgotten Heritage: Christaller's Antecedents." In Brian J. L. Berry (ed.), *Perspectives in Geography 3, Ideas in Evolution.* DeKalb, Ill.: Northern Illinois University Press.

Mulvihill, D. J., M. M. Tumin, and L. A. Curtis

1969 *Crimes of Violence.* Staff Report to the National Commission on the Causes and Prevention of Violence. Washington, D.C.: U.S. Government Printing Office.

Muth, Richard F.

1961 "The Spatial Structure of the Housing Market." *Papers and Proceedings of the Regional Science Association* 7:207–220.

1969 *Cities and Housing.* Chicago: The University of Chicago Press.

Nagel, Ernest

1961   *The Structure of Science.* New York: Harcourt, Brace and World.

Naroll, Raoul

1956   "A Preliminary Index of Social Development." *American Anthropologist* 58 (August):687–715.

National Academy of Sciences

1972   "Freedom of Choice in Housing: Opportunities and Constraints." Report of the Social Science Panel, National Academy of Sciences—National Academy of Engineering. Washington, D.C.

National Advisory Commission on Rural Poverty

1967   *The People Left Behind.* Washington, D.C.: U.S. Government Printing Office.

National Bureau of Standards

1973   *Standard Metropolitan Statistical Areas.* Federal Information Processing Standards Publication 8-3. Washington, D.C.: U.S. Government Printing Office.

National Commission on Civil Disorders

1968   *Report on U.S. Riots.* New York: Bantam Books.

National Committee Against Discrimination in Housing

1970   *Jobs and Housing.* New York: National Committee Against Discrimination in Housing.

Neenan, William

1970   "The Suburban–Central City Exploitation Thesis: One City's Tale." *National Tax Journal* 23:117–139.

Nelson, Howard J.

1955   "A Service Classification of American Cities." *Economic Geography* 31 (3):189–210.

Newling, Bruce E.

1960   "Urban Population Densities: A Comment on Colin Clark's Paper with Special Reference to Kingston, Jamaica." Unpublished manuscript, Northwestern University.

1962   "The Growth and Spatial Structure of Kingston, Jamaica." Ph.D. dissertation, Northwestern University.

1965   "Urban Growth and Spatial Structure: Mathematical Models and Empirical Evidence." Unpublished manuscript, City College of New York.

Newman, D. K.

1967   "Decentralization of Jobs." *Monthly Labor Review* 90:7–13.

Northwood, L. K., and E. A. T. Barth

1965   *Urban Desegregation, Negro Pioneers, and Their White Neighbors.* Seattle: University of Washington Press.

Odum, H. T., J. E. Cantlon, and L. S. Kornicker

1960   "An Organizational Hierarchy Postulate for the Interpretation of Species-Individual Distributions, Species Entropy, Ecosystem Evolution, and the Meaning of a Species-Variety Index." *Ecology* 41:395–399.

Olcott, George C.

1973   *Annual Blue Book of Land Values in Chicago.*

Packard, Vance O.

1972   *A Nation of Strangers.* New York: David McKay.

Pahl, R. E.

1970   *Patterns of Urban Life.* New York: Humanities Press.

Pappenfort, Donnell M.

1959   "The Ecological Field and the Metropolitan Community." *American Journal of Sociology* 64:380–385.

Park, Robert E.

1916   "The City: Suggestions for the Investigation of Human Behavior in an Urban Environment." *The American Journal of Sociology* 20:577–612.

1936   "Human Ecology." *The American Journal of Sociology* 42:1–15.

1950   *Race and Culture.* Glencoe, Ill.: The Free Press.

1967   "The Urban Community as a Spatial Pattern and a Moral Order." In Ralph H. Turner (ed.), *Robert Park on Social Control and Collective Behavior,* pp. 55–68. Chicago: The University of Chicago Press.

Park, R. E., and E. W. Burgess

1921   *Introduction to the Science of Sociology.* Chicago: The University of Chicago Press.

1925   *The City.* Chicago: The University of Chicago Press.

Parsons, Talcott

1937   *The Structure of Social Action.* New York: McGraw-Hill Book Company.

Pedersen, Poul O.

1967   *Modeller for Befolkningsstruktur og Befolkningssudvikling i Storbymorader Specielt med Henblik pa Storkobenhavn.* Copenhagen: State Urban Planning Institute.

Pendleton, W. C.

1962   "The Valuation of Accessibility." The University of Chicago, mimeographed.

Perevedentsev, V.

1972   Comments reported in *Current Digest of the Soviet Press*, 21 (9):8.

Perle, Sylvia M.

1964   "Factor Analysis of American Cities." M.A. thesis, The University of Chicago.

Perloff, Harvey S., E. S. Dunn, E. E. Lampard, and R. F. Muth

1960   *Regions, Resources and Economic Growth*. Baltimore: Johns Hopkins University Press.

Perloff, Harvey S., and Lowdon Wingo

1963   *Natural Resource Endowment and Regional Economic Growth*. Washington, D.C.: Resources for the Future, Inc.

Perroux, Françoise

1961   "Les Pôles de Croissance." In F. Perroux, *L'économie du XX siècle*, pp. 121–242. Paris: Presses Universitaires de France.

Perry, Clarence

1939   *Housing for the Machine Age*. New York: Russell Sage Foundation.

Peters, Philip

1968   "The Urban Ecology of Louisville, Kentucky." Unpublished paper. Center for Urban Studies, The University of Chicago.

Peterson, George L.

1965   "Subjective Measures of Housing Quality: An Investigation of Problems of Codification of Subjective Value for Urban Analysis." Unpublished Ph.D. dissertation, Northwestern University.

1967a   "Measuring Visual Preferences of Residential Neighborhoods." *Ekistics* 23:169–173.

1967b   "A Model of Preference: Quantitative Analysis of the Perception of the Visual Appearance of Residential Neighborhoods." *Journal of Regional Science* 7:19–31.

Pierce, J. R.

1961   *Symbols, Signals and Noise*. New York: Harper and Row.

Pinkerton, James

1969   "City-Suburban Residential Patterns by Social Class: A Review of the Literature." *Urban Affairs Quarterly* 4:499–519.

Pinkney, Alphonso

1969   *Black Americans*. Englewood Cliffs, N.J.: Prentice-Hall.

Pondy, Louis

1969   "Effects of Size, Complexity and Ownership on Administrative Intensity." *Administrative Science Quarterly* 14:47–60.

Population Research and Training Center

1962 "Comparative Urban Research: Progress Report to the Ford Foundation, May 1962." Chicago: Population Research and Training Center, The University of Chicago.

Pownall, L. L.

1953 "The Functions of New Zealand Towns." *Annals of the Association of American Geographers* 45 (4):332–350.

Pred, Allan

1966 *The Spatial Dynamics of U.S. Urban-Industrial Growth, 1800–1914.* Cambridge, Mass.: The M.I.T. Press.

1967 *Behavior and Location.* Lund Studies in Geography, Series B, No. 27. Lund, Sweden.

Price, Daniel O.

1942 "Factor Analysis in the Study of Metropolitan Centers." *Social Forces* 20 (4):449–455.

Proshansky, H. M., et al.

1970 *Environmental Psychology.* New York: Holt, Rinehart and Winston.

Rainwater, Lee

1966 "Fear and the House-as-haven in the Lower Class." *Journal of the American Institute of Planners* 22 (1):23–31.

Rapkin, Chester, and William G. Grigsby

1960 *The Demand for Housing in Racially Mixed Areas.* Berkeley: University of California Press.

Rashevsky, Nicolas

1953a "Outline of a Mathematical Approach to History." *Bulletin of Mathematical Biophysics* 15:197–234.

1953b "Some Quantitative Aspects of History." *Bulletin of Mathematical Biophysics* 15:339–359.

Ratzel, Friedrich

1882 *Anthropogeographie 1: Grundzüge der Anwendung der Geographie auf die Geschichte.* Stuttgart: J. Engelhorn.

1891 *Anthropogeographie 2: Die Verbreitung des Menschen.* Stuttgart: J. Engelhorn.

1903 "Die geographische Lage der grossen Städte." *Jahrbuch der Gehe-Stiftung*, 31–72.

Ray, D. Michael, and Brian J. L. Berry

1965 "Multivariate Socioeconomic Regionalization: A Pilot Study in Central Canada." In S. Ostry and T. Rymes (eds.), *Papers on Regional Statistical Studies*, pp. 75–123. Toronto: University of Toronto Press.

Ray, D. Michael, et al.

1968   "The Socio-economic Dimensions and Spatial Structure of Canadian Cities." Unpublished paper, University of Waterloo.

Redfield, Robert

1941   *Folk Culture of Yucatan*. Chicago: The University of Chicago Press.

Rees, Philip H.

1968   "The Factorial Ecology of Metropolitan Chicago, 1960." M.A. thesis, The University of Chicago.

1970   "The Factorial Ecology of Metropolitan Chicago." In Brian J. L. Berry and Frank E. Horton (eds.), *Geographic Perspectives on Urban Systems*, pp. 276–290 and 306–397. Englewood Cliffs, N.J.: Prentice-Hall.

Reissman, Leonard

1964   *The Urban Process*. Glencoe, Ill.: The Free Press.

Robson, B. T.

1969   *Urban Analysis*. Cambridge: Cambridge University Press.

Rodwin, Lloyd

1970   *Nations and Cities*. Boston: Houghton Mifflin.

Rose, Arnold, F. J. Atelsek, and L. R. McDonald

1953   "Neighborhood Reaction to Negro Residents: An Alternative to Invasion and Succession." *American Sociological Review* 18:497–507.

Rose, H. M.

1970   "The Development of an Urban Subsystem: The Case of the Negro Ghetto." *Annals of the Association of American Geographers* 60:1–7.

Rosen, Harry, and David Rosen

1962   *But Not Next Door*. New York: Ivan Obolensky.

Rosenwaike, Ira

1970   "Critical Examination of the Designations of SMSA's." *Social Forces* 48:322–333.

Rossi, Peter

1955   *Why Families Move*. Glencoe, Ill.: The Free Press.

Royal Commission on Local Government in England

1969   *Community Attitudes Survey: England*. London: Her Majesty's Stationary Office.

Royko, M.

1971   *Boss*. New York: New American Library.

Rummel, Rudolph

1967   "Understanding Factor Analysis." *Journal of Conflict Resolution* 11:440–480.

Rushing, William

1967   "The Effects of Industry Size and Division of Labor on Administration." *Administrative Science Quarterly* 12:273–295.

Saarinen, Thomas F.

1969   *Perception of Environment.* Commission on College Geography Resource Paper No. 5. Washington, D.C.: Association of American Geographers.

Schlüter, Ott

1899   "Bemerkungen zur Siedelungsgeographie." *Geographische Zeitschrift* 5:65–84.

1906a   *Die Ziele der Geographie des Menschen.* München: Bruckmann.

1906b   "Die Leitenden Gesichtspunkte der Anthropogeographie, insbesondere der Lehre F. Ratzels." *Archiv der Sozialwissenschaften und Sozialpolitik* 22:10–29.

Schnore, Leo F.

1957   "Metropolitan Growth and Decentralization." *American Journal of Sociology* 63:171–180.

1958   "Social Morphology and Human Ecology." *American Journal of Sociology* 63:620–634.

1959   "The Timing of Metropolitan Decentralization." *Journal of the American Institute of Planners* 25:200–206.

1961   "The Myth of Human Ecology." *Sociological Inquiry* 31 (2):128–139.

1962   "Municipal Annexations and Decentralization, 1950–1960." *American Journal of Sociology* 67:406–417.

1972   *Class and Race in Cities and Suburbs.* Chicago: Markham Publications.

Schorr, Alvin L.

1970   "Housing and Its Effects." In H. M. Proshansky et. al. (eds.), *Environmental Psychology*, pp. 319–333. New York: Holt, Rinehart and Winston.

Schwind, Paul J.

1971   *Migration and Regional Development in the United States, 1950–1960.* Department of Geography Research Paper No. 133. Chicago: The University of Chicago Press.

Schwirian, Kent P.

1974   "Some Recent Trends and Methodological Problems in Urban Ecological Research." In Kent P. Schwirian (ed.), *Comparative Urban Structure: Studies in the Ecology of Cities*, pp. 3–31. Lexington, Mass.: D. C. Heath and Company.

Scott, Mellier

1969   *American City Planning Since 1890.* Berkeley: The University of California Press.

Sears, D. O., and R. P. Abeles

1969   "Attitudes and Opinions." *Annual Review of Psychology* 20:253–288.

Seidman, D. E.

1964   *An Operational Model of the Residential Land Market.*

Sen, S. N.

1960   *The City of Calcutta: A Socio-Economic Survey.* Calcutta: Bookland Private Ltd.

Shaw, Clifford, F. M. Zorbaugh, Henry McKay, and Leonard Cottrell

1929   *Delinquency Areas.* Chicago: The University of Chicago Press.

Sherratt, G. G.

1960   "A Model for General Urban Growth." *Management Science: Models and Techniques* 2:147–159.

Shevky, Eshref, and Wendell Bell

1955   *Social Area Analysis: Theory, Illustrative Application and Computational Procedures.* Stanford: Stanford University Press.

Shevky, Eshref, and Marianne Williams

1949   *The Social Areas of Los Angeles: Analysis and Typology.* Berkeley: The University of California Press.

Short, James F.

1971   *The Social Fabric of the Metropolis.* Chicago: The University of Chicago Press.

Simmel, Georg

1902   "The Number of Members as Determining the Sociological Form of Groups I." *American Journal of Sociology* 8:1–46.

1903a   "The Number of Members as Determining the Sociological Form of Groups II." *American Journal of Sociology* 8:138–196.

1903b   "Die Grossstädte und das Geistesleben." In T. Petermann (ed.), *Die Grossstadt*, pp. 185–206. Dresden: Zahn and Jaensch.

1957   "The Metropolis and Mental Life." In Paul K. Hatt and Albert J. Reiss,
(1905) Jr. (eds.), *Cities and Society*, pp. 635–646. New York: The Free Press.

Simmons, James W.

1962   "Relationships Between the Population Density Pattern and Site of Cities." Unpublished M.A. thesis, The University of Chicago.

1968   "Changing Residence in the City: A Review of Intra-Urban Mobility."

Geographical Review 58 (4):622–651. Reprinted, in Brian J. L. Berry and Frank Horton (eds.), *Geographic Perspectives on Urban Systems*, pp. 395–413. Englewood Cliffs, N.J.: Prentice-Hall (1970).

Simon, H. A.

1955   "On a Class of Skew Distribution Functions." *Biometrika* 42:425–440.

Sjoberg, Gideon

1960   *The Pre-Industrial City, Past and Present*. Glencoe, Ill.: The Free Press.

Slayton, Donald

1963   "Property Values in Changing Neighborhoods." In A. Avine (ed.), *Open Occupancy vs. Forced Housing Under the Fourteenth Amendment: A Symposium on Anti-Discrimination Legislation, Freedom of Choice, and Property Rights in Housing*, pp. 28–33. New York: Bookmailer.

Sly, David F.

1972   "Migration and the Ecological Complex." *American Sociological Review* 37:615–628.

Smelser, N. J.

1963   *Theory of Collective Behavior*. New York: The Free Press.

Smith, Buckeley, Jr.

1959   "The Reshuffling Phenomena: A Pattern of Residence of Unsegregated Negroes." *American Sociological Review* 24:77–79.

Smith, Joel

1970   "Another Look at Socioeconomic Status Distributions in Urbanized Areas." *Urban Affairs Quarterly* 5:423–453.

Smith, R. H. T.

1965a   "The Functions of Australian Towns." *Tijdschrift voor Economische en Sociale Geografie* 56 (3):81–92.

1965b   "Method and Purpose in Functional Town Classification." *Annals of the Association of American Geographers* 55 (3):539–548.

Smith, Wilbur, and associates

1961   *Future Highways and Urban Growth*. New Haven.

Spencer, Herbert

1877   *The Principles of Sociology*. Vol. 1. New York: D. Appleton and Company.

Spencer, Robert F.

1959   *The North Alaskan Eskimo: A Study in Ecology and Society*. Smithsonian Institute Bureau of American Ethnology, Bulletin 171.

Spiegel, H. B. C.

1960   "Tenants' Intergroup Attitudes in a Public Housing Project With Declining White Population." *Phylon* 21:30–39.

Spodek, Michael A.

1968   "The Urban Ecology of Shreveport, Louisiana." Unpublished paper. Center for Urban Studies, The University of Chicago.

Srole, Leo

1972   "Urbanization and Mental Health: Some Reformulations." *The American Scientist* 60:576–583.

Srole, Leo, et al.

1962   *Mental Health in the Metropolis.* New York: McGraw-Hill Book Company.

Stanback, Thomas M., Jr., and Richard V. Knight

1970   *The Metropolitan Economy.* New York: Columbia University Press.

Stanford Research Institute

1968   *Costs of Urban Infrastructure for Industry as Related to City Size in Developing Countries.* Menlo Park, Calif.: Stanford Research Institute.

Stegman, Michael A.

1969   "Accessibility Models and Residential Location." *Journal of the American Institute of Planners* 35 (1):22–29.

Steigenga, William

1955   "A Comparative Analysis and Classification of Netherlands Towns." *Tijdschrift voor Economische en Sociale Geografie* 46 (6–7):105–119.

Stein, Maurice

1964   *The Eclipse of Community.* New York: Harper Torchbooks.

Stewart, John Q., and William Warntz

1958   "Physics of Population Distribution." *Journal of Regional Science* 1:99–123.

Sumner, William G.

1906   *Folkways.* Boston: Ginn.

Sussman, Marvin

1957   "The Role of Neighborhood Associations in Private Housing for Racial Minorities." *Journal of Social Issues* 13:31–37.

Suttles, Gerald D.

1968   *The Social Order of the Slum.* Chicago: The University of Chicago Press.

1972a   "Community Design." Paper prepared for the National Research Council, National Academy of Sciences.

1972b *The Social Construction of Communities.* Chicago: The University of Chicago Press.

Sweetser, Frank L.

1965 "Factor Structure as Ecological Structure in Helsinki and Boston." *Acta Sociologica* 26:205–225.

Taeuber, Irene

1972 "The Changing Distribution of the Population of the U.S. in the 20th Century." In Sara Mills Mazie (ed.), *U.S. Commission on Population Growth and the American Future.* Vol. 5, pp. 31–108.

Taeuber, Karl E., and Alma F. Taeuber

1965 *Negroes in Cities: Residential Segregation and Neighborhood Change.* Chicago: Aldine Publishing Company.

Tarr, Joel A.

1972 *Transportation Innovation and Changing Spatial Patterns: Pittsburgh, 1850–1910.* Pittsburgh: University of Pittsburgh Press.

Tennant, Robert J.

1961 "Population Density Patterns in Eight Asian Cities." Paper read at West Lake Division meeting of the Association of American Geographers.

Theodorson, George A. (ed.)

1961 *Studies in Human Ecology.* New York: Harper and Row.

Thomas, E. N.

1970 "Additional Comments on Population-Size Relationships for Sets of Cities." In W. L. Garrison (ed.), *Quantitative Geography,* pp. 167–190. New York: Atherton Press.

Thomas, Ray

1969 *London's New Towns.* Landau: P.E.P.

Thomas, W. I.

1967 *On Social Organization and Social Personality* (Morris Janowitz, ed.). Chicago: The University of Chicago Press.

Thompson, D'Arcy

1917 *On Growth and Form.* Cambridge: Cambridge University Press.

Thompson, Warren S.

1947 *The Growth of Metropolitan Districts in the U.S., 1900–1940.* Washington, D.C.: U.S. Government Printing Office.

Thompson, Wilbur

1965 *A Preface to Urban Economics.* New York: John Wiley and Sons, Inc.

1968   "Internal and External Factors in the Development of Urban Economics." In Harvey S. Perloff and Lowdon Wingo, Jr. (eds.), *Issues in Urban Economics*, pp. 43–62. Baltimore: Johns Hopkins University Press.

Thorndike, E. L.

1939   *Your City*. New York: Harcourt Brace Jovanovich.

Tidemann, T. Nicolaus

1968   "The Theoretical Efficacy of 'Potential' and Transport Cost Models of Location." Center for Urban Studies, The University of Chicago. Mimeographed.

Tillman, James A.

1961a   "Morningtown, U.S.A.—A Composite Case History of Neighborhood Change." *Journal of Intergroup Relations* 2:156–166.

1961b   "The Quest for Identity and Status: Facets of the Desegregation Process in the Upper Midwest." *Phylon* 22 (4):329–339.

1962   "Rationalization, Residential Mobility, and Social Change." *Journal of Intergroup Relations* 3 (1):28–37.

Tisdale, Hope

1942   "The Process of Urbanization." *Social Forces* 20:311–316.

Toennies, Ferdinand

1887   *Gemeinschaft und Gesellschaft*. Leipzig: Fues's Verlag.

Tropman, John E.

1969   "Critical Dimensions of Community Structure: A Re-examination of the Hadden-Borgatta Findings." *Urban Affairs Quarterly* 5 (2):215–232.

Turner, J. F. C.

1968   *Uncontrolled Urban Settlement: Problems and Policies*. United Nations, New York, Department of Economics and Social Affairs.

Turnham, D., and T. Jaeger

1971   *The Employment Problem in Less Developed Countries*. Paris: Development Center of OECD.

Ullman, Edward L.

1954   "Amenities as a Factor in Regional Growth." *Geographical Review* 44:119–132.

1957   *American Commodity Flows*. Seattle: University of Washington Press.

United Nations

1962   *Demographic Aspects of Manpower*. Report 1: "Sex and Age Patterns of Participation in Economic Activities." New York.

U.S. Bureau of the Census

1960   *Historical Statistics of the United States, Colonial Times to 1957*. Washington, D.C.

1963   *U.S. Census of Population, 1960.* Vol. 1. Characteristics of the Population, Chapter D, Detailed Characteristics. Washington, D.C.

1970   *City Government Finances in 1968–69.* Washington, D.C.: U.S. Government Printing Office.

1972   *1970 Census of Population Summary Tapes.* Fourth count, file B, Minor Civil Divisions, State of Wisconsin.

U.S. Department of Agriculture

1971a   *The Economic and Social Conditions of Rural America in the 1970's.* Economic Development Division, Economic Research Service, U.S. Department of Agriculture. Prepared for the Committee on Government Operation, U.S. Senate, 92nd Congress, 1st Session. Washington, D.C.: U.S. Government Printing Office.

1971b   *Table of Nonmetro Places with 1970 Population Between 25,000 and 50,000, Intercensal Growth Rate of 13.3 Percent or Greater, and 75 Miles or More from an SMSA, and Table of Nonmetro Places with 1970 Population Between 10,000 and 24,999, Intercensal Growth Rate of 13.3 Percent or Greater, and 75 Miles or More from an SMSA.* Populations Growth Studies, Human Resource Branch, Economic Development Division, U.S. Department of Agriculture. Mimeographed.

U.S. Department of Commerce

1965   *Growth Patterns in Employment by County, 1940–1950 and 1950–1960.* Office of Business Economics, U.S. Department of Commerce. Washington, D.C.: U.S. Government Printing Office.

U.S. District Court, Northern District of Illinois, Eastern Division

1972   Civil Action No. 72C 1197.

Valentine, Charles A.

1968   *Culture and Poverty. Critique and Counterproposals.* Chicago: The University of Chicago Press.

Van Arsdol, Maurice, Santo F. Camilleri, and Calvin F. Schmid

1958   "The Generality of Urban Social Area Indices." *American Sociological Review* 23:277–284.

Van den Berghe, Pierre L.

1960   "Distance Mechanisms of Stratification." *Sociology and Social Research* 44:155–164.

Vance, Rupert B., and Sara Smith Sutker

1954   "Metropolitan Dominance and Integration in the Urban South." In Rupert Vance and Nicholas J. Demerath (eds.), *The Urban South,* pp. 114–134. Chapel Hill, N.C.: University of North Carolina Press.

Vaughn, Robert

1843   *The Age of Great Cities.* Jackson and Walford.

472   References

Vernon, Raymond

1960   *Metropolis 1985*. Cambridge, Mass.: Harvard University Press.

Vidich, Arthur J., and Joseph Bensman

1958   *Small Town in Mass Society: Class, Power, and Religion in a Rural Community*. Princeton: Princeton University Press.

Voss, Harwin, and David Petersen (eds.)

1971   *Ecology, Crime and Delinquency*. New York: Appleton-Century-Crofts.

Ward, Benjamin

1962   *Greek Regional Development*. Athens: Center for Economic Research.

Warner, Sam Bass

1962   *Streetcar Suburbs*. Cambridge, Mass.: Harvard University Press.

Warren, Roland L.

1972   *The Community in America*. Chicago: Rand McNally and Co.

Wax, Murray L.

1971   *Indian Americans*. Englewood Cliffs, N.J.: Prentice-Hall.

Weaver, Robert

1948   *The Negro Ghetto*. New York: Harcourt Brace Jovanovich.

Webber, Melvin

1964   "Culture, Territoriality and the Elastic Mile." *Papers and Proceedings of the Regional Science Association* 13:59–69.

Weber, Adna F.

1899   *The Growth of Cities in the Nineteenth Century*. New York: Macmillan.

Weber, Alfred

1909   *Über den Standort der Industrien 1: Reine Theorie des Standorts*. Tübingen: Mohr.

Weber, Max

1922   *Wirtschaft und Gesellschaft*. Tübingen: Mohr-Siebeck.

1958   *The City*. Glencoe, Ill.: The Free Press.

Weiss, Herbert K.

1961   "The Distribution of Urban Population and an Application to a Servicing Problem." *Operations Research* 9:860–874.

Wells, H. G.

1902   *Anticipations: The Reaction of Mechanical and Scientific Progress on Human Life and Thought*. London: Harper and Row.

Westie, Frank

1952   "Negro-White Status Differentials and Social Distance." *American Sociological Review* 17 (5):550–558.

Wiener, Norbert, Arturo Rosenblueth, and Julian Bigelow

1943   "Behavior, Purpose and Teleogy." *Philosophy of Science* 10:18–24.

Wilcox, D. F.

1904   *The American City. A Problem in Democracy.* New York: Macmillan.

Wilensky, Harold L.

1961   "Orderly Careers and Social Participation: The Impact of Work History on Social Integration in the Middle Mass." *American Sociological Review* 26:521–539.

Williams, Robin M.

1964   *Strangers Next Door.* Englewood Cliffs, N.J.: Prentice-Hall.

Wilner, D. M., R. P. Walkley, and S. W. Cook

1955   *Human Relations in Interracial Housing.* Minneapolis: University of Minnesota Press.

Winder, Alvin

1951   "Residential Invasion and Racial Antagonism in Chicago." *Phylon* 12 (3):239–241.

Wingo, Lowdon, Jr.

1961   *Transportation and Urban Land.* Washington, D.C.: Resources for the Future, Inc.

1962   "An Economic Model of the Utilization of Urban Land for Residential Purposes." *Papers and Proceedings of the Regional Science Association* 7:191–205.

Winsborough, Halliman H.

1961   "A Comparative Study of Urban Population Densities." Unpublished Ph.D. dissertation, Department of Sociology, The University of Chicago.

1963a   "City Growth and City Structure." *Journal of Regional Science* 4 (2):35–49.

1963b   "An Ecological Approach to the Theory of Suburbanization." *American Journal of Sociology* 68:565–570.

Wirth, Louis

1928   *The Ghetto.* Chicago: The University of Chicago Press.

1938   "Urbanism as a Way of Life." *American Journal of Sociology* 44:1–24.

Withey, Stephen B.

1962   "Reaction to Uncertain Threat." In G. W. Baker and D. W. Chapman (eds.), *Man and Society in Disaster*, pp. 93–123. New York: Basic Books.

Wolf, Eleanor

1957   "The Invasion-Succession Sequence as a Self-Fulfilling Prophecy."
        Journal of Social Issues 13:7–20.

1963   "The Tipping Point in Racially Changing Neighborhoods." *Journal of the
        American Institute of Planners* 29 (3):217–222.

1965   "The Baxter Area: A New Trend in Neighborhood Change?" *Phylon*
        26 (4):344–354.

Wolf, Eleanor, and Charles N. Lebeaux

1967   "Class and Race in the Changing City: Searching for New Approaches to
        Old Problems." In Leo F. Schnore and Henry Farin (eds.), *Urban
        Research and Policy Planning, Urban Affairs Annual Reviews*. Vol. 1, pp.
        99–129. Beverly Hills, Calif.: Sage Publications.

Wolpert, Julian

1965   "Behavioral Aspects of the Decision to Migrate." *Papers of the Regional
        Science Association* 15:159–169.

1966   "Migration as an Adjustment to Environmental Stress." *The Journal of
        Social Issues* 22 (4):92–102.

Wood, Robert C.

1964   *1400 Governments*. New York: Doubleday-Anchor.

World Bank

1972   *Urbanization*. Washington, D.C.: I.B.R.D.

Young, David

1972   "Industry Flees Decaying City." *Chicago Tribune*, May 7, 1972.

Young, Michael, and Peter Willmott

1957   *Family and Kinship in East London*. Baltimore: Penguin Books.

Zehner, R. B.

1972   "Neighborhood and Community Satisfaction in New Towns and Less
        Planned Suburbs." *Journal of the American Institute of Planners* 37:379–
        385.

Zimmer, Basil G.

1955   "Participation of Migrants in Urban Structure." *American Sociological
        Review* 20:218–224.

Zimmer, Basil G., and Amos H. Hawley

1959   "The Significance of Memberships in Associations." *American Journal of
        Sociology* 64:196–201.

Zipf, George K.

1941   *National Unity and Disunity: The Nation as a Bio-Social Organism*.
        Bloomington, Ind.: The Principia Press.

1949   *Human Behavior and the Principle of Least Effort.* Cambridge: Addison-Wesley.

Zorbaugh, Harvey W.

1926   "The Natural Areas of the City." *Publications of the American Sociological Society* 20:188–197. Reprinted, in George A. Theodorson (ed.), *Studies in Human Ecology*, pp. 45–49. New York: Harper and Row. 1961.

1929   *The Gold Coast and the Slum.* Chicago: The University of Chicago Press.

Author Index

477

Subject Index

DATE DUE

DEMCO, INC. 38-2931